Making Sense of Evolution

Making Sense of Evolution

The Conceptual Foundations of Evolutionary Biology

MASSIMO PIGLIUCCI AND
JONATHAN KAPLAN

THE UNIVERSITY OF CHICAGO PRESS
CHICAGO AND LONDON

Massimo Pigliucci is associate professor of ecology and evolution at SUNY Stony Brook. He is the author of *Phenotypic Evolution, Denying Evolution, Phenotypic Plasticity,* and *Tales of the Rational: Skeptical Essays about Nature and Science.* Jonathan Kaplan is assistant professor of philosophy at Oregon State University. He is the author of *The Limits and Lies of Human Genetic Research.*

The University of Chicago Press, Chicago 60637
The University of Chicago Press, Ltd., London

Printed in the United States of America
15 14 13 12 11 10 09 08 07 06   1 2 3 4 5

ISBN-13: 978-0-226-66836-9 (cloth)
ISBN-13: 978-0-226-66837-6 (paper)
ISBN-10: 0-226-66836-3 (cloth)
ISBN-10: 0-226-66837-1 (paper)

Library of Congress Cataloging-in-Publication Data

Pigliucci, Massimo, 1964–
Making sense of evolution : the conceptual foundations of evolutionary biology / Massimo Pigliucci and Jonathan Kaplan.
p. cm.
Includes bibliographical references and index.
ISBN-13: 978-0-226-66836-9 (cloth : alk. paper)
ISBN-13: 978-0-226-66837-6 (pbk. : alk. paper)
ISBN-10: 0-226-66836-3 (cloth : alk. paper)
ISBN-10: 0-226-66837-1 (pbk. : alk. paper)
1. Evolution (Biology) I. Kaplan, Jonathan. II. Title.
QH366.2.P545 2006
576.8—dc22

2006011168

∞ The paper used in this publication meets the minimum requirements of the American National Standard for Information Sciences—Permanence of Paper for Printed Library Materials, ANSI Z39.48-1992.

Contents

Acknowledgments

The writing of a book is the sort of thing that would never get done unless many people contributed to it in a variety of ways. As inadequate as the following list is, we would nevertheless like to thank the following friends and colleagues: Richard Aquila, André Ariew, Josh Banta, Kathy Bohstedt, David Buller, Sharyn Clough, John Dupré, Sergey Gavrilets, Paula Kover, Mohan Matthen, Roberta Milstein, Norris Muth, John Nolt, Anya Plutynski, Kim Sterelny, Gunter Wagner, and Jason Wolf. Jonathan Kaplan's Philosophy of Biology seminar read drafts of several chapters and made many helpful suggestions; we would especially like to thank Jacob Tennesson, Seth White, Cecily Bishop, Katherine Lantz, Katharine Morris, and Jenny Moser. In addition, several anonymous reviewers read the entire manuscript at different stages and made a variety of helpful suggestions. Massimo Pigliucci also wishes to acknowledge the continuous support of the National Science Foundation, part of which has gone toward freeing enough of his time to allow him to co-author this book.

Some parts of this work were published previously in different versions: parts of chapter 5 appeared in *Trends in Ecology and Evolution* (Pigliucci and Kaplan 2000), parts of chapter 7 appeared in *Philosophy of Science* (Kaplan 2002), and parts of chapter 9 appeared in *BioEssays* (Pigliucci 2003d).

PRELUDE Evolutionary Biology and Conceptual Analysis

Biological evolution is, as has often been noted, both fact and theory. It is a fact that all extant organisms came to exist in their current forms through a process of descent with modification from ancestral forms. The overwhelming evidence for this empirical claim was recognized relatively soon after Darwin published *On the Origin of Species* in 1859, and support for it has grown to the point where it is as well established as any historical claim might be. In this sense, biological evolution is no more a theory than it is a "theory" that Napoleon Bonaparte commanded the French army in the late eighteenth century. Of course, the details of how extant and extinct organisms are related to one another, and of what descended from what and when, are still being worked out, and will probably never be known in their entirety. The same is true of the details of Napoleon's life and military campaigns. However, this lack of complete knowledge certainly does not alter the fundamental nature of the claims made, either by historians or by evolutionary biologists.

On the other hand, evolutionary biology is also a rich patchwork of theories seeking to *explain* the patterns observed in the changes in populations of organisms over time. These theories range in scope from "natural selection," which is evoked extensively at many different levels, to finer-grained explanations involving particular mechanisms (e.g., reproductive isolation induced by geographic barriers leading to speciation events).

In this book, we focus on some of the patchwork of theories that make up the explanatory side of evolutionary biology. In particular, our interest is in those aspects of evolutionary biology that are conceptually rich and complex—those that force us to confront questions about the nature of selection, adaptation, species, and the like. There are a number of ways in which these questions have been addressed, and a number of different accounts of these areas of evolutionary biology. These different accounts, we will maintain, are not always compatible, either with one another or with other accepted practices in evolutionary biology. The goal of this work is to uncover some of these conceptual difficulties and present alternative interpretations that remove or at least lessen them. We approached this project from the perspectives of a practicing evolutionary biologist and a philosopher of science, and the resulting analysis intertwines, we hope effectively, the scientific and philosophical levels of investigation.

While we strive to present a coherent picture of a portion of evolutionary biology, we do not wish to pretend that the vision we articulate is the only possible coherent picture. Rather, we hope that by drawing attention to some of the ways in which the relationships between particular problems can be profitably approached, our work will encourage others, both practicing evolutionary biologists and philosophers, to take the problems we identify seriously and to work to address the difficulties raised.

THE GENERAL FRAMEWORK OF THE BOOK: TWO METAPHORS AND THEIR IMPLICATIONS

Because we will be making some potentially controversial claims throughout this volume, it is crucial for the reader to understand two basic ideas underlying most of what we say, as well as exactly what we think are the implications of our views for the general theory of evolutionary quantitative genetics, which we discuss repeatedly in critical fashion in several chapters. We are most willing to receive criticisms based on genuine disagreement with what we say, but there is no sense in ruffling feathers because of a misunderstanding of the scope of our analysis.

The first central idea we wish to put forth as part of the framework of this book will be readily familiar to biologists, although some of its consequences may not be. The idea can be expressed by the use of a metaphor proposed by Bill Shipley (2000) and illustrated in figure 1a: the shadow theater popular in Southeast Asia. In one form, the *wayang golek* of Bali and other parts of Indonesia, three-dimensional wooden puppets are used to project two-dimensional shadows on a screen, where the action is presented to the spectator. Shipley's idea is that quantitative biologists find themselves very much in the position of wayang golek's spectators: we have access to only the "statistical shadows" projected by a set of underlying causal factors. Unlike the wayang golek's patrons, however, biologists want to peek around the screen and infer the position of the light source as well as the actual three-dimensional shapes of the puppets. This, of course, is the familiar problem of the relationship between causation and correlation, and, as any undergraduate science major soon learns, correlation is not causation (although a popular joke among scientists is that the two are nevertheless often correlated).

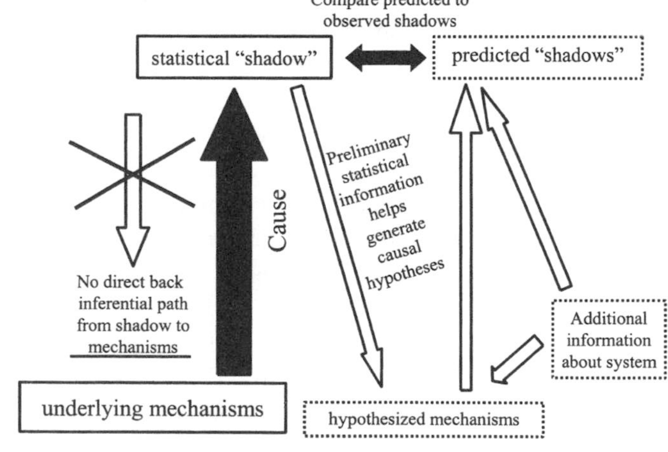

Figure 1. An example of Indonesian "shadow" theater. (a) In the *wayang golek* form, three-dimensional wooden puppets are used to project two-dimensional shadows on a screen. Bill Shipley (2000) used the wayang golek as a metaphor to illustrate the proper relationship between causation and correlation in biological research, a central idea underlying much of what is proposed in this book. (Photograph © Trisanna.com; used with permission.) (b) While no direct inference of underlying mechanisms based on observed statistical shadows is possible, a more complex process of hypothesis generation and testing does include a study of statistical shadows.

The loose relationship between causation and correlation has two consequences that are crucial to much of what we say. On the one hand, there is the problem that, strictly speaking, it makes no sense to attempt to infer mechanisms directly from patterns. That is because, to continue the shadow theater metaphor, there are too many possible combinations of three-dimensional shapes and light sources that result in the same two-dimensional shadows. On the other hand, as Shipley elegantly shows in his book, there is an alternative route that gets (most of) the job done, albeit in a more circuitous and painful way. What one can do (fig. 1b) is to produce a series of alternative hypotheses about the causal pathways underlying a given set of observations; these hypotheses can then be used to "project" the expected statistical shadows, which can be compared with the observed one. If the projected and actual shadows do not match, one can discard the corresponding causal hypothesis and move on to the next one; if the two shadows do match (within statistical margins of error, of course), then one has identified at least one causal explanation *compatible* with the observations. As any philosopher or scientist worth her salt knows, of course, this cannot be the end of the process, for more than one causal model may be compatible with the observations, which means that one needs additional observations or refinements of the causal models to be able to discard more wrong explanations and continue to narrow the field. A crucial point here is that the causal models to be tested against the observed statistical shadow can be suggested by the observations themselves, especially if coupled with further knowledge about the system under study (such as details of the ecology, developmental biology, genetics, or past evolutionary history of the populations in question). But the statistical shadows cannot be used as direct supporting evidence for any particular causal model.

The second central idea underlying much of what we will say in this book has been best articulated by John Dupré (1993), and it deals with the proper way to think about reductionism. The term "reductionism" has a complex history, and it evokes strong feelings in both scientists and philosophers (often, though not always, with scientists hailing reductionism as fundamental to the successes of science and some philosophers dismissing it as a hopeless epistemic dream). Dupré introduces a useful distinction that acknowledges the power of reductionism in science while at the same time sharply curtailing its scope. His idea is summarized in figure 2 as two possible scenarios: In one case, reductionism allows one to explain and predict higher-level phenomena (say, development in living organisms) entirely in terms of lower-level processes (say, genetic switches throughout development). In the most extreme case, one can also infer

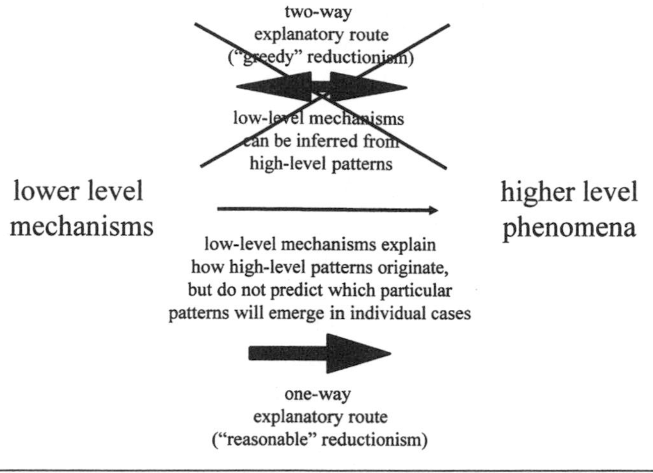

Figure 2. Dupré's concept of proper and improper uses of reductionism. *Top:* "Greedy" reductionism assumes that somehow it is possible not only to explain higher-level phenomena based on lower-level mechanisms, but also to follow the causal arrow backward and infer the mechanisms that produced given higher-level patterns. This kind of reductionism will not work for sufficiently complex systems, such as those of interest to evolutionary biologists. *Bottom:* A more reasonable form of reductionism assumes that lower-level phenomena can explain *how* certain higher-level patterns are generated, but they cannot predict *what* specific patterns will be observed because so many causal factors are interacting. A fortiori, reasonable reductionism denies the possibility of inferring mechanisms directly from patterns because of the many-to-many relationship between the two.

the details of the lower-level processes from the higher-level patterns produced (something we have just seen is highly unlikely in the case of any complex biological phenomenon because of Shipley's "statistical shadow" effect). This form of "greedy" reductionism, diagrammed in the upper part of figure 2, is bound to fail in most (though not all) cases for two reasons. The first is that the relationships between levels of manifestation of reality (e.g., genetic machinery vs. development, or population genetics vs. evolutionary pathways) are many-to-many (again, as pointed out above in our discussion of the shadow theater). The second is the genuine existence of "emergent properties" (i.e., properties of higher-level phenomena that arise from the nonadditive interaction among lower-level processes). It is, for example, currently impossible to predict the physicochemical properties of water from the simple properties of individual atoms of hydrogen and oxygen, or, for that matter, from the properties of H_2O molecules and the smattering of necessary impurities.

A second, more reasonable, form of reductionism, which is presented in the lower part of figure 2, essentially states that lower-level phenomena can be effectively used to explain, in broad outline, *how* higher-level outcomes materialize, but cannot usually be pushed to the point of making accurate quantitative predictions about specific instances of higher-level phenomena—that is, to predict *what* actually happens in individual instances. A fortiori, one cannot reason backward from the higher-level patterns and infer the specific underlying causal processes. For example, modern genetics and molecular biology provide a convincing (if as yet incomplete, as we shall see) account of how development can take place in living organisms, but even with complete knowledge of the genome, one could not predict developmental outcomes in specific human beings. Analogously, population genetics theory affords us a general framework for beginning to understand how genotypic frequencies change in populations during the course of evolution, but it is not generally sufficient for making specific predictions about real populations whose evolution is simultaneously affected by a variety of interacting processes. As a side note, it is irrelevant whether these limitations of reductionism are epistemic (i.e., whether they depend on our inability to obtain or process all the necessary information—most people would accept this, we suspect) or ontological (i.e., whether they are the result of fundamental properties of the world as it is, as Dupré suggests), because in either case practicing scientists will not be able to get around them.

The distinction between "greedy" reductionism (in which lower levels are sufficient to explain and predict anything we might wish to know about higher levels) and the more reasonable explanatory reductionism (in which explanations in terms of the features of lower levels are often necessary to explain how the properties observed at higher levels are possible) is somewhat independent of a third strand of reductionism; namely, methodological reductionism. Methodological reductionism is not a thesis about what kinds of explanations or predictions of higher-level phenomena the features of lower levels will tend to produce; rather, it is a thesis about what kinds of methods will tend to make us most successful in understanding complex systems. Methodological reductionism, roughly, claims that when one wishes to understand a complex system, attempting to understand the parts of that system and the interactions among those parts is generally a good (and often the best) place to start. There are very few (if any) contemporary philosophers *or* scientists who would object to this form of reductionism, and indeed, some would claim that, historically, the acceptance of methodological reductionism was part of what became known as the scientific revolution.

Finally, we get to the crucial point of what all this means in terms of evolutionary quantitative genetics, the theory that takes the worst beating throughout the rest of this book. We think there are three different roles that quantitative genetics theory can play within the broader context

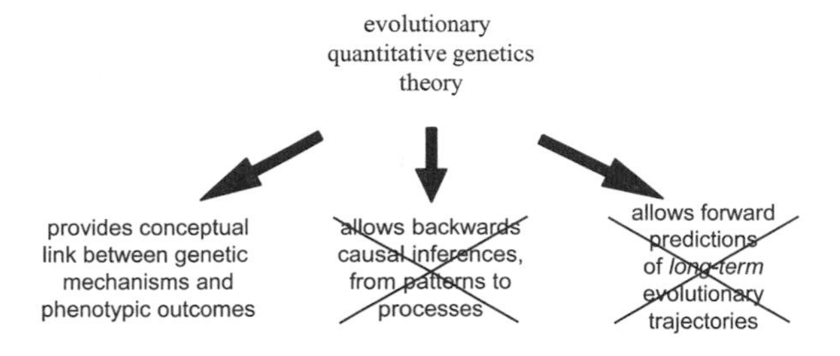

Figure 3. How we see the proper and improper role of quantitative genetics theory throughout this book, as a direct consequence of Shipley's "shadow theater" metaphor and of Dupré's ideas on the proper role of reductionism.

of evolutionary theory: one that is legitimate, and two others that represent conceptual and empirical dead ends. Figure 3 summarizes these roles. First, one can conceive of quantitative genetics theory as providing a conceptual link between Mendelian and molecular genetics on the one hand and the evolution of complex phenotypes on the other; in this sense, the theory tells us how the elements of the lower levels of explanation (chiefly genes and their direct products in individual organisms) can causally join together to produce the types of patterns that are observed at higher levels (evolving phenotypes in populations). Alternatively (though not mutually exclusively), one can see quantitative genetics as providing the conceptual tools to engage in backward causal inference—that is, to infer the actual genetic architecture (the number of genes and their interactions) underlying specific phenotypes. Finally (and again, not exclusively of the other two alternatives), one can think of the quantitative genetics framework as affording researchers the possibility of making quantitative predictions on the (long-term) evolutionary future of a particular population (it is obvious that very short-term predictions are possible, since they have been used for a century in plant and animal breeding; the really interesting question hinges on whether the conceptual and statistical tools of quantitative genetics can be extrapolated over evolutionarily significant time frames).

It should be clear from our preceding discussion of Shipley's and Dupré's ideas that we think that the first use of quantitative genetics theory is perfectly viable, but that the second and the third are not. Hence, we do see a proper—indeed, a crucial—role for quantitative genetics theory in modern evolutionary biology. Moreover, this role extends to empirical practice, because the tools available to quantitative geneticists (which we will discuss in due course) for quantifying natural selection and constraints are necessary starting points for theorizing about causal mechanisms. Indeed, these tools are the very same ones that provide researchers with the all-important "statistical shadow" to be used as a reference during the quest for causal understanding. But we also see troubling signs in much of the recent literature of language consistent with the second and third uses of quantitative genetics theory. It is on these uses that our criticism will focus.

CONCEPTUAL ANALYSIS, PHILOSOPHICALLY INFORMED CRITICISM, AND THE PHILOSOPHY OF SCIENCE

The approach we adopt in this book puts our work firmly in the tradition of philosophy of science that attempts to produce a particular kind of "science criticism"; that is, a critical meta-evaluation of the concepts employed, the claims made, and the methodologies adopted within specific fields of science. Our work is therefore meant to be *prescriptive* in that, unlike descriptive work, it is concerned not only with how terms are used, but also with how they ought to be used; we discuss not just how things are done, but how (we think) they should be done.

Our conceptual analysis of how scientists think and talk about their subject matter is the less controversial aspect of this project. Conceptual analysis has a long tradition in philosophy and has probably produced the bulk of the recent literature in the field. Among the examples that come to mind in the philosophy of biology are investigations into the nature of natural selection (e.g., Sober 1984; Waters 1991; Matthen and Ariew 2002; Millstein 2002), the use of the concept of causality in science (e.g., Sober and Lewontin 1982; Rieppel 1990; Pigliucci 2003c), and the never-ending discussions on what biological species are (Pigliucci 2003d; see also chapter 9 and references therein).

Examples of philosophical work critical of particular areas of science tend to be more controversial, but are hardly rare. Here, the focus is on critiques of particular areas of research, particular research projects, or even individual studies. Examples include the rigorous criticism of sociobiology and, more recently, evolutionary psychology that aims to show that there is too often a serious disconnect in these fields between the claims made and the empirical evidence adduced to back up those claims (see chapter 7 and, e.g., Lewontin 1998; Kaplan 2002; Buller 2005). These critical projects can have a decidedly "applied" component. Most obviously, of course, those people engaged in the critical projects are recommending a change of practices to the scientists involved, and those changes—if implemented—will have practical implications for funding and publishing research. Moreover, since historically some sloppy scientific research has had profound social and political implications (think of the successes of the eugenics movement in the United States during the first part of the twentieth century), these critical projects will tend to have social and political consequences as well. For example, talk of genes "for" certain traits (see chapter 6 and references therein) can lead to heated debates affecting social policy and societal attitudes.

Science criticism can be irritating for the scientist, and indeed, some scientists are even annoyed by descriptive philosophical projects focused on the sciences. For example, physicist Steven Weinberg (1992) has argued, in an essay provocatively titled "Against Philosophy," that philosophers are not only useless to science, but are at times positively harmful. Weinberg maintains that philosophy has never solved a scientific problem, and that furthermore, some philosophical schools of thought have temporarily hindered the progress of science. His favorite example is the alleged slow acceptance of quantum mechanical theory in the early part of the twentieth century, which he attributes at least in part to the fact that some scientists were buying into the logical positivists' contention that science should never make use of unobservable quantities (of which quantum mechanical theory is full).

While there are many avenues available to question and criticize Weinberg's examples, the more fundamental problem seems to us to be that it is strange to expect philosophers to "solve" scientific problems. It seems equally wrong to assume that only projects that solve particular scientific problems are useful to scientists. Rather, there are a number of valuable roles that science criticism can play, even if scientists sometimes find that criticism frustrating. If properly carried out, we believe, critical analyses of the concepts employed, the methodologies used, and the evidence arrayed in support of particular claims can actually benefit both science itself (by, for example, keeping scientists from making claims that find little justification in their science, and by pointing toward different possible avenues of research) and society at large (by providing a system of checks on the increasingly prominent and powerful role of science in people's lives). The point here is not to see whether philosophers are better than scientists at pursuing particular scientific research programs, and hence at solving scientific problems. Rather, it is to put the essence of philosophical inquiry—the ability to conduct rigorous analyses of the meaning of concepts, of how various concepts relate to one another, and of how they are applied—at the service of scientists (and society at large) in order to help them cut through the conceptual fog and get to questions that are actually answerable on empirical grounds (or, sometimes, to show why this is not feasible).

One important caveat needs to be added here: Some scientists have interpreted critical descriptions of science or criticism of particular research projects (see, e.g., Latour and Woolgar 1979; Pickering 1984; Collins and Pinch 1998) as claims that there is no objective reality out there, that scientists just "make stuff up" and then defend their claims for political reasons. This interpretation of science studies has led to what have come to be known as the "science wars" (see Hacking 1999 for a particularly lucid discussion). To be clear, we do not support any of the radically relativist positions sometimes attributed to those people engaged in science studies. However one thinks of the notion of an objective reality, we presume here both that science is the best tool available to us for investigating the nature of that reality, and that doubts about the validity of research programs are best addressed by particular critiques of the techniques used and the data gathered—that is, by competing research programs rather than vague epistemological worries (see Clough 2003).

TOWARD A MORE COHERENT PICTURE OF EVOLUTIONARY THEORY: AN OVERVIEW

This volume, then, is a combination of conceptual analysis and science criticism applied to a specific set of questions that emerge from particular areas of evolutionary biology. Our focus is on areas in which certain ways of conceptualizing the fields in question influence, and are influenced by, particular kinds of methodological practices; that is, on the way a given conceptual understanding informs lines of inquiry in biology. Often, our claim will be that conceptual mistakes—views that are, strictly speaking, incoherent or in conflict with other, more reasonable interpretations—have resulted in the deployment of methods that are unlikely to yield satisfying results, and that refocusing on the underlying conceptual issues can help draw attention to what are likely to be more profitable avenues of research. Let us be clear that we do not mean to suggest that biologists are particularly lazy or sloppy in their thinking. We all have to take certain things for granted in order to get the work done. However, assumptions and metaphors can become internalized habits of thought to which we no longer pay conscious attention. It is our aim to identify and discuss some of these habits and their consequences, present more coherent interpretations of some of the basic background assumptions, and thereby direct attention toward potentially more fruitful alternative methodologies. Many of these alternatives are already present in the biological literature, but have been treated as minority or exceptional views. Our goal is not so much to develop new methods as to situate underutilized methods and underrepresented views in a more compelling overall picture of evolutionary biology.

Our analysis focuses on several interrelated areas of concern to organismal biologists, and ranges from the treatment of problems with some very general aspects of the theory of evolution to very specific examples illustrating what we think reasonable solutions to some of those general problems might look like. We start the book with a focus on natural selection and fitness, two of the most important concepts in evolutionary biology. We argue that a critical examination of some recent ideas on the nature of natural selection (chapter 1) should affect our understanding of what natural selection is and how it can (and cannot) be measured (chapter 2), as well as the debate over what it is that is "selected" by natural selection (genes, individuals, groups, etc.: chapter 3). The next several chapters are focused on the idea of "adaptation" and the relationship of our understanding of that concept to how we think about what constitutes "natural selection," what (if anything) counts as a "constraint" on adaptive evolution, and how adaptive hypotheses can (and cannot) be tested (chapters 4–7). This leads us naturally into questions regarding the role these reconceptualizations have for our view of macroevolution, speciation, and the nature of species (chapters 8 and 9). We end with a more general discussion of the use of null models and hypotheses in biology (a theme that recurs in several of the previous chapters) and some suggestions for alternative ways of thinking about experimental tests and presenting the results of empirical research.

We choose to focus on these issues because they are all still current and active areas of study in biology and have already proved to be superb examples of some very fruitful cross-pollination between biologists and philosophers. There are obviously other issues that are equally ripe for the sort of analysis we provide, and the issues we do discuss have additional implications. Even those issues we address could not have been treated in anything like an exhaustive manner within the scope of a reasonably sized book. In any case, our goal is not to provide the final answer to the questions we address; rather, we hope that both philosophers and scientists will find plenty of food for thought in this book and that it will encourage more conversations about the kinds of issues we address. If so, we will have accomplished a large part of our goal.

1 Natural Selection and Fitness

After the "Force" Metaphor

I HAVE CALLED THIS PRINCIPLE, BY WHICH EACH SLIGHT VARIATION, IF USEFUL, IS PRESERVED, BY THE TERM OF NATURAL SELECTION.
—CHARLES DARWIN, *ON THE ORIGIN OF SPECIES*

Natural selection is often conceived of as a "force" that acts on populations through the differences in relative fitness among individuals within each population. Recent work by such authors as Matthen and Ariew points out difficulties with the force metaphor and the attendant concept of fitness. This is far from being a trivial semantic quibble; the way in which one conceives of natural selection and fitness strongly influences what one will take to be evidence for natural selection having occurred, and how one thinks of the difference between drift and selection. Following other authors, we suggest that one must distinguish between informal and formal fitness, and between informal and formal selection. Doing so not only makes clear why the force metaphor is misleading, but also has implications for our ability to distinguish selection from drift, for measurements of selection in natural populations (chapter 2), for the levels of selection debate (chapter 3), for our conception of constraints on evolutionary change (chapter 4), and for the way that the metaphor of adaptive landscapes is used (chapter 8).

WHAT IS NATURAL SELECTION?

When Darwin (1859, 80ff) coined the term "natural selection," he did so in hopes that an appeal to a process that was well known—artificial selection—would make the process to which he was referring clearer. Just as animal or plant breeders can change the distribution of traits in a population by selectively breeding those animals or plants with traits deemed desirable, Darwin argued, the distribution of traits in natural populations of organisms changes over time as those organisms with traits better fit to the environment succeed in reproducing at a higher rate than those with traits less well fit, at least where the differential fit between organism and environment is shared between parents and offspring (box 1.1). It is the increasing "fit" between the organisms and the environment in popula-

Box 1.1 Natural Selection and Evolution by Natural Selection

In order for evolution by natural selection to occur, it is generally agreed that the following three conditions must be met:

1. Organisms in a population differ from one another with respect to some trait; say, *a*.
2. Differences in that trait, *a*, are associated with differences in the organisms' reproductive success (average number of offspring).
3. The differences in that trait, *a*, are *heritable*–offspring are, on average, more like their parents than they are like the population average with respect to trait *a*.

Conditions 1 and 2 are necessary for natural selection to occur. Condition 3 is necessary for natural selection to have evolutionary consequences.

tions that explains "the extraordinarily complex and intricate organization of living things" (Bell 1997). Darwin reasoned that just as, over time, artificial selection can produce radically different breeds from the same ancestral population, natural selection, given the enormously greater amount of time available, can produce different species, adapted to different local conditions or to different ways of living, from the same ancestral populations. Natural selection, then, is the explanatory principle accounting for the fit between organism and environment, as well as being an integral part of the explanation of the variety of organisms.

Still, there are important dis-analogies between artificial and natural selection. Artificial selection is undeniably a process, the result of which may be change in a population (given certain assumptions, such as the heritability of the trait in question); there is someone doing the selecting, and it is this that permits us to say unequivocally that selection is taking place. Contrast this with Futuyma's (1998) definition of natural selection as "*any consistent difference in fitness* (i.e., survival and reproduction) *among phenotypically different biological entities*" (349, emphasis in original). Here it seems as if natural selection is not a process, but rather a fact about the population in question—that is, the existence of an average difference in fitness between certain entities. Of course, Futuyma is quick to point out that natural selection can be decoupled from evolution, both because it is possible for a population to evolve through nonselective mechanisms, and because natural selection can occur without evolution (if, for example, the variation in the traits under selection is not heritable) (1998, 365). But the idea that natural selection is a fact about populations still sits awkwardly with Futuyma's claim that natural selection "is one agent of change in the pattern of variation" in populations, and hence one of the causes of evolution (365); *facts* about populations and *agents acting* on populations are generally conceived of as two different things.

Sober's (1984) detailed development of a force metaphor for evolutionary change is one way of trying to make the view of natural selection as a process cohere with the view that takes it to be a fact about populations. If fitness is defined as average differences in reproductive success among members of a population, then, Sober suggests, natural selection can be viewed as the force resulting from those fitness differences (1984, 46). This "force" is then conceived of as one of the causes of evolution; it is the only such force that can result in the evolution of adaptations. However, as evolution in this view is conceived of as heritable changes in the composition of a population,[1] other forces need to be evoked to explain evolution more generally. For example, since genetic drift can change the frequencies of alleles within a population, it must be a force as well, as must, say, migration events, genetic linkage, and mutations (Sober 1984, 38). The actual changes in a population are the result of the combination of these forces, just as the actual changes in the velocity of physical objects are the result of the combination of forces in Newtonian physics (1984, 50). This talk of combining forces in a Newtonian way is not supposed to be just metaphorical; Sober and Lewontin (1982) state that "[s]election for *X*, and against *Y*, and so on, are component forces that combine vectorially to determine the dynamics of the population" (160) (fig. 1.1).

This interpretation of population changes being the result of various forces also fits in very well with one of the standard visual metaphors used to understand the relationship between fitness and evolution; namely, the fitness landscape.[2] It is natural to think that in order to explain the movement of a population on the landscape, one must appeal to forces pushing it in various directions; selection pushes the population up "hills" of higher fitness, drift pushes it randomly, and so forth (fig. 1.2). In the absence of any force, the population is assumed to remain in place on the landscape.

The force metaphor, however, is misleading in important ways. It is not just that the metaphor is not perfect (something everyone agrees with), nor is it just that there are ways in which the comparison between "forces" in evolutionary biology and Newtonian forces breaks down. If the only places where the comparison failed were not significant to understanding selection, nor to empirical work in evolutionary biology, and

1. It is often assumed that the only variation that is heritable is variation that is associated with genetic differences. However, as we argue in chapter 3, this position is empirically inadequate, as we now know of several different ways in which phenotypic variation can be reliably inherited through nongenetic pathways (see, e.g., Sollars et al. 2003; Oyama, Griffiths, and Gray 2001; Jablonka and Lamb 2005).

2. See chapter 8 for further discussion of the problematic nature of fitness landscapes as they are usually conceived.

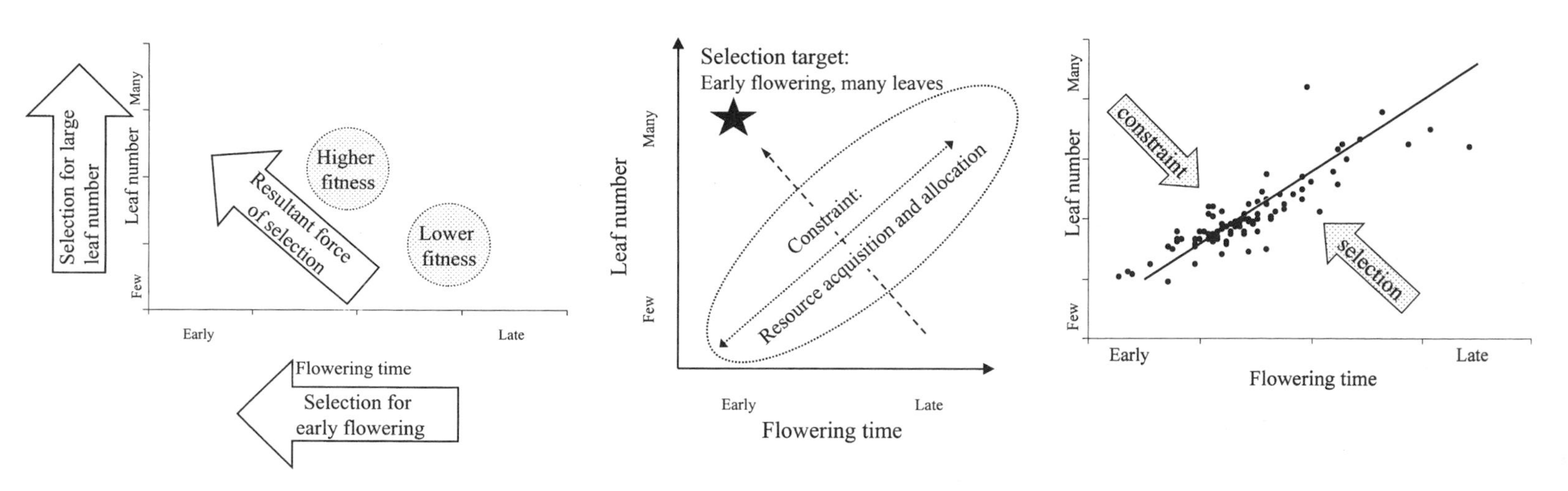

Figure 1.1. The force metaphor in action. The following kind of analysis is generally what is meant when natural selection is conceived of as a force. (a) In the plant *Arabidopsis thaliana*, there is directional selection both for early flowering and for large leaf number at flowering. The result of these two selective "forces" interacting is a net selective force for a combination of both large leaf number and early flowering time. But, while for any particular leaf number selection favors earlier flowering, and for any particular flowering time selection favors larger leaf numbers, there is a limit on how many resources *A. thaliana* can acquire and devote to growth and reproduction. In other words, both leaf number and flowering time are constrained by the ability of the plant to gather and partition the resources necessary for growing leaves and for flowering. This "constraint" directs selection along some particular paths rather than others, and results in another "force." (b) The "selection target" here is the "optimum" of both early flowering and high leaf number at flowering; selection for this target is the result of the net "force" from panel a. But the evolution of the two traits is constrained by fundamental limits on resource acquisition and allocation; this is a "force" that channels phenotypic development along particular directions (the double-headed arrow). (c) The resulting force from panel a (directional selection both for early flowering and for large leaf number at flowering) and the constraint from panel b (limits on resource acquisition and allocation) combine to result in a strong correlation between flowering time and leaf number.

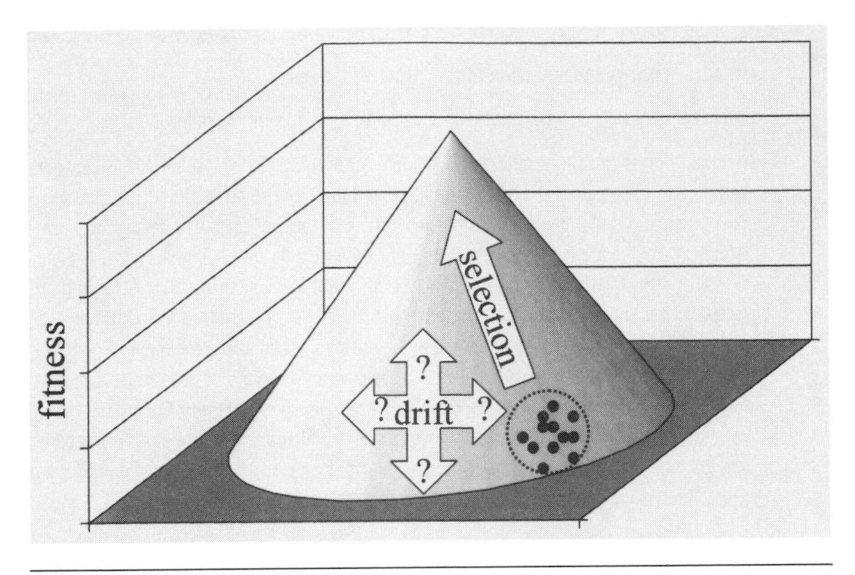

Figure 1.2. The fitness landscape and the force metaphor. A population near a fitness peak is pushed "up" the peak by the "force" of selection, and pushed in random directions by the "force" of drift.

if the comparison were intellectually fruitful where it did not break down, then the failures of the metaphor would not matter. We argue here that this is not the case; rather, the places where the metaphor fails and where the comparisons break down are conceptually important, and the study of selection and evolution is hampered by the (often implicit) use of the force metaphor. Further, even where the metaphor is not obviously inadequate, other ways of looking at the relationship between fitness, natural selection, and evolution are able to do all the work done by the force metaphor without engendering the conceptual problems it does. The force metaphor simply is not more "fruitful" than other ways of understanding selection and fitness. Moreover, carefully distinguishing different uses of "selection" and "fitness" permits one to avoid the force metaphor altogether, while simultaneously pointing toward new (and more fecund) solutions to philosophical problems as well as more productive avenues of empirical research in biology.

TWO WAYS OF THINKING ABOUT FITNESS AND NATURAL SELECTION, REVISITED

Natural selection is supposed to result whenever there are heritable differences in fitness between biological entities in a population (see box 1.1). Matthen and Ariew (2002) make an important distinction between informal fitness (box 1.2) and predictive (or formal) fitness. In their charac-

Box 1.2 Fitness and Selection in the "Informal" Sense

We use the terms "informal fitness" and "informal selection" in the following way:

Informal fitness (also referred to as individual fitness) is about the relationship between variation in some particular trait (or small set of traits) and some particular physical process (or small set of processes). Organisms differ in informal fitness with respect to a particular trait and a particular process if that process interacts with the organisms' different versions of the trait such that organisms with different versions of the trait differ in their expected reproductive success because of that process. A physical process that interacts with different versions of a trait such that organisms with the different versions differ in their expected reproductive success may be said to be *discriminate* with respect to that trait. A particular process can be discriminate with respect to some traits, but not to others.

We restrict the term *vernacular* fitness to the overall propensity of a particular organism to be reproductively successful; vernacular fitness, we argue, is epistemologically inaccessible, and of no significance to evolutionary theory more generally.

Informal selection is what takes place when a particular physical process interacts with the variation in a particular trait in a "discriminate" way. However, because of the stochastic nature of reproductive success or failure through discriminate processes, even informal selection is at heart a statistical, rather than an individual, phenomenon.

terization, *vernacular* fitness refers to a biological entity's overall relative ability to survive and reproduce; it is thus linked to the idea of there being a "fit" between the organism and the environment. In this view, those organisms in a population that are, overall, better fit to the conditions of their existence may be expected to do better (be more reproductively successful), on average, than those organisms that are less well fit to those conditions (see also Ariew and Lewontin 2004). In this interpretation, the "fit" between the organism and the environment emerges from the continuous interactions between the two; that is, particular organisms interact with particular aspects of the environment in particular ways, and those organisms better suited to succeed in their interactions with those aspects of the environment are better fit to that environmental regime.

This sense of vernacular fitness can play little or no role in actual research programs because it is difficult or impossible to determine the (overall) vernacular fitness of an organism. There are too many environmental variables and too many traits, and they interact in ways that are too complex. This problem is not just epistemic; the vernacular fitness of an organism is supposed to be about how well it is fit to its environment, not about how it actually fares. That is, vernacular fitness is about the *propensity* of the organism to succeed in environments like the ones it finds itself in. The *actual* reproductive success of the organism, which emerges from its interaction with the environment in all its details, does not tell us how fit the organism is in the vernacular sense, because random factors that do not reflect the organism-environment fit also contribute to reproductive success (e.g., an individual may be potentially very fit given its typical environment, but extremely unlucky in being hit by lightning). But neither is it obvious how to interpret the idea of the organism's *average* reproductive success in relevantly similar environments: what is to count as an environment that is "relevantly similar" for the purpose of evaluating vernacular fitness? If we demand that many of the details be maintained, the idea of fitness is in danger of collapsing back into actual reproductive success; if we permit too coarse an analysis, on the other hand, we will miss too much of what makes one organism in a population more or less fit than another. Nor is the overall fitness of individuals particularly valuable for thinking about changes in populations over time—the features of the population that change over time are not obtainable in any obvious way by summing over the reproductive success of individual organisms; instead, they must be obtained by comparing the success of individuals with and without particular variants of particular traits (or clusters of traits).

In a biologically more realistic, but less satisfying, fashion, one might think instead about an entity's relative ability to survive and reproduce given some particular trait (or some combination of a relatively small number of traits) and some set of physical processes that interact with the trait in question. One organism can be more fit than another in this, which we dub the *informal* sense of fitness, if the physical processes that interact with the traits of interest are *discriminate* with respect to the variations in those traits; that is, if organisms with one variant of a trait are more likely to survive and reproduce than are organisms with another variant of that trait, given the interaction of a particular physical process with the trait. If, on the other hand, the processes that interact with the trait in question are *indiscriminate* with respect to variation in that trait, there will be no informal fitness differences between the organisms when those processes and traits are considered. Notice that this way of thinking about fitness does not take into account how frequently organisms interact with the physical processes in question; it fails to take into account any differences in the mean frequencies with which organisms with different variants of the trait might interact with the relevant physical processes. This version of informal fitness is only about the expected success of an organism given a particular relationship between a trait and some particular physical process. It is focused on the "individual" nature of fitness, but does not address the likely reproductive success of the organism overall. In this way, this version of informal (or individual) fitness differs from the usual vernacular view of fitness, and is much more tractable than is the vernacular version.

In order to make clear the sense in which this informal version of fitness is tractable, we need to be clear about what "discriminate" and "indiscriminate" mean in this context. If, given a particular kind of physical process the organisms will face, the variations in a trait among those organisms make a difference to their abilities to survive and reproduce, the process is discriminate with respect to that trait, and there is an informal fitness difference between the organisms (with respect to that process and that trait). On the other hand, if the variations in the trait do not make a difference given the process, there will be no difference in the informal fitness of the variants. Note that whether a trait is relevant to informal fitness differences depends on the physical processes that trait is interacting with. For example, with respect to death by predation, running speed may be a trait in which differences result in differential success; however, with respect to death by lightning strikes, running speed is very likely not a trait in which variations matter to survival and reproduction. So, in this example, with respect to running speed, predation is a discriminate physical process, whereas lightning is an indiscriminate process. Similarly, mitosis is a process that is usually indiscriminate with respect to particular alleles; however, in the case of segregation distortion (discussed in chapter 3), it can be discriminate with respect to different alleles.

This much more limited conception of individual fitness—the fit between a given trait and a physical process—suggests that the idea of an organism's *overall* fitness (its *vernacular* fitness in the above sense; its overall propensity for reproductive success) could be given by something like the net result of the interactions of all the discriminate and indiscriminate processes with one another and with all of the organism's traits, given something akin to the average expected environment for that organism. But again, there are good reasons to reject the notion of "overall fitness" in this context. Consider, for example, how difficult it is to make sense of the idea of an average expected environment for an organism, especially when we consider that the environments faced by organisms with different variants of a trait may well be different because of that very same variation (as in the case of "niche-constructing" traits). Even if this idea of overall fitness could be made conceptually coherent, it is certainly intractable in practice, and can play little or no part in our understanding of either the actual or the mean population dynamics in any real instance. That is, the concept of an overall (vernacular) fitness can play no role in either our ability to understand what has actually transpired in some particular population or our ability to understand what is likely to transpire in populations of that kind.

This account of informal fitness in terms of particular traits and physical processes leads to another epistemic difficulty; namely, how it can be determined whether a physical process is discriminate or indiscriminate with respect to some particular trait. In general, of course, physical processes like those being considered here are probably stochastic with respect to their influence on organisms. For example, whether an organism succeeds in fleeing from a predator at any given time no doubt depends on a vast number of variables, and even if the outcome of each encounter is determinate given the antecedent conditions,[3] there is no reason to think that observing individual encounters of this kind would permit us to make accurate predictions about outcomes of relevantly similar (but not identical) encounters. Just because a process is discriminate with respect to a particular trait, there is no guarantee that any particular interaction of that process with that trait will reflect the general propensity of that process vis-à-vis that trait. So, just because organisms with a particular variant of some trait in a particular (finite) population succeed in fleeing predators more often than those with a different variant, that does not guarantee that they are, in the informal sense, fitter (and still less, of course, that they would be fitter in the overall, vernacular, sense). Put another way, since whether or not a given process is discriminate with respect to some trait is itself a statistical matter, the finite size of real populations would seem to make determining with certainty the status of given processes vis-à-vis particular traits impossible.

Insofar as one hopes to derive informal fitness from some kind of straightforward census of a population and the physical processes that interact with the traits of each member of the population, this might be a serious problem. But of course evolutionary biology is not primarily about such counting procedures; insofar as we are trying to uncover actual casual processes (rather than engaged in mathematical modeling), ecologically informed fieldwork will be a necessary part of our work. Observations of the interactions of processes and traits combined with careful "engineering" analyses[4] will, we suspect, permit relatively sound judg-

3. This is, of course, itself a metaphysical assumption we might well question (see, e.g., Cartwright 1983 and Dupré 1993).

4. An engineering analysis in this sense is simply an analysis of the way the trait under consideration is likely to interact with particular kinds of physical processes (for instance, escaping predators); such an analysis can be as sophisticated as careful optimization modeling designed to yield quantitative predictions (see, e.g., Seger and Stubblefield 1996), or it can be a less formal analysis, the aim of which is more qualitative (see, e.g., Lauder 1996).

ments to be made regarding which processes are discriminate and which are indiscriminate with respect to certain traits. This kind of biologically informed common sense is difficult to formalize but easy to recognize, and its importance to practicing biologists should not be underestimated.

A similar point may be made regarding the identification of traits and processes. There is nothing trivial about the question "What constitutes a trait?"; however, the problem of identifying traits is not wholly intractable, nor are we completely lacking criteria by which to judge whether the purported traits identified are likely to be biologically meaningful (see Wagner 2001). While we may lack a straightforward algorithm for determining what counts as a trait, the task is not thereby rendered impossible, nor are the traits we pick out to study entirely arbitrary. Similarly, while there is no formal algorithm available that will permit us to say what physical processes are of biological significance, or whether those processes are discriminate or indiscriminate with respect to particular traits, the task of determining these kinds of facts is not impossible, nor are the answers we derive arbitrary.

At the informal level, one is tempted to think of natural selection as a process akin to artificial selection (following Darwin's original intuition): discriminate physical processes tend to "select" those organisms with the more (informally) fit traits. In this conception of informal natural selection, there are a number of ways in which it can fail to have evolutionary consequences. The discriminate physical processes selecting on the basis of one trait variant may be opposed by other discriminate physical processes selecting on the basis of another trait variant; the two processes may then balance each other. More trivially, the variation in the trait in question may not be heritable; under these conditions, of course, while the trait may be selected in the informal sense, the composition of the population, vis-à-vis that trait, will not be altered. In either case, while informal selection is occurring, the structure of the population is not changing, and hence evolution is not taking place.

This interpretation of informal (a.k.a. individual) natural selection and fitness (see box 1.2) can be usefully contrasted with what Matthen and Ariew (2002) refer to as formal (or *predictive*) fitness and selection (box 1.3). Predictive fitness and formal selection are statistical facts about the average reproductive success achieved by organisms (or biological entities more generally) possessing one variant of a trait rather than another, all things considered. These are not facts about actual outcomes, nor, a fortiori, about the particular physical processes that led to the actual outcomes. In this view, the predictive fitness of some variant of a trait is the *expected* mean ratio of the growth rate (or a similar measure of relative suc-

Box 1.3 Fitness and Selection in the Formal Sense

We use the terms "formal fitness" and "formal selection" in the following way:

Formal fitness (also referred to as predictive fitness) can be defined as the difference from one in the mean ratio of the growth rates of parts of a population divided on the basis of variation in some trait. Formal (predictive) fitness is a statistical property of populations. When a population can be divided into a number of groups based on variations among the members of the population with respect to some trait, the average growth rates of these groups can be compared. When the ratio of the growth rates of two groups is not equal to one, there is a difference in formal fitness, and, if the trait is heritable, one is justified in expecting an average change in the makeup of the population. It is in this sense that formal fitness is predictive. Note, however, that formal fitness is defined as mean change in populations; therefore, without additional evidence, we cannot legitimately expect the behavior of any actual population to reflect this average. Rather, we would need a number of replicates of the population in question to generate a reasonable estimate of the average through experimental means.

Formal selection is the distribution of the ratios of growth rates of "trait populations" picked out by differences in trait variants within the ensemble; it describes the way in which particular kinds of outcomes are statistically distributed around the mean outcome of growth rates of particular trait populations for populations of that type.

cess) of the portion of the population sharing that variant to that of the portion of the population lacking that variant. Let us call the segments of the population so apportioned (by virtue of sharing a trait variant) "trait populations." For traits for which the mean ratio of the growth rates of the trait populations is one, there is no fitness difference in this formal sense; for traits for which the mean ratio of the growth rates of the trait populations is different from one, there is a fitness difference.

It is important to note that the notion of an average difference is being used in a nested sense here. On the one hand, in a particular population divided on the basis of some trait, the actual reproductive success of the trait populations so identified will be the result of summing over the individual entities with and without the trait variant in question. In that sense, the ratio of the growth rates of the trait populations emerges from an averaging of the actual reproductive success of entities with and without the given traits. But predictive fitness is not about the actual outcome within a single population; rather, it is about the (statistically) expected outcome given a particular kind of population interacting with a particular kind of environmental context. For this reason, formal fitness and selection are about features of *ensembles* of populations—collections of relevantly similar populations in relevantly similar environments.

To derive the predictive fitness of trait variants in a population, then, requires that one consider the likely distribution of ratios of growth rates of the trait populations so identified; that is, one is required to consider the distribution of outcomes of entire populations. This is easiest to understand in the case in which one has access to multiple replicates of a population. For example, in the case of bacteria, it is relatively easy to set up replicates of a population and determine which variants do best; this is logistically feasible in the case of some more complex and longer-lived organisms as well. Even where one cannot determine the distribution of population-level outcomes for epistemic reasons, however, it is the mean difference in the reproductive success of the populations so picked out that is what is meant by predictive (formal) fitness. At the formal or predictive level, natural selection is a fact about the statistical distribution of outcomes surrounding the differences in mean growth rates of the trait populations (see box 1.3).

In this formal view, natural selection has little in common with artificial selection. Since the formal level is concerned about the mean differences in fitness of trait variants in collections of similar populations, it ought to be clear that there is nothing doing the selecting at that level. Fitness differences in this formal sense are statistical facts about ensembles of relevantly similar populations; while, given these facts, particular patterns in the success or failure of individual trait populations may be predicted, the facts themselves do not necessarily reflect causal pathways. Differences in formal selection are about the expected patterns in ensembles of populations; because many different causal processes can result in the same kinds of patterns, the patterns do not reliably reflect the underlying causal structures (see Shipley 2000). For example, fitness differences at the predictive level may depend on any of a number of traits that are statistically correlated, only one (or some) of which are causally implicated in fitness and selection at the informal level.

Traditionally, differences in (overall) vernacular fitness have been considered to be the cause of differences in predictive fitness, but as Matthen and Ariew (2002) note, that position glosses over too many of the difficulties in moving from events at the individual level to the statistical trends that the predictive level is concerned with. The most obvious of these difficulties is epistemic: the interaction of physical processes with the traits of individual organisms is, naturally, extraordinarily complex, and determining how, on average, subpopulations divided by the traits of interest will respond is an intractable problem.

Even if the vernacular fitness of an entity were conceptually coherent and estimable in practice, it would still (in general) tell us little about what would happen to other entities that happen to share some (but not all) traits with that entity. We would, therefore, be unable to predict the average success of organisms with one variant of a trait rather than another, and similarly unable to predict the mean changes in trait (including allele) frequencies. The same would hold under most conditions even if we had access to the vernacular fitness of every organism in the population. The predictive fitness of particular traits (including alleles) cannot be derived from vernacular fitness measures of the idiosyncratic collections of traits that constitute individual organisms.[5]

While the informal fitness of particular variants of traits vis-à-vis certain physical processes may be accessible, in order to move to predictive fitness, we would need to be able to find the average informal fitness of those variants of the traits, given all the physical processes they interact with, and we would need to know the average frequency with which organisms with those variants interacted with those physical processes throughout their life cycles. Again, given the number of potentially interacting variables and the finite size of real populations, it might be impossible to generate the information necessary to move from observations at the individual level to the statistical facts that make up the predictive level. This problem is especially acute given that predictive fitness (and formal natural selection, the statistical distribution of outcomes that emerges at the ensemble level) is not about particular populations, but rather about the average differences that emerge from collections of relevantly similar populations; facts about what happens in particular individual cases will not permit one to make generalizations about the statistical trends that will emerge in ensembles of cases.[6]

The stochastic nature of many of the physical processes that interact with individual organisms make it impossible to predict, given a particular set of antecedent conditions, what the final outcome will be at the level of any individual population. The realized relative growth rates of trait populations of any particular population may or may not reflect the predictive fitness differences of those trait variants, for example, because relatively many of the organisms with the less fit (in the predictive or formal

5. This point is related to an objection Dupré (1993) levies against Salmon's statistical relevance model of explanation. Dupré argues that for many of the cases we are interested in, the number of causally relevant interacting factors is high enough that most of the classes of relevantly similar cases will have zero members (they will be empty), and even those that are not empty will generally have only one member (or at most a very few members). Under such conditions, he suggests, our ability to identify the effects of particular sets of causal factors using Salmon's approach will be, at best, severely limited.

6. Of course, for some specific instances, many of these problems will be tractable, and for a few, they will not emerge as problems at all—consider, for example, the case in which one variant of the trait in question is always lethal. But these kinds of cases are relatively rare and do not represent the most general form of the problem.

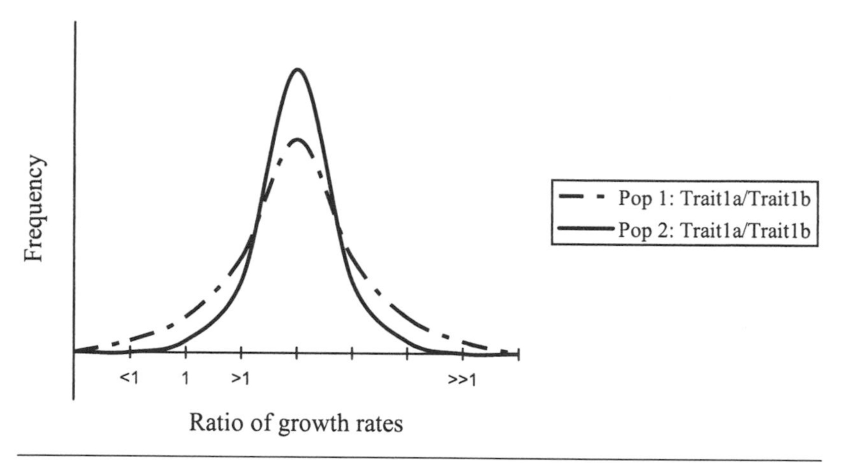

Figure 1.3. Hypothetical distribution of ratios of growth rates of trait populations: Two cases of selection. Here, the ratio of growth rates for trait populations is graphed against the frequency with which particular ratios are expected to be observed. The mean ratio of growth rates of trait populations based on the trait of interest (which has two variants, *a* and *b*) is different from one in both of two hypothetical populations, so in each hypothetical population there is a difference in (predictive) fitness between the two variants, and there is (formal) selection going on. Here we can say that there is at least one discriminate process going on at the individual level that is statistically correlated with the variation in the trait used to assign members of the population to subpopulations (trait populations) in both instances. Note that in some cases, the trait population picked out by the "trait1*a*" variant of both populations grows less quickly than does the trait population picked out by the "trait1*b*" variant, despite the fact that predictive fitness favors the "trait1*a*" variant in both cases. In other cases, the "trait1*a*" variant grows much more rapidly compared with the "trait1*b*" variant than the difference in predictive fitness would suggest. These kinds of cases may be called instances of "drift" if we are thinking of drift as an outcome. Note that while the mean difference in fitness between "trait1*a*" and "trait1*b*" is the same in both populations, population 2 has a lower variance than does population 1—so "drift," we might say, is a more frequent expectation in the case of population 1 and correspondingly a less common expectation in the case of population 2.

sense) variant of the trait in question may have gotten lucky, and relatively many organisms with the more fit variant may have gotten unlucky. This implies that at the level of ensembles of populations, there may often be substantial scatter around the mean ratio of the growth rates of the subpopulations in question: in some actual (realized) populations, the ratio of the growth rates of the subpopulations will fall below the mean, and in others it will fall above the mean. The shape of the distribution of ratios of growth rates of the subpopulations around the mean will be influenced by the kinds of processes involved, as well as by factors such as the effective population size (fig. 1.3).

But, while both discriminate and indiscriminate processes will influence the shape of the distribution of populations around the mean (the scatter), only discriminate processes can result in the mean ratio of the

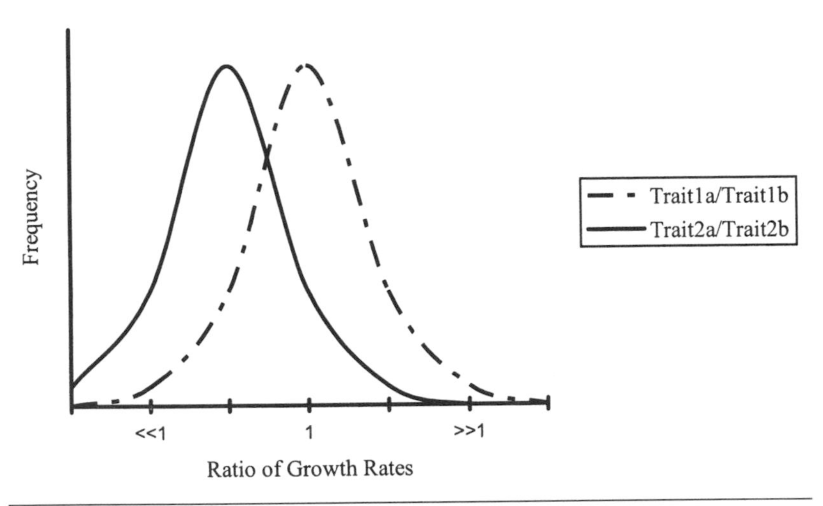

Figure 1.4. Hypothetical distribution of ratios of growth rates of trait populations: No selection versus selection. The mean ratio of growth rates of the trait populations picked out by variations in "trait 1" (with variants *a* and *b*) is equal to one, so there is no difference in predictive fitness between them, and no formal selection at work. This, however, does not imply that there are no discriminate processes going on at the individual level related to the trait in question. It may be that the particulars of the physical instantiation of the traits prevent the discriminate processes in question from influencing the frequencies of the traits in the ways one might expect (e.g., it might be a case of heterosis, or of pleiotropic effects). Note that the same population can, of course, show differences in predictive fitness when divided on the basis of another trait (trait 2 in this example, which also has variants *a* and *b*, trait 2*b* here having the higher predictive fitness).

growth rates being different from one (fig. 1.4). So, given a difference in predictive fitness (a statistical fact), one can say something about the processes going on at the physical level; namely, that at least one of them is a discriminate process with respect to a trait that is (statistically) associated with the trait used to identify the "trait populations" of interest. The opposite, unfortunately, is not the case: having identified a discriminate physical process with respect to some trait at the informal level, one cannot be sure that there will be a difference in the predictive fitness of the subpopulations picked out by that trait. The reason is that the relevant discriminate process may be opposed by other discriminate processes, or the variation in question may not be heritable, or the developmental pathways that produce the trait may prevent the trait's spread toward fixation for other reasons. For example, in the case of heterozygotic superiority, the way that some traits are generated genetically and developmentally can result in a stable equilibrium of phenotypes despite there being a "superior" variant vis-à-vis the discriminate physical process involved (if the superior genotype is the heterozygote, the two homozygotes will keep being

Box 1.4 The Relationship between Informal and Formal Fitness

Formal fitness differences imply discriminate processes. Given the existence of a formal (predictive) difference in fitness between organisms within a population grouped by the particular form of a trait, there exists at least one physical process that is discriminate with respect to either that trait itself, or to some trait that reliably covaries with the trait in question. Given a particular result at the level of ensembles of populations, then, we can say something about individual-level events.

Discriminate processes *do not* imply formal fitness differences. The existence of processes at the individual level that are discriminate with respect to particular variation in some trait does not imply that this trait will be associated with differences in formal (predictive) fitness. There may be other physical processes that interact with the variation in the trait in ways that prevent the actions of the first process from having an average effect in the population, or the variation in the trait may not be heritable.

produced simply by virtue of the mechanics of meiosis). It is only at the statistical level that the average results of the particular constellation of processes at work can be seen. Similarly, lack of a difference in predictive fitness in subpopulations picked out by a trait does not imply that there are no discriminate processes operating at the individual level with respect to that trait (box 1.4). Note too that discriminate processes can alter the distribution of outcomes *without* altering the formal fitness of the traits in question; failure to find a formal fitness difference, then, does not even imply the lack of formal selection (fig. 1.5).

It is in part for reasons like this that Matthen and Ariew (2002) argue that it makes no sense to ask, of a particular population, whether the actual outcome is due to "selection" or "drift"—all one can ask is how likely any particular outcome is, given the statistical distribution of outcomes around the mean fitness difference between trait populations in the ensemble of populations (2002, 78; see also Walsh, Lewens, and Ariew 2002). In fact, in this view, drift is not a process at all; the best sense that can be made of the concept appeals not to some property that particular populations share, but rather to the relative frequency of the kinds of changes that the populations have experienced. In a case in which the changes observed are close to the changes statistically expected for relevantly similar populations, we might say that the outcome reflects our expectations from predictive selection; in a case in which the changes are more distant from the mean, we might say that the outcome does not reflect those expectations—that is, we might choose to call it an example of drift. But that does not imply that any kind of process took place in the latter population that did not take place in the former (box 1.5).

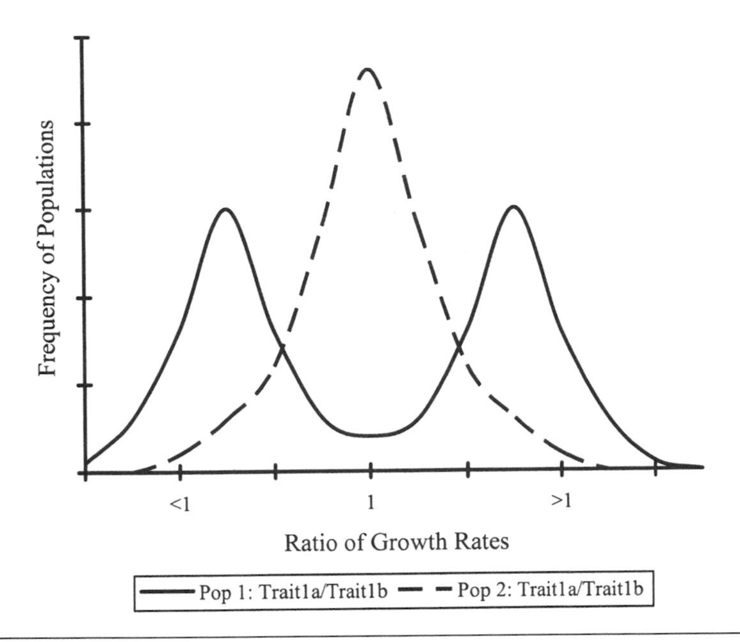

Figure 1.5. Differences in formal selection without differences in formal fitness. In the population in which there is disruptive selection, as well as in the population in which there is stabilizing selection, there is no difference in *formal fitness* between the two variants of the trait used to divide the population. However, the two populations experience very different selection (in the formal sense); note the radical difference in the *distribution* of outcomes.

It is worth stressing the conclusion that drift is not a process in any meaningful sense, not least because many biologists have lavished substantial attention and energy on trying to test particular hypotheses regarding selection versus drift (some recent examples include Gavrilets and Hastings 1995; Travisano et al. 1995; Orr 1998; Roff 2000; Philips, Whitlock, and Fowler 2001). Insofar as "drift" is simply a name we give to certain outcomes that are at a particular place in the statistical distribution of likely outcomes, one cannot meaningfully ask the question, "Is this outcome the result of selection or of drift?" Further, it follows that the related problem of distinguishing drift from selection in natural populations (see chapters 2 and 4) has no formal solution, and indeed, that there can be no automatic or foolproof method of identifying selection in natural populations through statistical analyses. There is, in other words, no way to tell by a strictly quantitative analysis of a single population, even one resampled repeatedly over time, whether there is a formal fitness difference associated with an identifiable trait. The only way to approach questions like this is by gathering evidence of many different kinds, including especially evidence from ecologically informed fieldwork.

Box 1.5 Why Drift Is Not a Process (or an Error)

Imagine tossing a fair coin 10 times and recording the ratio of heads to tails (H/T). While the "expected" outcome is 5/5, one is not very likely to get that exact ratio. If the trial of 10 tosses is repeated, and the outcomes are recorded each time, one expects to derive a distribution of outcomes—very few cases of 0/10 or 10/0, relatively many cases of 5/5, and so forth. The outcomes will follow a normal distribution, and a graph of the outcomes, with frequency plotted on the *y* axis and the H/T ratio on the *x* axis, will resemble, famously, a bell curve.

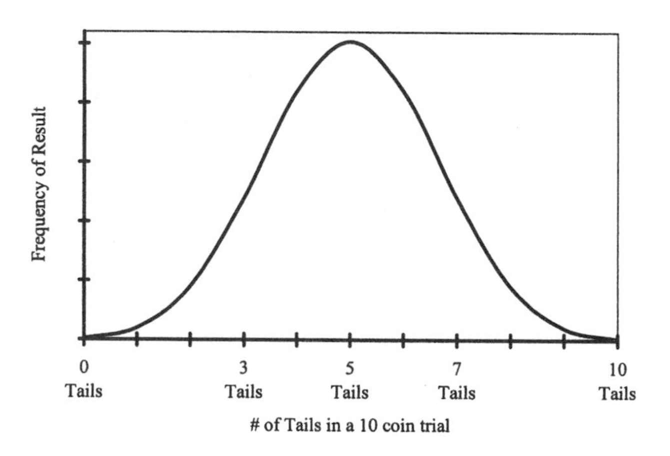

Consider all the trials (the sets of 10 tosses of the coin) in which the ratio was 7 heads to 3 tails (7/3). What explains why the ratio was 7/3 in these cases? Was there some process in play in the case of the 7/3 ratio that was not in play in the cases that produced a 5/5 ratio? Was there any "force" acting on the coin in the case of the trials that resulted in a 7/3 ratio that was not acting on the coin in the case of the trials that resulted in a 5/5 ratio? Clearly, the answer is no—the only forces acting on the coin are straightforwardly *physical* forces (gravity, air, friction, etc.). If coin tosses are deterministic, then if one knew *all* the physical forces, one could say why the coin landed heads-up or heads-down on any particular toss, but one would still not be able to find anything shared by all the trials that resulted in a 7/3 ratio rather than a 5/5 ratio.

If one were to ask why these trials resulted in a ratio of 7/3 rather than 5/5, the answer would have to be something like "there is no reason—that is just part of the expected distribution of outcomes—we expect a 7/3 ratio in about 12 percent of the cases." One may call the 7/3 outcome the result of "drift," but again, there is nothing shared by the 7/3 trials that is not also shared by the 5/5 trials.

Drift is often referred to as a kind of "sampling error" (see, e.g., Futuyma 1998, 297), but this seems misleading as well. If one is taking a small sample of a large population in order to estimate the frequency of some trait, and the sample differs markedly from the population at large, this, it seems, is a "sampling error." But is a trial of 10 coin tosses that has resulted in a 7/3 ratio a case of "sampling error"? What was being sampled? And, therefore, where is the error?

(*continued*)

Box 1.5 (*continued*)

Similarly, in finite populations, organisms with particular variants of a trait will sometimes do better or worse than the mean expected outcome; this is neither a "sampling error" nor the result of some special process. Rather, the likelihood of any particular outcome can be derived from the frequency with which it appears in the distribution of outcomes.

THE TROUBLE WITH FORCES

If one accepts these analyses of fitness and selection, the difficulties with the force metaphor become clear. The idea that a theory of forces can explain evolutionary change confuses what stands in need of explanation. Insofar as we are trying to explain changes in the distribution of heritable traits in a single population, the only "forces" at issue are straightforwardly physical ones (see box 1.5). Individual populations change over time because of the way physical processes interact with particular organisms—organisms that live or die, or reproduce or not. How successful any particular organism is will be determined by the physical processes that organism happens to interact with, and by how those interactions happen to go. None of these facts is given by predictive fitness, nor by any reasonable notion of informal fitness.

If one interprets informal fitness in its overall (vernacular) sense, the interactions an organism has with physical processes cannot be determined, even in part, by the overall individual fitness of the organism, since the overall fitness is a statistical summation of all of the possible interactions, their probabilities, and their possible outcomes. Nothing in this description would permit one to say what particular physical processes an organism in fact interacted with, and hence none of these facts, even when summed over the entire population, could explain the changes that actually took place in any particular population. If one interprets informal fitness to be about the relationship an organism with a given trait variant has to certain discriminate physical processes, again, this cannot determine which physical processes the organism in question actually interacts with, nor how often it does so, nor the details of those interactions. The actual success or failure of an organism, then, is not given by the restricted view of individual fitness (that is, the relationship of some trait to some particular discriminate process). By the same reasoning as above, informal fitness in this more limited sense, even if summed over the population as a whole, cannot be used to explain the changes that take place within any particular population. Informal fitness is simply not an explanatorily salient factor when applied to particular individuals, and hence is not a relevant factor when applied to particular populations—to a single population consisting, as it does, of individuals.

There is another rather similar reason why, if one is trying to understand what happens to a particular population, "selection" in this informal sense is not explanatory. To say that a certain physical process is discriminate with respect to some trait is not to say that the trait in question is being selected in the population in any way that would permit us to explain the dynamics of the relevant population. Overall, when considered from the standpoint of formal or predictive fitness, the trait in question may be neutral, or even detrimental. This kind of fact, however, emerges not at the level of individual organisms, or even of a single population; rather, it is a fact about the mean fitness of the trait—a statistical trend most appropriately conceived of at the ensemble level (that is, as a fact about the distribution of outcomes of relevantly similar populations).

In the formal sense, natural selection can be explanatory at the level of mean changes in frequencies of heritable features in populations—that is, at the level of ensembles of populations. But even at the formal level, it does not make sense to think of natural selection as a "force," generated by formal fitness differences, that is acting on populations. At the formal level, after all, fitness differences are statistical facts about particular correlations, not facts about causally efficacious pathways. Of course, if there are differences in predictive fitness, there must be discriminate processes at the individual level. However, the particular differences in predictive fitness we find at the formal level do not necessarily reflect those discriminate physical processes in any straightforward way.

Note too that there is nothing (no force) that can be said to be acting on the population as a whole. The idea that selection "pushes" populations "up" fitness landscapes confuses the effect of selection with the causes. Insofar as we understand points on a landscape to correspond to individuals with particular genotypes (as in Wright's original formulation; see chapter 8), populations move up landscapes because the reproductive success of the individuals is correlated with their location on the landscape. If, as in some contemporary versions of the fitness landscape (see chapter 8 for criticism of those versions), points correspond to populations with particular genotypic frequencies, the movement of populations on the landscape just *is* evolution, and the "hills" on the landscape are defined by the statistical tendencies of the population in question. There is no force that pushes populations uphill; all it means for a population to be on a "hill" is that the population is more likely to undergo some kinds of transformations than others.

The formal failures of the comparison between Newtonian forces and evolutionary processes should, by this time, be clear, as the metaphor breaks down in key places. Matthen and Ariew (2002) note several of these failures, including difficulties with resolving the different forces into a single force responsible for the actual evolutionary trajectory of the population (2002, 66ff), and the absence of a specifiable zero-force law (2002, 60ff).[7] If the difficulty with the force metaphor were simply that metaphorical forces in evolutionary theory were not perfectly isomorphic to Newtonian forces, this would hardly come as a surprise. But the metaphor as a whole seems misguided—it does not make sense of what we are trying to explain in evolutionary theory, nor are the problems it generates fruitful. Attempting, for example, to distinguish between the actions of natural selection and drift on a population is hopeless, insofar as drift is not a process that can act on a population at all. Since drift is better thought of as the expected statistical scatter around the mean predictive fitness, it is a conceptual mistake to attempt to distinguish it from selection. It is, therefore, no wonder that the formal tools developed to attempt to do so are inadequate to the task (see chapter 2 and especially the discussion on G-matrices in chapter 4). We should expect that it will often be impossible to detect selection in a straightforward manner through statistical analyses, and often impossible to show, even to a relatively high degree of confidence, that selection is taking place (but see fig. 1.6). But this is not cause for despair: biologically informed ecological research will often be able to link the weak statistical associations that can (at least sometimes) be found in studies of natural populations to particular traits and discriminate physical processes, such that it makes sense to think in terms of selection.

The metaphor of natural selection, constraints, and drift as forces is tempting; however, Matthen and Ariew (2002) argue convincingly that it is a poor one, and that much conceptual confusion in evolutionary biology and the philosophy of biology can be traced back to inadequacies in the way of thinking about evolution that this metaphor encourages. Making a sharp distinction between the individual level and the statistical level in evolutionary biology has profound implications for many areas of biology and philosophy of biology.

Matthen and Ariew (2002) focus on the implications of the force metaphor for thinking about the causal structure of evolutionary theory, and here we have followed up on those implications in a broader context. For

7. As Matthen and Ariew (2002) point out, populations evolve in ways that are stochastic; the Hardy-Weinberg law of population genetics not only fails to apply to any real population, but fails even to describe the behavior of ideal finite populations (see chapter 10). Populations change composition in fundamentally unpredictable ways without any force being applied to them at all.

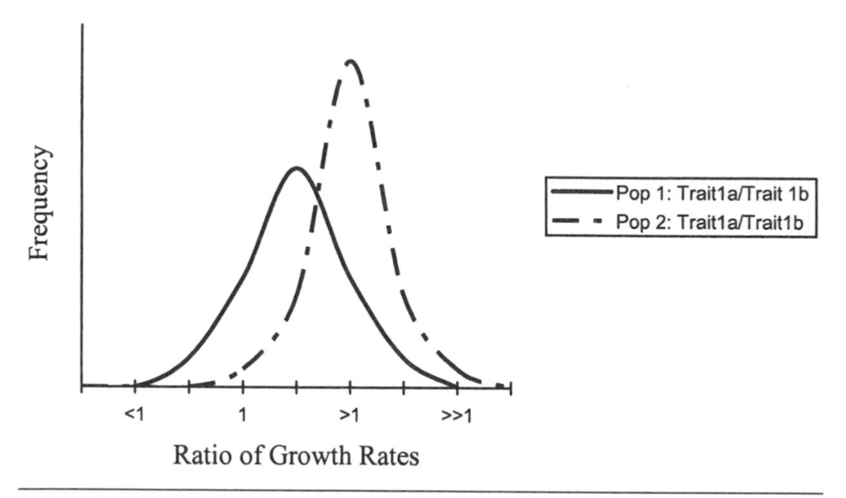

Figure 1.6. Hypothetical distribution of ratios of growth rates of trait populations: Detecting selection. In this case, there is a mean difference in fitness between trait populations divided on the basis of variation in trait 1 (variants *a* and *b*) in both populations. However, the chance that one will observe a difference in growth rates that reflects the direction of selection in population 2 is much higher than in population 1. If most traits in natural populations that are under selective pressure result in cases like population 1, field studies of natural selection will be very difficult, as many observed changes in trait frequencies will not reflect the direction of selection, even where the trait is in fact being selected at the ensemble level (even where there is a difference in predictive fitness).

example, the arguments above suggest that one way of rearticulating the difference between "neutralists" and "selectionists" (see chapter 10 and, e.g., Hey 1999) is as a difference about a particular empirical question. The key (though often epistemically unobtainable) information necessary to answer that question is the distribution of ratios of growth rates of identified trait populations around the mean ratio of the growth rates (the predictive fitness of the traits and the standard deviation of the ensemble of populations). If, in the case of directional selection in the formal sense, the standard deviation of ratios of growth rates is small compared with the distance of the mean ratio of the growth rates from one, most of the actual changes that we observe in particular populations of this sort can be attributed to the expected direction of change given by the formal difference in fitness. On the other hand, if the standard deviation is large compared with the distance of the mean ratio of the growth rates from one, then many of the actual changes that we observe in particular populations of this sort will not reflect the expected direction of change (see fig. 1.6). Obviously, which situation obtains more often, and under what conditions we can expect each kind of situation to obtain, is an empirical question, but one that may be difficult to answer. But again, attempting to find formal ways of identifying selection and drift in particular populations is unlikely to be helpful.

In the next two chapters, we focus on the implications of these ways of conceiving of fitness and natural selection at the formal and informal levels for thinking through methods that attempt to detect selection in natural populations (chapter 2), and for various issues in the levels of selection debate (chapter 3). We first argue that some contemporary methods that purport to be useful for detecting the presence of selection in natural populations cannot deliver what they promise, and discuss some of the reasons we believe these methods continue to be employed despite their widely acknowledged weaknesses. Next, we turn our attention to the levels of selection debate, arguing that traditional arguments regarding the invisibility of genes to many kinds of selection processes, and the ability of facts at the phenotypic level to "screen off" information about the genotypic level from selection, are irrelevant when applied to the statistical level. Likewise, we suggest that claims about genic selection's ability to represent evolutionary change are irrelevant to the individual level. Since no single conception of selection fits both levels, this inability to reconcile the different intuitions regarding the levels of selection should hardly come as a surprise. But again, this way of analyzing the levels of selection debate is unavailable if selection is to be conceived of as a force of some sort. Rather, one will tend to get hung up on two related problems: the question "What is the force acting on?" seems to demand one kind of answer, whereas the question "What is the currency of evolutionary forces?" seems to demand a very different kind of answer. The force metaphor, in other words, is part of what makes the levels of selection debate seem like it is addressing a real problem.

While the multilevel approach to selection and fitness articulated here demands a rethinking of many of the relationships among the basic concepts of evolutionary biology, and indeed, of those concepts themselves, the payoff, we think, will be worth it. This more careful way of distinguishing between senses of selection and fitness points toward solutions to many long-standing problems and, just as important, nudges us away from long-standing conceptual confusion.

2 How (Not) to Measure Natural Selection

IT IS NOT THE SCIENTIST'S FAULT THAT NATURE HAS MADE SOME OF ITS SECRETS RELATIVELY OPAQUE TO HUMAN SCRUTINY.
—ELLIOTT SOBER, *THE NATURE OF SELECTION*

While natural selection may be the most important concept in evolutionary biology, empirical research designed to measure it in natural populations is (perhaps surprisingly) a relatively recent phenomenon. In this chapter, we argue that what is now the standard approach to measuring natural selection in wild populations (though by no means the only approach to *studying* selection)—namely, multiple regression analyses—in fact falls far short of that goal. Furthermore, despite the claims made by some researchers, these techniques cannot produce long-term predictions of evolutionary change. We suggest that our real goal ought to be to determine the underlying causal structure of the changes in populations, including selection, and that this requires a different set of tools and a different mind-set—that of the detective more than the statistician.

THE MOST IMPORTANT CONCEPT IN EVOLUTIONARY BIOLOGY

Ever since Darwin (1859) proposed natural selection as the fundamental process explaining (adaptive) evolutionary change, selection has been at the center of both intense empirical study and heated debate. Entire books have been written on what selection is (Sober 1984), how it works (Bell 1997), and how it can be measured (Manly 1985; Endler 1986). It is fair to say that both conceptually (see chapters 1 and 3) and empirically (this chapter and chapter 4), natural selection unquestionably captures center stage in evolutionary biology. Selection is, after all, what makes biological adaptation possible, and so it is, we think, not too much of a stretch to say that everything else studied by evolutionary biologists is an (admittedly elaborate) footnote to Darwin's original insight.

It is odd, then, that attempts to measure natural selection in a way that is useful to such areas of evolutionary theory as quantitative population genetics are a somewhat recent addition to the literature (Lande and Arnold 1983). This late start is all the more remarkable when one considers that one of the hallmarks of modern evolutionary biology and genetics has been the quantification of its objects of study (think of Mendel's peas [Stern and Sherwood 1966] or of contemporary approaches to morphometrics [Rohlf and Marcus 1993]). The modern quantitative study of selection has been characterized by repeated attempts at conceptual and methodological improvements during the past two decades, which in turn have spurred a host of empirical research. And yet, as we argue in this chapter, a critical analysis of these methods shows that they do not deliver what they claim (or at least, what most practitioners take them to claim, for the originators of these methods are often much more savvy than their followers). Indeed, partly for the reasons we discussed in the previous chapter, we conclude that fitness trait covariances do not really measure natural selection, and that it is simply not conceptually sound to attempt to distinguish drift from selection via any strictly quantitative analyses of individual populations.

Rather than being able to measure natural selection (or identify "constraints"; see chapter 4), the best we can do, we argue, is to obtain quantitative estimates of the statistical relationships among certain traits and use them as the basis to formulate causal hypotheses about selection to be tested by other means. Some evolutionary biologists, we suggest, have been focused too much on developing and deploying statistical methods of analysis, ending up glossing over the conceptual underpinnings of the techniques they employ. This emphasis has led to something of a disconnect between empirical approaches and the concept they refer to. We argue that a redirection of the current efforts would probably yield far more significant insights into the nature of selection and evolutionary change than the methods currently in widespread use.

Our investigation is articulated around the discussion of a series of crucial papers dealing with the measurement of natural selection and constraints, all of which have appeared since 1983. While our choice is, by necessity, limited to a handful of works, these works have been carefully selected from among the most influential and conceptually important papers in this field. We analyze the claims and contributions of these papers, and—more important—how they are related to one another to form the complex conceptual web that currently drives the empirical measurement of selection and constraints in the wild. We focus not only on the reasons why the things that many researchers imply these methods can do are not in fact what they can accomplish, but also on what each of the pivotal papers in this field has actually contributed to our understanding of evolution and natural selection. We conclude this chapter with an analysis of the current state of the field, including a brief analysis of the directions we think would be less useful for researchers to pursue, as well as those we think would be more promising avenues of inquiry. Of course, this is very much an open area of investigation for both philosophers of science and theoretically inclined biologists, and we certainly do not pretend to provide the final word on the subject.

IN THE BEGINNING: LANDE-ARNOLD AND MULTIVARIATE ANALYSES

Selection on single traits was a constant preoccupation of animal and plant breeders throughout the twentieth century, and both the theory and empirical study of this topic were well represented in the literature (see, e.g., Falconer 1981). However, until the landmark paper published in 1983 by Russ Lande and Steve Arnold in *Evolution,* this kind of analysis had not been generalized to complex, multivariate phenotypic traits. Lande and Arnold's goal was to find a technique for measuring selection coefficients that would yield results suitable for input into the standard equations of multivariate evolution as elaborated by quantitative genetics theory. If this could be accomplished, it would allow evolutionary quantitative geneticists to predict—given certain assumptions (see chapter 4)—the evolutionary trajectory of natural populations. This would bring evolutionary biology closer to being a predictive science, and hence (though this certainly was not Lande and Arnold's stated goal) closer to being a "hard" science like physics (Cleland 2001).

Before we tackle some of the details necessary to understand both Lande and Arnold's important contribution and the many papers that followed it, it is important to keep in mind that the goal of finding a way to quantify the action of natural selection on complex, multivariate phenotypes is conceptually independent of the goal of generating results that can be used in the standard equations of quantitative genetics. One could, for example, devise several statistical procedures to measure natural selection (indeed, see Manly 1985 and Endler 1986 for a review of a variety of approaches), yet the resulting parameter estimates might not be suitable to be plugged into the standard equations describing evolutionary change because of the specific assumptions underlying the mathematical methods used to derive such equations. Conversely, the theory of evolutionary quantitative genetics had obviously developed until 1983 (and, one could reasonably argue, ever since) largely independently of the problem of empirically measuring the necessary parameters.

The promise of the Lande-Arnold approach to multivariate analysis, however, was enormous. If successful, it could be used to measure natural selection directly in wild populations, to distinguish between selection and drift, to identify constraints on multi-trait evolution, and thereby to predict the likely path of multivariate evolution in the populations in question (and even to reconstruct past selective pressures, given the current status of populations). But the coupling of the goals of quantifying multivariate selection and predicting the response of a population, which is what Lande and Arnold were explicitly aiming at, presents a serious problem. First, note that "selection" can be thought of in a variety of ways (see chapter 1), some of which cannot be measured in the way Lande and Arnold wish; even if it could be, it would, of course, be a very local measure, labile to change with changes in the population. Given this, we argue that the goal of predicting long-term evolutionary trajectories is hopeless (see also chapter 4; note that short-term predictions are perfectly possible, and in fact were being used by plant and animal breeders for decades before modern evolutionary quantitative genetics). Second, there are severe conceptual limitations to the Lande-Arnold approach, which emerge from the structure of some of the key concepts used in the equations (especially that of the genetic variance-covariance matrix, **G**—see chapter 4). Because of these limitations, these kinds of statistical approaches do not provide any information about causal pathways or the underlying causal structure of the populations being studied—which, we think, is the ultimate goal of selection studies.

Because the Lande-Arnold approach claims to produce measurements of selection that can be used to predict evolutionary change, it has been used extensively in place of other methods of assessing quantitative relationships between aspects of complex phenotypes. Some of these methods—for example, path analysis and structural equation modeling (see Shipley 2000)—are more suitable for yielding the sort of information that biologists are actually interested in when it comes to understanding selection, but so far have played a relatively minor role in the evolutionary literature.[1]

Lande and Arnold start out by reiterating the fundamental conceptual distinction between natural selection (which acts on phenotypes regardless of their genetic basis) and the response to it (which does depend on the inheritance and developmental systems underlying those phenotypes). This distinction is nicely captured by the simple "breeder's equation" for the case of one character:

$$R = h^2 S, \tag{2.1}$$

1. Techniques such as path analysis and SEM will not, of course, permit one to determine formal fitness differences or identify formal selection through purely quantitative means; rather, as we make clear below (and in chapter 10), we think these methods are to be preferred because they make more perspicuous the kinds of further tests and additional evidence necessary to make these kinds of claims.

where S is the selective pressure on a trait, h^2 is its heritability, and R is the response to selection.

We noted above that the problem of how a population will respond to selection on a single trait (which is captured by the breeder's equation) has been extensively studied, both conceptually and empirically, but it should be remarked that the problems with such work are also well established in the literature. Many of these problems can be traced back to h^2, the heritability term. As similar problems emerge in the related term in the multivariate version of the equation (**G**, the genetic variance-covariance matrix), it is worth pausing here to reflect on the limitations of h^2 and what those limitations mean for the usefulness of the breeder's equation to evolutionary biologists.[2]

The basic idea behind the univariate breeder's equation is straightforward: the phenotypic mean of a trait in a population will change in response to selection to a degree proportional to the variance of the trait associated with heritable differences in the phenotype and the strength of selection. The heritability of a trait, h^2, is defined as the fraction of phenotypic variance in the population that is associated with genetic variation in that population, and is given by the following equation:

$$h^2 = \frac{V_G}{(V_G + V_E + V_{G\times E} + V_{err})} = \frac{V_G}{V_P}, \tag{2.2}$$

where V stands for various components of the phenotypic variance (V_P) of a trait within a given population (in a particular set of environments), with the subscripts indicating the "genetic" variance (G, the component associated with differences among genotypes; see Lewontin 1974), the "environmental" variance (E, the component associated with variation in the environment, or phenotypic plasticity; see Pigliucci 2001 for a review), the genotype-environment interaction variance (G × E, a measure of how different genotypes react to distinct environments), and the "error" variance (err, which subsumes any other factor not explicitly considered by the model, including unique environmental variation). Often, heritabilities are calculated by further subdividing V_G into additive (abbreviated with the subscript A) and nonadditive (e.g., dominance, epistatic, dominance-by-dominance interaction, etc.) components. In largely outbreeding species, the additive genetic component is the only part of the

2. We are not arguing that the breeder's equation is useless. Indeed, it has been successfully applied in plenty of research projects in plant and animal breeding. However, the limitations we discuss become crucial whenever one wishes to generalize results to multiple populations, or extend such results to long periods of time (many generations). The latter are, of course, the dominion of interest of evolutionary biology, not crop enhancement.

total phenotypic variance that results in a predictable response to selection; phenotypic variation in the population that is associated with nonadditive genetic variation is not available for straightforward selection, since the response of the population will not necessarily be in the expected direction. For this reason, the additive genetic variance is the only genetic variance of immediate interest to selection theory (though in the long term, nonadditive genetic variance can be significant), and it is, of course, the factor of primary concern to plant and animal breeders (who are interested in the expected, and hence predictable, response of the population to selection in the relatively short term).

The idea behind the breeder's equation can be clearly understood by considering the two extreme situations: if the heritability is zero, then there is no response to selection at all; there can still be selection if there is phenotypic variation, but the population's mean does not move from one generation to the next in response to the applied selection. If there is no heritable variation associated with the variation in the trait—that is, if offspring fail to resemble their parents more than they resemble the population at large—then any variation in the trait can be expected to be recreated in the next generation, no matter what portion of the population is bred. On the other hand, if the heritability is 100 percent, then the response to selection is exactly equivalent to the selective pressure: if, for example, the upper quartile of the population is selected for breeding, the character's mean will be correspondingly shifted in the following generation. One can therefore think of heritability as an estimate of the (short-term!) "fuel" that makes phenotypic evolution possible (box 2.1).

There are several well-known problems with the idea of heritability. First, the heritability of a trait is a result (among other things) of the particular genes and gene frequencies that are present in a given population. This means that heritabilities are not specific to a trait, but are more narrowly defined for a particular trait in a particular population at a particular time (Lewontin 1974). A corollary of this definition is that evolution itself will change the heritability of a trait, since processes such as selection, mutation, and recombination change the gene frequencies of an evolving population.[3] So using the breeder's equation to predict long-term responses to selection is simply out of the proper scope of the quantitative genetic approach—selection on a trait changes the trait's heri-

3. Note the absence of drift from this list. Because, as we discussed in chapter 1, drift is not a process at all, we are not considering it, as would normally be done. Rather, the stochastic nature of all the processes enumerated here means that in finite populations, there will be alterations in gene frequencies as a function of population size—without the action of any distinct "process."

Box 2.1 Using the Breeder's Equation

The breeder's equation (eq. 2.1) includes three variables: the response to selection R, the heritability h^2, and the selection differential S. Given the value of any two of these variables, the third, of course, can be calculated. If a trait has a known heritability h^2 (within a particular population), and the population is subject to a known intensity of selection S on the variation in that trait, we can then calculate the population's response to selection—or how much the mean of the trait is expected to shift from one generation to the next. Similarly, if the value of h^2 for a trait in the population before selection is known, and the shift in the mean of that trait is also known, the intensity of selection can be calculated.

For example, consider a trait in a population (say, the height of the organism) with a mean value of 10 cm, a heritability in that population of 0.6, and a selection regime such that the mean value of the trait in those individuals that successfully breed is 14 cm. Then R, the response to selection, will be 2.4, because $R = 0.6 \times (14 - 10) = 2.4$; the population mean will have shifted upward by 2.4 cm to a mean of 12.4 cm.

Similarly, the response to selection R can be used to discover S, the intensity of selection, when the heritability of the trait in the population in question is known. In the above example, if we know that $h^2 = 0.6$ in the population before selection, and that the measured mean before selection is 10 cm and after selection is 12.4 cm, we can calculate that in this case S is 4.

Note that, unless various assumptions are made about the population under study, R is not a measure of the predictive fitness (see chapter 1 and box 1.3) of the character trait. Rather, R is a measure of how much the character changed over the course of a generation. Only if the population is assumed to be of infinite size, or assumptions are made about the typicality of the change being considered, can R be thought of as a measure of predictive fitness.

tability, and hence affects further response to selection. The only way around this problem is to devise a way to calculate the continuous alteration of the heritability itself as selection proceeds. This, however, is not feasible, because heritabilities are actually a statistical summary of the underlying genetics and may correspond to a variety of possible patterns of genetic architecture at the molecular level. Depending on the exact causal makeup of the statistical genetic variance, heritabilities could change in different ways in response to the same selective pressures. Sidestepping this problem would involve knowing how the underlying genetic and molecular architecture would respond to particular kinds of selection in these particular cases; it seems clear that any technique that revealed this kind of information would also eliminate the need to estimate the changes in heritability to begin with.

The second major difficulty is that heritabilities change—sometimes dramatically—with the environment in which they are measured. This was pointed out on theoretical grounds by Lewontin (1974) and has been repeatedly demonstrated empirically in a variety of systems and circumstances (e.g., Vela-Cardenas and Frey 1972; Daday et al. 1973; Paulson, Roberts, and Staley 1973; Mazer and Schick 1991; Ebert, Yampolsky, and Stearns 1993; Simons and Roff 1994; Bennington and McGraw 1996; Hoffmann and Schiffer 1998; Sgrò and Hoffmann 1998; Loeschcke, Bundgaard, and Barker 1999). Because different genotypes respond differently to distinct environments, the heritability of a trait can vary with the environment of the population under study (fig. 2.1a).

Finally, note that heritability does not provide significant information about the causal pathways involved in the development of the trait itself. That is, heritability is simply a measure of a particular covariation (which is a statistical attribute) between phenotype and genotype, and is not an indication of specific causal pathways. It is in part for this reason that heritability must not be interpreted as a measure of genetic determinism (contrary, alas, to much of what is assumed or explicitly stated in, for example, all too many entries in the vast literature on IQ and other cognitive traits in humans [e.g., Bouchard et al. 1990]; see Feldman and Lewontin 1990) (see figs. 2.1a and 2.1b).

Recall that what Lande and Arnold (1983) were trying to provide was a method to quantify the type and intensity of selection for the case of several (possibly intercorrelated) characters in natural populations, not just the evolution of single characters as described by the breeder's equation. That is, they needed a way to quantify the multivariate equivalent of S, the selection intensity. R, the response to selection, is then obtained from the multivariate version of the breeder's equation, once we know the intensity of selection and the genetic variance-covariance matrix (**G**, see chapter 4).

Lande and Arnold elegantly showed that the multivariate version of the breeder's equation is

$$\mathbf{\Delta} z = \mathbf{G}\mathbf{P}^{-1}\mathbf{s}, \quad (2.3)$$

where $\mathbf{\Delta} z$ is a vector summarizing the difference in the phenotypic traits before and after selection (the multivariate response to selection), **G** is the genetic variance-covariance matrix, **P** is the phenotypic variance-covariance matrix (a statistical summary of the correlations between the character traits in question), and **s** is a vector measuring selection intensities on all traits being considered. The vector **s** is defined as

$$\mathbf{s} = \mathrm{cov}[w,z]. \quad (2.4)$$

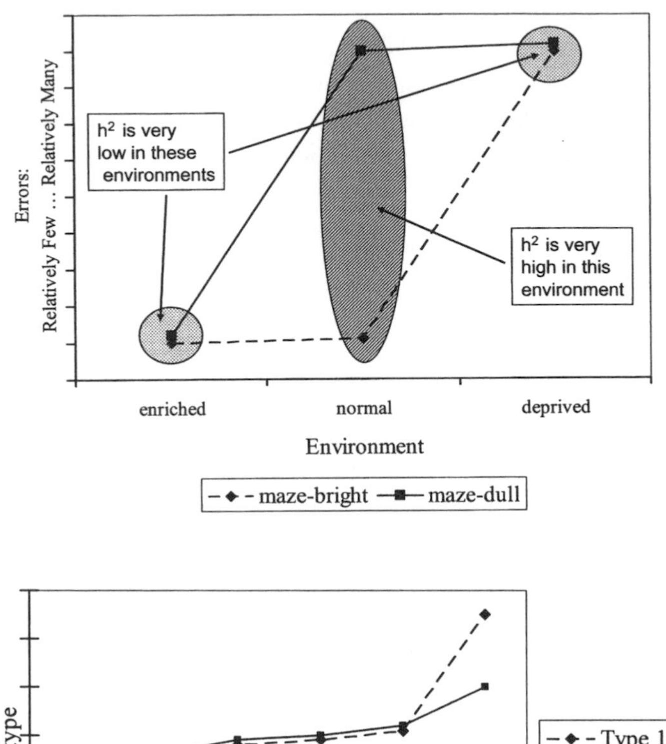

That is, **s** is the set of covariances between variations in the traits (z_i) and variations in fitness (w). At this point, the similarity between the multivariate and the univariate versions should be clear.

Lande and Arnold then introduced a crucial quantity:

$$\boldsymbol{\beta} = \mathbf{P}^{-1}\mathbf{s}, \tag{2.5}$$

where $\boldsymbol{\beta}$ is the vector of "selection gradients" (as opposed to the vector of "selection differentials," **s**; box 2.2). That is, $\boldsymbol{\beta}$ measures the intensity of selection on a trait when the effect of selection on the trait due to its correlation with other selected traits has been statistically removed. So, in

Figure 2.1. Norms of reaction and heritability (h^2). (a) A norm of reaction is the response of a particular genotype to variation in some aspect of the environment. For example, Cooper and Zubek (1958) raised two strains of rats that had been bred to be very good and very bad (respectively) at running mazes. In the environments in which selection took place, maze-running ability was highly heritable—much of the variation in maze-running ability was associated with differences in the genotype. However, when Cooper and Zubek raised the rats in "enriched" environments (environments with lots of toys, colors, and stimulation) the "maze-dull" rats performed as well as the "maze-bright" rats. So, in that environment, differences in the genotype were not associated with maze-running ability, and the heritability of maze-running ability was very low (effectively zero). Similarly, when the rats were raised in "deprived" environments (gray cages with no moving parts or visual stimulation), the "maze-bright" rats did just as poorly as the "maze-dull" rats, and again, in that environment, heritability was very low.

So what is the heritability of maze-running ability in these rats? Obviously it depends critically on the environment in which they are raised. For populations of rats raised in a variety of environments, the proportion of the relevant kinds of environments will determine the heritability. Note too that variation in the performance of the rats cannot be simply divided into a "genetic" and an "environmental" component because rats with different genotypes respond differently to different environments (i.e., there is variation in phenotypic plasticity or genotype-environment interactions). Heritability is, then, a *local* measure, and the heritability of a trait tells us little or nothing about the extent to which either the trait itself or differences in the trait are "determined." A trait can have either a very high or a very low heritability and yet be changed radically by a change in the environment or the population structure. Measures of heritability are simply unrelated to any reasonable definition of "genetic determinism" and cannot be used either to predict what will happen in different environments or to explain the causes of the variation that exist in the current range of environments.

(b) A graphic representation of one meaning of the phenomenon of canalization (considered at the population level—there are additional meanings at the level of individuals) that is crucial to understanding why measures of heritability tell us nothing about genetic determination. In environments b through e (the range of canalization), the phenotype is developmentally buffered from environmental variation. In this example, the different genotypes respond to changes in the environment in the same way, and similar phenotypes are produced whatever the genotype within the environmental range of canalization. However, in environments outside the range of canalization, the different genotypes produce radically different phenotypes. These variations are heritable, but, of course, if heritability is estimated only on populations in environments b through e, the heritability of the trait will seem very small. Just because differences in the genotype do not make a difference to differences in the phenotype in those environments, it does not follow that the genetic differences between organisms of type 1 and type 2 are not significant, nor does it imply that any particular genes do or do not play a role in the development of the trait in question.

fact, $\boldsymbol{\beta}$ is what Lande and Arnold had set out to estimate: they needed a statistical tool that would allow biologists to measure the intensity of selection on traits in real populations, in which those traits are probably correlated with numerous other traits through a variety of different causal pathways. According to the theory developed by Lande and Arnold, one can also interpret $\boldsymbol{\beta}$ as the average gradient of the relative fitness surface, weighted by the phenotype distribution (1213). In other words, $\boldsymbol{\beta}$ can be thought of as approximating an aspect of the surface of the phenotypic adaptive landscape (see chapter 8) for a given population (box 2.3).

The tool suggested by Lande and Arnold (after an original idea by Pearson, proposed as early as 1903) in order to quantify $\boldsymbol{\beta}$ is a standard statistical technique codified in countless commercial software: multiple

Box 2.2 Multiple Versions of the Multivariate Breeder's Equation

Equation 2.3 represents one version of the multivariate breeder's equation, where $\mathbf{\Delta z}$ is a vector of the differences in phenotypic means for several traits before and after selection, **P** is the phenotypic variance-covariance matrix (inverted, as the -1 superscript implies), and **s** is the vector of selection differentials. However, another version of the equation, which we will be using primarily in chapter 4, reads:

$$\Delta z = \mathbf{G}\boldsymbol{\beta}, \tag{2.6}$$

where the vector $\boldsymbol{\beta}$ quantifies the selection gradients, and **G** is the genetic variance-covariance matrix.

There is often confusion between the terms "selection gradients" and "selection differentials," as used in these equations. Selection differentials, **s** (used in eq. 2.3), are the overall covariances between each trait and a measure of fitness; differentials, then, measure the total selection on a trait, including selection directly on that trait and indirect selection through the fact that the focal trait is correlated with others that may in turn be under selection. Making certain assumptions about the typicality of the population under selection (or that the population is effectively of infinite size, etc.), one can go in principle from measures of **s** to measures of predictive (or formal) fitness (as defined in chapter 1).

On the other hand, selection gradients, $\boldsymbol{\beta}$ (used in eq. 2.5), are the standardized regression coefficients between trait means and fitness; they are, in other words, the intensity of selection corrected for the statistical interdependence among traits (something that is common in multivariate analyses). Selection gradients, then, measure specifically the intensity of direct selection on the focal trait—that is, they measure the intensity of selection that remains once the additional effects of selection on other traits have been statistically removed. $\boldsymbol{\beta}$, therefore, cannot be interpreted as a measure of predictive fitness as defined in chapter 1, because predictive fitness may be in the opposite direction of the selection gradient suggested by $\boldsymbol{\beta}$. Neither, however, is $\boldsymbol{\beta}$ a measure of fitness in the vernacular (overall) sense, since it is supposed to be a measure of the fitness of a trait given all the environmental interactions with that trait, but with the correlation with other traits removed. For reasons made clear in the body of this chapter, we think that it is difficult to achieve a conceptually coherent interpretation of $\boldsymbol{\beta}$.

While the two versions (eqq. 2.3 and 2.5) of the multivariate breeder's equation are, of course, mathematically equivalent, they are used in different contexts depending on the interest of the investigator (a focus on total or partial selection). It is, however, important to keep them conceptually distinct from each other, as—because of their emphasis—they convey very different pieces of information about selection in natural populations.

regression analysis (fig. 2.2). This technique is a way to estimate the statistical relationship between a "dependent" variable (in our case, fitness) and a set of "independent" (but potentially intercorrelated) variables (the individual traits). If the analysis is carried out by using raw data (i.e., without applying the standard transformations aimed at improving statistical attributes such as normality of the data distribution) and a measure of relative fitness (such as standardized reproductive output), the multiple re-

Box 2.3 Interpreting and Visualizing s and $\boldsymbol{\beta}$

In a 1989 paper, Steve Arnold and Patrick Phillips sought to provide methods to visualize selection analyses of the Lande-Arnold type. Much of the Phillips-Arnold paper is concerned with statistical techniques for reducing the dimensionality of data sets in order to make the analyses more tractable. Their methods, while mathematically straightforward, produce variables that tend to be difficult to interpret biologically, and hence the methods they suggest have not been used extensively by biologists. In the process of developing these techniques, however, Phillips and Arnold clarified several of the key concepts used in multivariate selection analyses in general, and it is on these clarifications that we will concentrate here. Phillips and Arnold provide a table that is extremely useful for discussions of selection analyses (see their table 1, 1212). In it, they summarize the meanings of all proposed Lande-Arnold-style selection coefficients. For simplicity, we focus here only on the linear selection coefficients (representing directional selection).

According to Phillips and Arnold, the selection differential, **s**, can be interpreted in two ways:

1. Dynamically: as a shift in character means due to direct and indirect (i.e., correlated through other characters) effects of selection
2. Statistically: as the covariance between relative fitness (however measured) and the characters in question

The selection gradient, $\boldsymbol{\beta}$, can be interpreted in three different ways:

1. Dynamically: as the strength of the direct "force" of selection (see chapter 1) on each character when correlations with other characters are statistically controlled for
2. Statistically: as the set of partial regression coefficients of relative fitness on each character, holding the other characters constant
3. Geometrically: as the direction of the steepest uphill slope from the population mean on the selection surface

The geometric interpretation of $\boldsymbol{\beta}$, as a direction on a fitness surface, itself admits to three different interpretations, depending on the type of fitness surface used:

1. **Individual:** In Phillips and Arnold's words (1213), $\boldsymbol{\beta}$ is the association between the expected fitness of an individual and its phenotypic values for various traits. Ecologically, it describes the fitness consequences of phenotypes interacting with their environment.
2. **Best quadratic approximation to the individual surface:** This value is given by the Lande-Arnold regression coefficients, since they are the coefficients of a quadratic regression of fitness on the measured traits.
3. **Adaptive landscapes:** The third kind of fitness surface is, of course, Wright's adaptive landscape, which Simpson applied (not without serious conceptual problems, discussed in chapter 8) to phenotypic traits.

As Phillips and Arnold point out, the fitness surfaces of interpretations 1 and 2 are *not* Wright's or Simpson's adaptive landscapes. However, a connection between interpretations 1 and 2 and Simpson's version (the phenotypic one) of the adaptive landscape can be found.

(*continued*)

Box 2.3 (*continued*)

Following Phillips and Arnold, $\boldsymbol{\beta}$ can also be interpreted as the slope of the adaptive landscape, *if* the distribution of the phenotypic traits is multivariate normal. How often this condition actually holds is a rarely tested empirical matter, as we have seen.

Phillips and Arnold highlight some of the problems and assumptions that go into these interpretations. One of the most serious, in our view, is that many distinct individual selection surfaces can yield the same selection gradients (i.e., the same quadratic approximations), since the linear and quadratic multiple regression coefficients of Lande and Arnold estimate the average slope (linear) and curvature (quadratic) of the individual surface. In addition, Phillips and Arnold note that the individual selection surface represents the relationship between individual traits and individual fitness, while the selection gradients are attributes of the population being studied, and they describe changes in population characteristics such as means and variances. We have argued in chapter 1 that the bridge between individual and population levels of analysis is problematic, to say the least, but assuming the kind of evolutionary quantitative theory discussed here, one can, in principle, go from one to the other.

gression coefficients will estimate the linear $\boldsymbol{\beta}$, which Lande and Arnold interpret as a coefficient of directional selection. They then go on to extend the technique to a series of quadratic coefficients, which are supposed to estimate stabilizing, disruptive, and correlational selection. For the sake of our discussion, we focus here on the linear coefficients—those that are interpreted as revealing directional selection—as the same reasoning applies to the other cases.

In their seminal 1983 paper, Lande and Arnold went on to anticipate many of the limitations and objections to their proposed approach; this discussion probably directly spurred some of the papers to which we turn next. However, the main points so far are worth summarizing before we proceed:

1. Lande and Arnold articulated a multivariate extension of the breeder's equation.
2. They proposed to use multiple regression analysis to estimate selection gradients.
3. They interpreted selection gradients (i.e., standardized multiple regression coefficients) as approximating the fitness surface.
4. They showed how the coefficients obtained from multiple regression analyses of fitness against traits can be plugged into the multivariate breeder's equation to predict the population's response to selection (if one knows the genetic variance-covariance matrix, **G**; see chapter 4).

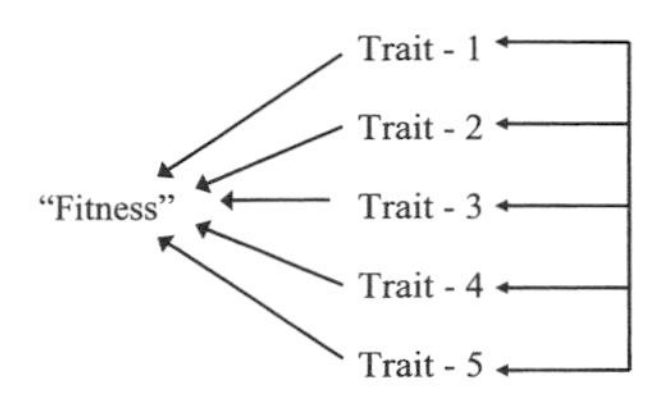

Figure 2.2. A visualization of multiple regression. Several "independent" variables (*right*) are statistically related to a "dependent" variable (*left*). The "independent" variables can be, and often are, correlated among themselves. A multiple regression analysis finds the degree to which each independent variable is correlated with the dependent variable, when variation in the other independent variables is statistically controlled for. Notice how the language of such analyses implies the adoption of a particular causal theory, even though nothing in the analysis directly substantiates any particular causal scenario.

All of this makes for a neat and powerfully attractive package (box 2.4). But, as we shall see, it is a package that—upon closer examination—reveals itself to be fraught with conceptual as well as practical problems that make its usefulness much more limited in scope than originally hoped for.

STATISTICAL AND CONCEPTUAL DIFFICULTIES IN MEASURING SELECTION

The central challenge for these multivariate approaches to understanding natural selection is moving from the statistical analyses to scientific interpretations of the results, a problem that is, of course, of much wider import in the biological and social sciences than, say, in some areas of physics or chemistry. We noted above that in interpreting the results of regression analyses as estimates of selection, Lande and Arnold made a number of critical assumptions, three of which are especially problematic: that the traits in question will be distributed in a way that is multivariate normal, that the traits will not be so strongly correlated with one another that the resulting correlations matrices will end up being singular (thereby yielding an output that is highly unstable), and that no unmeasured factors will be causally significant. The first two assumptions are primarily statistical in nature, and the chief difficulty with them is that when they fail, the standard techniques for solving the multiple regression analyses do not deliver. The third assumption is in some ways much more serious and can be dealt with only by moving beyond the realm of statistical inference; it is also a matter of general concern across many scientific disciplines, not limited to the study of selection or even to evolutionary biology.

Let us examine each of these problems in a bit more detail. The assumption of multivariate normality is an extension of the standard

Box 2.4 Measuring Selection on Multivariate Phenotypic Traits

It is instructive to see in some detail how the Lande-Arnold selection analyses work. Consider the following case (adapted from Lande and Arnold 1983): A rainstorm kills off a significant portion of a local population of flying bugs. In order to discover what phenotypic traits were selected for by this event, several traits (width of head, width of thorax, length of scutellum/upper thorax, length of forewing) are measured in a small sample of the population (about a hundred bugs) before and after the storm:

Mean values of traits before and after selection

Trait	Standardized mean value before selection	Standard deviation before selection	Standardized mean value after selection	Change (**s**)	Is change statistically significant?
Head	0.88	0.03	0.88	−0.04	No
Thorax	2	0.05	1.997	−0.03	No
Scutellum	1.5	0.06	1.484	−0.02	Yes
Wing	2.3	0.04	2.28	−0.02	Yes

Based on **s**, the selection differentials, one would guess that selection acted to reduce the mean size of the scutellum and wing, but had no significant effect on the head and thorax. However, this line of reasoning ignores the fact that the traits in question are correlated with one another:

Correlation between traits before selection

	Head	Thorax	Scutellum	Wing
Head	1	0.72	0.5	0.6
Thorax		1	0.59	0.71
Scutellum			1	0.62

Given these correlations, Lande and Arnold calculated $\boldsymbol{\beta}$, the directional selection gradient. They interpreted this value as the intensity of selection on a trait when the effects of its correlations with other traits are removed.

Directional selection gradient ($\boldsymbol{\beta}$)

Trait	$\boldsymbol{\beta}$	Significant?
Head	−0.7	No
Thorax	11.6	Yes
Scutellum	−2.8	No
Wing	−16.6	Yes

Based on this analysis, we can see that there is selection for a larger thorax, but even stronger selection for a shorter wing. The net result of these selective pressures is that wing length decreases slightly relative to thorax size. The apparent selection on the scutellum

(*continued*)

Box 2.4 (*continued*)

size is revealed to be an artifact of the strong selection for reduced wing length, as scutellum size and wing length are positively correlated.

If the reasoning here is correct, then, simply by observing the changes in the phenotypes of organisms over the course of generations, the direct effects of selection can be disentangled from changes that are merely correlated with selection. It is worth recalling, however, that a conceptual analysis reveals that these methods measure neither the formal (predictive) nor the informal (and still less the vernacular or overall informal) fitness of the traits in question as defined in chapter 1. The interpretation of $\boldsymbol{\beta}$ coefficients, then, may not be as straightforward as most biologists seem to think.

assumption of normality for the distribution of individual trait values, common in statistical analyses of most data sets. Ordinarily, one can transform one's data to improve their fit to such a distribution (see Sokal and Rohlf 1995), but in the case of Lande-Arnold-style selection analyses, this move is blocked, as such transformations make it impossible to calculate selection gradients that can then be plugged into quantitative genetic equations predicting phenotypic evolution. Because, as we noted above, the ability to generate selection gradients was what made the Lande-Arnold approach so appealing, statistical techniques that produce results that cannot be so used must be rejected. While some techniques for skirting these problems have been suggested, none are without problems of their own (see Mitchell-Olds and Shaw 1987). In any event, how often the multivariate character traits under study are multivariate normal, or can be usefully transformed so that they are multivariate normal, is an empirical question to be dealt with case by case.

The second problem is that if too many of the traits in the analysis are highly correlated with one another, a situation known as multicollinearity will be obtained. In this case, one will not be able to invert the phenotypic correlation matrix, **P**, as used in the multivariate breeder's equation above—that is, it will be impossible to calculate $\mathbf{P}^{-1}$. Because moving from **P** to $\mathbf{P}^{-1}$ is required for using the multivariate breeder's equation to generate selection gradients, this creates a serious obstacle. The usual solution to this problem is statistically straightforward, but can be biologically problematic: one needs to eliminate some of the correlated variables from the regression analysis to reduce the degree of collinearity—that is, to reduce the number of variables that are strongly correlated with one another. But it may be difficult to justify doing so in biological terms—for example, one sometimes loses the ability to include in the analysis characters that are biologically very important or obviously related to fitness. Therefore, while the difficulty can be dealt with from the standpoint of the statistical problem, doing so makes the biological interpretation of the results more difficult and more problematic.

We now come to what we regard as the most serious problem, the one created by unmeasured traits. This problem has, in essence, nothing to do with the statistics and everything to do with both the biology and, more broadly, the conceptual issues surrounding the measurement of selection—and it is a very well-known thorn in the side of every practitioner in the field, acknowledged by Lande and Arnold themselves in their initial paper. The basic idea is that the statistical accounting of the variance in fitness in terms of a series of independent variables is only as good as the completeness of the set of such variables. If one or more characters with important effects on fitness are missing from the list (because they could not be measured, or because biologists did not think of measuring them), the attempt to approximate the underlying fitness surface fails: the real fitness surface may be very different from the estimated one, because crucial pieces of information are missing.

Let us be perfectly clear about this: There is essentially no general solution (and certainly no statistical solution) to this problem. The best biologists can do is to justify their lists of characters and rack their brains about what they can possibly have missed and how to go about measuring it the next time around. The problem of unmeasured influences on fitness becomes more acute when one considers that what is missing from the statistical account may not be the influence of yet another phenotypic trait, but that of an environmental factor (see Mitchell-Olds and Shaw 1987). For example, if some plants in a population find themselves in a soil patch characterized by higher levels of nutrients than the surrounding patches, they may as a result have both larger size and higher fitness—through additional causal pathways connecting the three variables. This would create a measurable covariance between size and fitness that would actually be the result of a third common causal factor (nutrient availability). This problem is illustrated in figure 2.3 as a path diagram, an approach that we will repeatedly come back to and recommend in this discussion, as others have repeatedly done over the years. If it is the influence of unmeasured traits that actually produces the observed correlations (or lack thereof), it follows that the inferred selection gradients will be biologically misleading—what the analysis suggests is being selected may be no such thing.

Mitchell-Olds and Shaw addressed this problem explicitly in their classic 1987 paper in *Evolution*. As they noted, the goal of Lande-Arnold-style

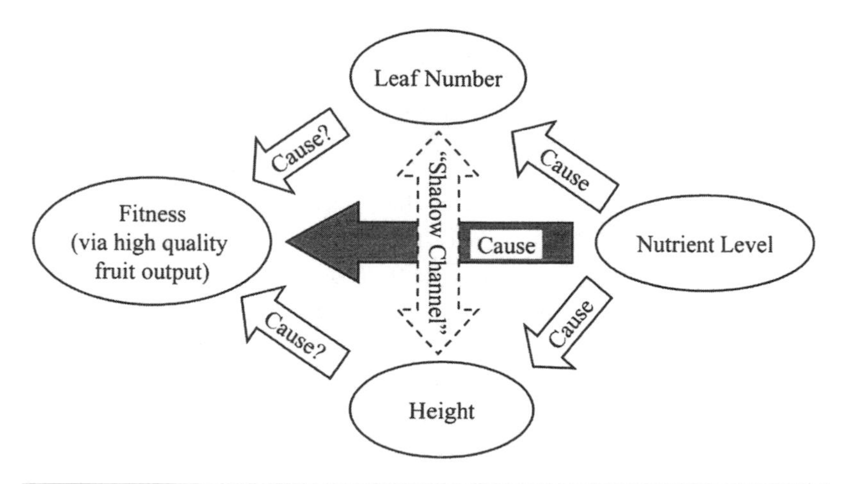

Figure 2.3. Modeling causal pathways. If changes in nutrient level cause both increased height and (say) increased fruit output in a plant population, then size and fitness will be correlated, but may or may not be causally related. If nutrient level is unmeasured, the analysis will miss an additional pathway that connects the variables in question, and will "reveal" pathways that may not in fact be causal, such as the "shadow channel" connecting height and leaf number.

regression analyses, and of the variations on those analyses that they suggested, is to improve our ability to make inferences about natural selection, not just to provide a statistical summary of the covariation of traits in a population. But they noted that the method in question is "an inherently observational method," and as such, while it can "reveal the effects of selection on characters," it cannot "unambiguously ascribe the cause of differences in fitness to the associated differences in characters" measured (Mitchell-Olds and Shaw 1987, 1152). This, indeed, is the problem: regression analysis is a descriptive statistical tool that simply cannot be employed to yield information about causal structures. If it is properly used, it can conveniently summarize information and *suggest* causal hypotheses to be tested by other means, but attempting to infer causal pathways by using multiple regression analysis is an improper use of the method. This should not come as a surprise—after all, Lewontin's primary complaint about the use of analyses of variance (ANOVAs) in the univariate case was just this (see Lewontin 1974), that often people confuse an analysis of (statistical) variance with an analysis of causes. There is nothing in the more complex multivariate case that would permit an easy move from statistical associations to causal structures. But what we wish to learn by studying selection is the causal pathways between traits and differences in fitness—not just the statistical associations!

The inferential problem caused by the fact that different underlying causal structures can have the same observational (in this case, statistical) consequences cannot be solved by any statistical approach. For example, while there is no way to guarantee that the characters chosen for the analysis are appropriate, Mitchell-Olds and Shaw note that "careful choice of characters, preferably on the basis of prior information regarding their functional significance, reduces the chance that selection apparently on one character is actually due to selection on some unmeasured character" (1987, 1154). Similarly, the problem of unmeasured environmental effects cannot be eliminated statistically (contrary to recent attempts by Rausher 1992 and Stinchcombe et al. 2002; box 2.5), but "the effect of environment on fitness and on the relationship between the characters and fitness can be inferred, if information regarding the environment is included in the analysis" (1155).

Information about the environment, according to Mitchell-Olds and Shaw, can be included in the analysis in at least three ways: by identifying and actually quantifying important environmental factors; by collecting data in spatially or temporally distinct blocks (to randomize the effect of unmeasured environmental factors); or by experimentally manipulating crucial aspects of the environment. Done well, these techniques can reveal the influence of specific environmental variables on the character traits in question and the correlations among them, and can suggest additional lines of research. But, of course, all of this requires significantly more work—both conceptually and empirically—than the simple act of measuring a bunch of traits and analyzing the data with a standard statistical package. Nonetheless, we do not see any available shortcut.

It is for this reason that Mitchell-Olds and Shaw suggest that the Lande-Arnold method of fitness regression be used "primarily as a tool for *suggesting hypotheses* about the forces of selection" (1987, 1150, our emphasis). But despite the fact that Mitchell-Olds and Shaw's paper is well known, few biologists studying selection using the multiple regression approach have adopted the methods suggested in that paper. While working biologists surely realize that there is, and can be, no straightforward way to measure selection in wild populations, the temptation to pursue the simple statistical methods, rather than engaging in the more difficult field and laboratory work necessary to understand the causal pathways actually in place, is apparently irresistible (see box 2.5).

So the assumptions built into the Lande-Arnold method make it very difficult to move from the results of regression analyses to reasonable biological interpretations of selection. But that, recall, was only part of what Lande-Arnold wanted to achieve. The hope was that the measurements of selection generated by these methods could be used in the standard

Box 2.5 The Problem of the Environment

Rausher (1992) focused on a problem pointed out by both Lande and Arnold (1983) and Mitchell-Olds and Shaw (1987): the possibility that selection analyses will be biased by unmeasured environmental effects (see fig. 2.3). Rausher proposed to avoid this problem by feeding the multiple regression analyses with "breeding values" rather than phenotypic data. In this context, breeding values are the genotypic trait means: the mean values of the traits associated with particular genotypes, given the environmental variation in the population. Genotypic means cannot be measured directly in wild populations, where all one has available is a collection of individuals and their particular phenotypes, but they can be estimated in a variety of ways if one can sort the organisms by genotype into environments in a controlled way. Given certain assumptions, breeding values will bypass the effect of unmeasured environmental factors.

One critical assumption demanded by Rausher's method is that the phenotypic covariance of a trait with fitness be understood as composed of a series of *independent* contributions, such as genetic effects, environmental effects, and so forth. If this were true, then one could partition these factors out statistically and extract the genetic component of the covariance, by definition freed of the unwanted environmental effects.

While empirical tests of Rausher's contention have been ambiguous (see Stinchcombe et al. 2002; Kingsolver et al. 2001), the logistical and conceptual problems with using breeding values are clear. The logistical problems include carrying out a carefully designed breeding program and transplanting the progeny into the field, something that is practical only with a few model organisms. Even where this is possible, such breeding programs and transplant studies have the potential of altering the measurement conditions to the extent that it is questionable that one is really measuring "natural" selection at all when using this approach.

However, there are two major conceptual problems with the use of breeding values that are more relevant here. Both Rausher's (1992) original proposal and the modification implemented by Stinchcombe et al. (2002) are based on the classic partition, in quantitative genetics theory, of phenotypic variance (and, by extension, covariance) into genetic and environmental components (i.e., portions of the phenotypic variance that are correlated with known differences among genotypes or environments). In terms of an analysis of variance, this looks like

$$V_P = V_G + V_E + V_{err}. \tag{2.7}$$

That is, the observed phenotypic variance (or covariance) can be partitioned into genetic, environmental, and so-called "error" components. Note that the equation as written here includes variation in phenotypic plasticity—the different ways in which different genotypes react to environmental variation—only as part of the error term. But genotype-environment interactions (G $\times$ E) are both ubiquitous and of enormous evolutionary importance—hardly something to be dismissed as "error." Rausher's correction, however, relies on removing an *overall* effect of the environment: if there is significant G $\times$ E interaction, this environmental effect may be swamped by the G $\times$ E term. This, of course, is an empirical matter, but as G $\times$ E interactions can be accurately determined only by dramatically altering the field settings (by carefully controlled breeding experiments and the manipulation of envi-

(*continued*)

Box 2.5 (*continued*)

ronmental variables), attempting to determine G × E in the field gets one even further away from measuring "natural" selection.

The second conceptual problem with Rausher's proposed improvement on the Lande-Arnold method is that it is actually based on a rather questionable causal model of environmental effects on phenotypes. If we refer back to figure 2.3, we see that the unmeasured environmental factor *simultaneously* affects a given trait and fitness, thereby creating a "spurious" covariance between the two, which the observer will interpret as directional selection on the trait. But how could an environmental factor affect fitness *directly*? Under any reasonable biological scenario (including those, very similar to each other, advanced by Mitchell-Olds and Shaw and by Rausher in this context), one would expect any environmental effect on fitness to be *mediated* by intervening phenotypic traits. For example, it sounds reasonable to say that a patch of high-nutrient soil will increase both, say, the number of leaves produced by a plant and any reasonable measure of reproductive fitness, such as fruit production. However, mechanistically speaking, the two effects are unlikely to be independent. Presumably, at least part of the reason that higher nutrient levels *cause* enhanced fruit production is that the plant is able to generate more leaves because it develops on a high-nutrient patch. More leaves mean a higher photosynthetic output, hence more energy to devote to growth and reproduction, eventually resulting in increased fitness. Hence, the problem of unmeasured environmental factors reduces to the problem of unmeasured, but relevant, phenotypic traits that mediate those environmental effects. Rather than eliminating environmental factors, the goal ought to be to include them explicitly as causal pathways in statistical models of the population in question.

If this general scenario for the origin of spurious phenotypic correlations induced by the environment is correct, then we certainly do not want to make any correction for environmental variation, because environmental variation here acts to augment the observable phenotypic range for both phenotypic traits and fitness, thereby making it more likely that we can identify the underlying causes of selection. Indeed, the evolutionary biology literature is rich with splendid examples of environmental, genetic, or artificial manipulation of phenotypes aimed at increasing the phenotypic variance observable in a population, and therefore the experimenter's chances of testing causal hypotheses about adaptation (e.g., Sinervo and Svensson 1998; Cipollini and Schultz 1999; Preziosi et al. 1999; Schmitt, Dudley, and Pigliucci 1999; Svensson and Sinervo 2000).

quantitative genetic equations in order to make predictions about the evolution of the traits under study. In addressing this aspect of the program, Mitchell-Olds and Shaw (1987) argued that using the resulting selection gradients in the multivariate breeder's equation in order to predict the long-term results of selection involves a number of assumptions; namely, that

> (1) many genes contribute to genetic variances and covariances; (2) genetic variance-covariance matrices remain approximately constant through time, which implies weak selection and large population size; (3) new or unmeasured

> environments do not alter **G**, **P**, or **z**; (4) there is no genotype-environment correlation or genotype by environment interaction. In addition, inbreeding and nonadditive gene action may greatly complicate prediction of response to selection. (1159)

Given that these assumptions are known generally not to be true, it is hard to imagine a clearer exposition of the strict limitations of the evolutionary quantitative genetic research program; indeed, these facts have been used repeatedly by several authors to criticize the overreach of evolutionary quantitative genetics (e.g., Turelli 1988; Barton and Turelli 1989; Shaw et al. 1995; Pigliucci and Schlichting 1997). And yet, Lande-Arnold-style selection analyses are by far the most common type of study of selection published in the recent and current literature.

At this point, the crux of the problem with studies of selection in the Lande-Arnold style should be increasingly clear: Evolutionary biologists are interested in testing causal hypotheses about selection, not simply in obtaining statistical descriptions of patterns of covariances among characters in natural populations. Yet multiple regression analyses, regardless of whatever improvement may be suggested on their basic template, simply are not a suitable tool for testing causal hypotheses (again, see Lewontin 1974 in the context of ANOVAs). The proper role of Lande-Arnold-style calculations is to summarize the statistical "shadow" of a population (see the prelude) and to suggest hypotheses to be tested, as Mitchell-Olds and Shaw (1987) remarked. But testing those hypotheses requires more, and more complex, work than do the regression analyses usually performed.

One well-understood method for testing causal hypotheses (not just in evolutionary biology) on data gathered at the population level relies on path analysis. Path analysis is, of course, a statistical approach, but the crucial difference between it and standard analyses of variance or multiple regression is that it can be used in a *confirmatory* rather than an exploratory mode. The tools to apply this method to the problem of testing selective hypotheses have been repeatedly proposed in the evolutionary literature. These suggestions have rarely been taken up for a number of reasons, chief among them the insistence—stemming from Lande and Arnold's original work—that selection analyses have to yield quantities that can be plugged into the equations describing the long-term evolution of quantitative characters (something that, at present at least, cannot be done with the results of path analysis).[4] If, however, as we have argued, these equa-

4. But see Scheiner and Callahan 1999.

tions are not very useful for the purposes of tracking or predicting the long-term evolution of wild populations, path analytical methods should look much more tempting, and it is to a consideration of such an alternative that we now turn.

IF NOT REGRESSION ANALYSIS, THEN WHAT? PATH ANALYSIS AND TESTING CAUSAL STORIES

In a particularly crisp analysis, Shipley (2000) discussed in detail the relationship between correlation (or covariance) analyses and analyses of causation in biology (see the prelude). The confusion in this area has always been great (which is why, for example, Lewontin had to publish a crucial paper on heritability and the concept of the norm of reaction in 1974, one that has influenced generations of researchers). The problem here is not the hackneyed "correlation does not equal causation" refrain, but rather the more subtle issues that result when multiple biologically plausible underlying causal structures can produce the same observable patterns of correlations (or covariances).

As we mentioned in the prelude, Shipley's apt metaphor for the relationship between causes and patterns is that of the wayang golek, an Indonesian form of theater in which characters in a story appearing on a screen are actually the shadows of wooden puppets behind the screen. All we can observe are the bidimensional projections on the screen, and we have only indirect evidence regarding what the three-dimensional shapes behind the screen—the causal mechanisms responsible for the shadows—actually are. But while it is true that a given shadow does not uniquely identify a specific shape—that is, one cannot go from correlation to causation—any particular shape can cast only certain shadows and not others—that is, causal hypotheses can be tested by the relationship between the correlation patterns they predict and those that are actually observed. While different casual stories involving, for example, past selection and past population structures can explain the correlations observed among phenotypic characters today, many possible causal stories can be excluded as not compatible with the correlation patterns actually observed. In this way, different classes of causal hypotheses can be tested.

Enter a seminal paper on measuring selection by Crespi and Bookstein (1989) that called for the use of path analysis in this context, and which has been followed by several other similar calls (e.g., Crespi 1990; Kingsolver and Schemske 1991; Mitchell 1992; Pigliucci and Schlichting 1996; Scheiner, Mitchell, and Callahan 2000; Scheiner et al. 2002). Path analysis is a statistical technique of which regression analysis can actually be

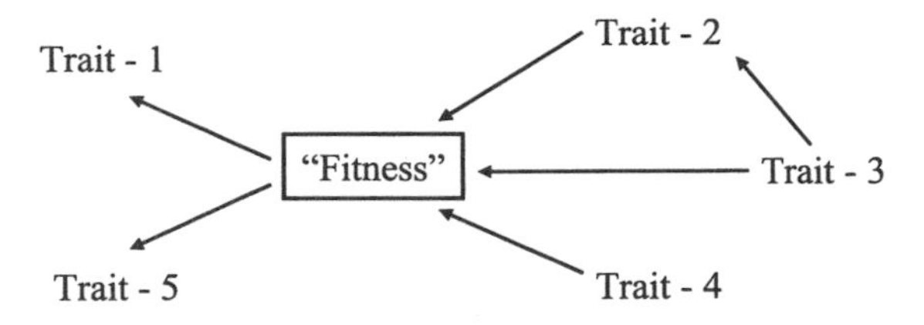

Figure 2.4. A simple structural equation model of the hypothesized relationships among a set of variables, some of which are unobserved directly (*box*). A researcher can construct several of these models based on previous knowledge of a given system, and then compare the variance-covariance matrices among variables predicted by the model(s) with the one actually observed. Models that accurately predict the observed covariance structure are retained for a further round of testing—triggered by additional observations or by experimental manipulation—while models that fail are discarded. (Additional effects due to unmeasured variables are not included in the diagram for simplicity, but they are part of the model, and can provide an important clue to the investigator as to the degree of completeness of the model itself.)

considered the simplest possible version. Path analysis itself is, in turn, a particular version of the broader category of structural equation modeling (SEM), although in the following discussion we refer to the latter two categories interchangeably. Path analysis allows the researcher to construct causal models of a given phenomenon based on the hypothesized hierarchical interaction of several variables, some of which may be unobservable directly (the so-called "latent" variables in SEM, such as "fitness" in fig. 2.4).

Crespi and Bookstein clearly understood and explicitly stated that Lande-Arnold-style multiple regressions and path analysis reflect two very different ways of thinking about measuring selection, and indeed, aim at two distinct goals (which they see as complementary): "The [Lande-Arnold] multiple-regression model provides a statistically optimal prediction, but the coefficients can be interpreted in causal biological terms only if the characters independently and completely determine survivorship. This assumption is unwarranted for virtually all selection studies" (20). This is contrasted with what SEM offers: "the explanation of the covariance among the characters, morphological factors, survivorship, and fitness, with the ultimate goal being to determine the causes and objects of selection" (27). Crespi and Bookstein clearly see path analysis as an interactive tool that allows investigators to set up a series of competing models of the causal structures underlying the covariances among phenotypic traits and other observable phenomena. These models can then be compared by their ability to predict the observed covariance structure. Models that fail can be eliminated, while those (not necessarily one) that survive the test are then used to suggest what additional information needs to be collected, for example, by measuring more characters or by experimental manipulation. The new data are then confronted with the surviving models (and possibly with additional ones suggested by new studies), and the cycle continues. Crespi and Bookstein agree with Endler (1986) and Mitchell-Olds and Shaw (1987) that "biological understanding of the causes of selection is ultimately the only way to determine how selection occurs on phenotypes" (26).

Notice also (see figs. 2.3 and 2.4) that path analysis aids the biologist enormously in three very difficult tasks, one of which has not even been mentioned so far. The first task, to which we have repeatedly referred, is the identification of additional, as yet unmeasured, factors affecting the system under study. Path analysis includes a series of tools to estimate not only the overall fit of the model to the data (as does multiple regression), but also the amount of variation *in each factor* that remains unexplained. The latter information can point the investigator toward areas of the causal model that may require further thought, thereby aiding in the search for additional variables that need to be measured on the following round. The second crucial conceptual advantage of SEM is that it can incorporate latent variables that cannot be directly measured. Clearly, "fitness" is one such variable. In all the preceding discussion, we have stayed clear of the Pandora's box of what exactly fitness is and how to measure it (see, e.g., Rosenberg 1978, 1983; Mills and Beatty 1979; Sober 1984; de Jong 1994; Matthen and Ariew 2002; Ariew and Lewontin 2004). Path analytical approaches avoid the problem of measuring fitness directly; rather, one must make explicit arguments (partly embedded in the path diagram) for considering particular observable variables (e.g., fruit production, biomass, survival rates, etc.) to be directly caused by the underlying fitness of an organism. This solution is simply not allowed in the context of multiple regression analyses, so the advantages of path analysis should be obvious. Third, recall the problem with potential environmental effects that may affect measured phenotypic traits, thereby biasing our conclusions on where selection is acting. Path analysis provides a simple and elegant way to include any number of environmental variables in the causal model (as opposed to attempting to "eliminate" them), provided, of course, that one has actually done the painful job of measuring such variables *at the individual level.* Any such environmental variable is simply added in the proper places in the path diagram, where it contributes to the prediction of the statistical shadow projected by the population.

Given the clear conceptual merits of path analysis and the equally great limitations of multiple regression analyses of selection, why do Crespi and Bookstein recommend that biologists conduct both, especially given that it is the latter, weaker program that is mostly practiced? Again, it is because of the promise of the Lande-Arnold approach to yield numbers that can be used in equations predicting long-term evolutionary change: as Crespi and Bookstein (27) put it, "the selection coefficients computed by the path model cannot be combined directly with genetic data to yield expected evolutionary trajectories, as can the partial coefficients of multiple regression." But as we have already noted, even if the parameters can be estimated accurately, the "expected" evolutionary trajectories will be accurate only over a very short time; they may therefore be useful for plant or animal breeding, but the long-term prediction of evolution in specific populations is simply not possible given the complexity and nonlinearity of the problem at hand.

We would like to suggest that this puzzling insistence on pretending that the regression analytical approach provides predictive power emerges from the well-recognized but underappreciated phenomenon informally referred to as "physics envy." Ever since Fisher (1930) likened his "fundamental theorem of natural selection" (see chapter 8) to the second principle of thermodynamics, evolutionary biologists—like many of their colleagues in other disciplines—have insisted on taking physics as the model of how science should proceed. The previous chapter explored some of the reasons why the "force" metaphor is tempting, but certainly its link to physics is one reason. And the hallmark of physics ever since Galileo and Newton has been the ability to make long-range predictions of natural phenomena—for example, the positions of planets at (almost) any given time in the past or future.[5] A theory of evolutionary forces, combined with a method for deriving values for those forces that could be plugged into the standard equations of population genetics, seems to imply that a similar project would be possible for biology. But historical sciences such as evolutionary biology are fundamentally different from largely ahistorical sciences such as (parts of) physics (see, e.g., Cleland 2002); the explanatory goals of evolutionary biology are quite different from those of, say, statistical mechanics. Just as contemporary physics does not take predicting specific instances of the decay of unstable isotopes of elements to be part of its goal (because it is impossible), neither should evolutionary biology take predicting the details of the particular evolutionary paths of specific populations to be part of its goal (since it is equally impossible, albeit for different reasons).

5. Of course, much of modern physics itself has moved away from this simplistic view of how "real" science works. Consider, for example, the field of nonequilibrium thermodynamics, in which long-term quantitative predictions are impossible because of the inherent nonlinearity and chaotic behavior of the relevant physical systems. Nobody, however, has suggested that this somehow makes thermodynamics less than a full-blown science.

The moral of this story, we suggest, is that one of the most influential papers in the recent history of evolutionary biology, the one published by Lande and Arnold in 1983, has led both theoretical and empirical investigators (including one of the authors of this book) down a less than optimally productive path for more than twenty years. This does not diminish the importance of Lande and Arnold's seminal contribution: without it, and without the several attempts at improving on it, people like Crespi and Bookstein (and, of course, us!) would have had a much more difficult time articulating what should be done in the study of natural selection. The progress of science cannot be likened to a single ever-ascending trajectory, and wrong turns and slowdowns are often crucial to finding our way forward.

The pivotal question here, as Crespi and Bookstein so clearly put it, is about what we want to accomplish in the quantitative study of natural selection: whether we want to obtain a set of statistical coefficients that can be plugged into equations describing the long-term evolution of populations, or whether we want to use methods that can aid us in testing causal hypotheses about the action of selection. The Lande-Arnold approach and its subsequent improvements attempt to pursue the first goal. Path analysis and SEM are clearly suited to go after the second goal. Crespi and Bookstein suggest that both goals are worthwhile, but we disagree: in our view, the Lande-Arnold approach should be restricted to providing the initial steps (the statistical summary of the population's "shadow") from which to branch into more sophisticated statistical and experimental approaches to testing causal hypotheses. It is hard to find papers in the literature in which the predictions of the first goal have ever been made, and essentially in no case have such predictions been subjected to empirical tests! Therefore, our efforts should instead be concentrated on the much more complex and scientifically rewarding task of dissecting the causal structure of selection in natural populations. As a bonus, properly understanding why the first goal is not worth pursuing may even help some historical scientists to get over their physics envy.

On a final note, we should, of course, observe that the above discussion is framed entirely in terms of statistical analyses, both those that are problematic when it comes to causal inference (analysis of variance, multiple regressions) and those that can be used (with much caution) in inferential investigations (path analysis, structural equation modeling). However, the scope of the literature on the causal study of natural selection is much broader, including combinations of genetic and phenotypic manipulations (e.g., Sinervo and Basolo 1996; Schmitt, Dudley, and Pigliucci 1999) as well as phylogenetic comparative studies (e.g., Larson and Losos

1996; Hansen 1997). Our discussion has been limited to inherently statistical approaches because they are the most popular and the most fraught with the kinds of conceptual ambiguities that interest us. But it is important to make full use of the toolbox available to evolutionary biologists in studying natural selection and, by extension, the process of adaptation (see chapters 5–8).

3 The Targets and Units of Selection

Individual Events and Population Dynamics

THE SAME MISTAKE WOULD BE MADE BY A CHILD WITNESSING THE MARCH-PAST OF A DIVISION, WHO, HAVING HAD POINTED OUT TO HIM SUCH AND SUCH BATTALIONS, BATTERIES, SQUADRONS, ETC., ASKED WHEN THE DIVISION WAS GOING TO APPEAR.
—GILBERT RYLE, *THE CONCEPT OF MIND*

Following our reasoning in chapter 1, we argue here that it is important to distinguish between two senses of fitness and natural selection: (1) as the result of individual interactions between organisms with particular traits and the physical processes that impinge on them, and (2) as the statistical distribution of changes in the makeup of populations. This distinction points toward a possible resolution of the units of selection problem in biology. When we are thinking of natural selection as a phenomenon at the individual level, the *targets* of selection are determined by the traits that the physical processes actually interact with. When, however, we are thinking of natural selection as a statistical phenomenon at the level of populations, the *units* of selection are determined by whatever kind of trait we use for "bookkeeping"—that is, to keep track of the statistical changes. While there is sometimes a best trait to use for these bookkeeping purposes given a particular population, in other instances several different kinds of traits are equally well suited to the bookkeeping task. Given this conclusion, both Dawkins's metaphor of the "selfish gene" and more recent defenses of gene-centric interpretations of selection are shown to be misguided; genes are neither the only entities that are causally relevant to selection, nor are they privileged for bookkeeping.

THE LEVELS OF SELECTION DEBATE

What kinds of entities does natural selection act on—individual organisms, genes, species? The so-called "levels of selection" debate is one of the longest and most acrimonious in modern evolutionary biology, and it has helped to define philosophically sharp distinctions among various schools of thought and the major figures that helped to establish them. We argue here that recognizing the different senses of fitness and selection (see chapter 1) can help to resolve the debate; once one stops thinking of selection, drift, and other (supposedly) causal explanations in evolution as analogous to forces and embraces the distinction between the informal and formal levels of analysis, it is easier to see how to accommodate the various intuitions at play in discussions of the level of selection question in evolutionary biology.

There are two major intuitions at work in the levels of selection debate in biology. The first is that, insofar as different physical processes are responsible for the differential survival and reproduction of organisms (as surely they are), those physical processes will interact with different traits of the organisms (or other entities) involved, and therefore natural selection will be best characterized by attention to the particular traits that the physical processes interact with. In this view, while it is usually phenotypes that are selected (because it is individual organisms that live or die, reproduce or not, and they do so because of particular interactions between particular physical processes and aspects of their phenotypes), other kinds of units can be selected under particular circumstances (box 3.1). The second intuition is that, insofar as evolution can be thought of as changes in gene frequencies in a population (the standard textbook definition of evolution), the best place to look in order to characterize natural selection is at the genic level.

Dawkins (1976) went even further in defense of the second, genecentric view, arguing that genes are the "fundamental" things that are replicated, and hence are the primal source of heritable fitness differences. It follows from this argument that the significance of the interaction between organismal traits and those physical processes associated with natural selection is only apparent, and the real interaction that matters to selection is always between those processes and genes. But of course, this aspect of Dawkins's line of argument will not be tempting if one accepts that the genetic system is just one of several systems of inheritance and that other, nongenetic systems can be evolutionarily important. For example, if one takes seriously the role of epigenetic developmental processes (e.g., Oyama, Griffiths, and Gray 2001) or research on epigenetic inheritance systems (e.g., Jablonka and Lamb 2005), the idea that the only entities of evolutionary importance are genes will seem misguided. Overall, neither view on levels of selection seems to be able to deal in a wholly satisfactory way with the intuitions that drive the alternate one.[1]

1. A comprehensive review of the literature devoted to the units of selection debate is, of course, beyond the scope of this chapter. For a good summary of the two main positions as outlined here, see Sterelny and Griffiths (1999) and citations therein; for a summary of some of the recent work on the relationship between the levels of selection debate and the units of selection problem, see Okasha (2003).

Box 3.1 Entities and Organisms

Part of the units of selection debate is about what kinds of entities can be the object of selection. Here we briefly discuss the kinds of entities that have been suggested as candidates and why they have been suggested (for a different perspective, reaching some of the same conclusions we arrive at, see Wagner and Laubichler 2001).

Individual organisms. Individual organisms are the most obvious candidates to be the objects of selection, as it is usually they that reproduce or fail to reproduce, that live or die, and so on. Indeed, most classic examples of (informal) selection involve individual organisms with either physical or behavioral traits that make them more or less likely to survive and reproduce than organisms with traits closer to the mean of the population. Selection takes place through physical processes that discriminate on the basis of different physical or behavioral traits, and what gets selected (in the informal sense) are individual organisms that possess particular variants of traits (or not). While one sometimes speaks of the traits being "selected," strictly speaking, it is the entire organism that is selected (in the informal sense) when a physical process interacts with a particular trait of that organism.

Genes. In Dawkins's (1976) original articulation of his "selfish gene" metaphor, genes were the only entities of evolutionary significance. The central idea started out as a popularization of earlier work by Hamilton (1963) and by Williams (1966). Both of these authors intended their work to be a refutation of the then rampant fuzzy thinking about group selection (Wynne-Edwards 1962), and both later retreated considerably from the extreme gene-centric view (see, e.g., Williams 1992).

As noted above, it is usually individual organisms that are either reproductively successful or not, not the particular alleles that those organisms happen to possess. In general, then, it makes little sense to speak of genes being selected. That said, it seems undeniable that genes can be the direct targets of selection, at least under certain restricted circumstances. Indeed, in some cases, even the "selfish gene" metaphor seems apt. The *SD* allele in *Drosophila* and the *t* allele in the house mouse are excellent examples of segregation distorters, and hence of truly selfish genes that increase their reproductive success even at the expense of their host organism. Other plausible candidates include highly repetitive DNA sequences, transposons (discussed below), and supernumerary chromosomes (which might be thought of as "selfish chromosomes" rather than selfish genes: Camacho, Sharbel, and Beukeboom 2000). The best-known mechanisms used by these selfish elements are transposition and meiotic drive (Hatcher 2000; Hurst and Werren 2001).

In these cases, it makes sense to think of genes as the objects of selection; the physical processes involved in their replication interact more or less directly with the particular sequences of nucleic acids that make up the genes themselves, and hence it is the genes that are being selected (in the informal sense) by the relevant physical processes. Note that, in order to be considered genic selection, the selection has to take place at the genic level, and the entity reproduced has to be the gene. For example, we would suggest that in *Drosophila*, the selection of the *SD* allele counts as genic selection because it is the structure of the allele itself that prevents gametes carrying that allele from being poisoned, not a downstream phenotype that is the result of complex developmental interactions.

Sub-gene elements. Highly repetitive DNA sequences, such as the human *Alu*, are very short (180 to 280 bases) noncoding regions that cannot be considered genes even under the most

(*continued*)

Box 3.1 (*continued*)

permissive definition (e.g., Waters 1994). Through errors of duplication, which their shortness and repetitiveness facilitate, these sequences can grow in number dramatically, up to the point of constituting a large fraction of an organism's genome: for example, millions of copies of *Alu* are present in the human genome (Kazazian and Moran 1998), amounting to about 10 percent of the total. Transposons are mobile genetic elements that are flanked by DNA sequences that make it particularly easy for them to excise themselves from a chromosomal location and insert themselves into another location, on either the same or another chromosome. Some transposons are capable of leaving copies of themselves behind; since often their DNA does not code for anything of use to the organism, they can rightly be thought of as selfish sequences.

Again, as in the case of genic selection, sub-gene elements are selected by physical processes that operate at or very near the level of nucleic acid sequences, and hence it is these stretches of DNA that are selected.

Groups and superorganisms. When selection "sees" features of a group in a way that makes the features of the individual organisms that make up the group invisible to the physical processes involved, it makes sense to think in terms of the group being selected. Wade's flour beetle experiments are an excellent example (Griesemer and Wade 1988; Wade 1977), as he set them up so that the features of the individual beetles would be irrelevant to whether a particular group was selected. In nature, it is often harder to distinguish cases of "true" group selection from cases in which individual selection results in group characteristics, but it seems likely that some of the purported cases of group selection are real in the sense described here (Sober and Wilson 1998).

Species and higher taxonomic levels. Finally, it has been suggested that species (or even higher taxonomic units) can be considered individuals in a sense that permits natural selection to sort intertaxonomic differences in a way analogous to the action of selection on individual organisms. Stanley (1975) articulated an early theory of species selection based on Eldredge and Gould's (1972) idea of punctuated equilibria. The most recent, and excruciatingly detailed, argument for the possibility of species selection was articulated by Gould (2002).

In order for species selection to occur, one needs to make a reasonable argument that (a) species can be considered individuals (e.g., because they are characterized by definite existence in time and space), and that (b) at least some of the characteristics of species are not directly reducible to those of the individuals that constitute them (e.g., speciation rates). The degree to which either of these conditions obtains in nature is still controversial and under intense debate.

In order to account for both sets of intuitions, one must make a distinction between the *individual*-level and the *statistical*-level (or ensemble-level) concepts of fitness and natural selection (as we argued one ought to do in chapter 1, following, e.g., Matthen and Ariew 2002). Insofar as one is concerned with the individual level, one's attention will be focused on the physical processes that impinge on the organisms (or other entities) themselves, and in particular, on the different ways in which those processes influence the survival and reproduction of organisms, given variation in the organisms' traits. Here, where one is concerned with *informal* fitness and informal selection (to continue the use of the terminology adopted in chapter 1), one will naturally focus on particular traits that interact with specific processes when identifying the level at which selection is taking place. These traits can be referred to as the *targets* of selection; it is these targets that particular physical processes interact with.[2]

On the other hand, when one is focused on the statistical level, one's attention will be directed not at any particular physical process, but rather at the statistical distribution of the ratios of the growth rates of subpopulations identified on the basis of some trait (the "trait populations" as we defined them in chapter 1). Where one is concerned with *predictive* fitness and *formal* selection, the trait one uses to pick out subpopulations for analysis will determine what kinds of evolutionary events one can detect, but there is, in principle, no right kind, or level, of trait to focus on (this point is made by Walsh, Lewens, and Ariew 2002). The question really is one of "bookkeeping"—some traits make for easier bookkeeping than others, and some evolutionary events will show up only when certain kinds of bookkeeping devices are used rather than others.[3] The traits one uses for bookkeeping may be called the *units* of selection—"units" having a nice noncausal sound to our ears. Insofar as one considers evolution to be changes in gene frequencies, and thinks of the technical work of population genetics as representative of evolutionary theory more generally, one will naturally choose to pick out subpopulations on the basis of genetic traits, and hence think of (formal) selection as being about the genic level. Of course, if one rejects this view in favor of, say, the more pluralistic approach endorsed by supporters of developmental systems theory

2. There are reasons, broadly following Ariew and Lewontin (2004) and Walsh (2003), why one might suggest that these traits are not in fact targets of selection, but rather simply targets of discriminate physical processes; in this view, functional adaptations are not to be explained as the result of selection, but rather as the result of the particular physical processes involved in certain reliable developmental processes that themselves are present because of the statistical phenomenon that *is* natural selection. In table 3.1, we note that, in our view, discriminate processes *may* be said to select on the basis of the traits they interact with. It is in this sense that the unit of selection depends on the particular physical processes that interact with particular traits, when one is focused on the individual level.

3. However, if in fact all evolution involved changes in gene frequencies (a claim that we reject; see below), then it is likely that any evolutionary event could be captured at the predictive level by dividing up the population of interest on the basis of genetic traits; that is, by forming trait populations on the basis of genetic markers. If this were so, then the genic level would in fact be the maximally adequate representation for bookkeeping purposes (Sterelny and Kitcher 1988).

(Oyama, Griffiths, and Gray 2001) or the students of epigenetic inheritance systems (Jablonka and Lamb 2005), one will think that a variety of different bookkeeping methods will be necessary to capture all that is interesting about evolutionary processes (box 3.2).

As we noted in the previous chapter, moving between claims about individual-level phenomena and claims about statistical-level phenomena is, unfortunately, often difficult. In general, the complexities inherent in natural populations prevent an easy transition between levels. However, even in realistically complex cases, it is sometimes possible to draw some conclusions about what might be happening at one level from information gleaned at another level. Careful attention to what is going on at the individual level can, for example, focus our interest on some possible subpopulations rather than others as potentially interesting objects of study at the statistical level. Similarly, how particular kinds of outcomes at the statistical level are possible can often best be explained by attention to the individual-level processes that occur in representative examples of specific populations.[4]

INDIVIDUAL INTERACTIONS, INFORMAL FITNESS, AND THE TARGETS OF SELECTION

When one is focused on the particular ways in which individual entities interact with the physical processes that influence their survival and reproduction, it is generally (but not always) possible to determine whether those processes are discriminate or indiscriminate with respect to those traits. In this case, the informal fitness of organisms with particular variants of a trait is understood to be about the relationship that the trait can be expected to have with a particular kind of physical process. For example, with respect to the physical processes involved in escaping predators, fast gazelles may be fitter than slow ones. Informal selection is the result of discriminate physical processes interacting with particular variants of traits in ways that influence the survival and reproduction of the organisms with those variants. To continue the example, we can observe that slow gazelles are more often killed by predators than are fast ones. Running speed, in this example, is a *target* of predation.

When we are looking at the individual entities involved, then, the

4. This feature—that attention to a lower-level phenomenon can often explain how what happens at the higher level is possible without being able to account for the details of the higher-level phenomenon—may be a quite general characteristic of the world. See Dupré (1993) for discussion and details, as well as our conceptual framework as laid out in the prelude.

Box 3.2 Is Evolution Just Changes in Gene Frequencies?

Evolution is often described as changes in the allelic frequencies in a population: where there is such a change, evolution has occurred; where there is no such change, there has been no evolution. Futuyma (1998), for example, claims that "natural selection can have no evolutionary effect unless phenotypes differ in genotypes" (365). The intuition that leads to this conclusion is, apparently, that the genetic system is the only system in which variation is reliably heritable, and so it is only changes in the genetic system that can be of evolutionary significance. The suggestion that phenotypic differences that are not reflected in genetic differences might be heritable and of evolutionary significance often meets the accusation of "Lamarckism" (see discussion in Jablonka and Lamb 2005). This view, however, is empirically inadequate and conceptually unclear.

It is first worth noting as a matter of empirical fact that there are changes of evolutionary significance that do not involve changes in gene frequencies, or indeed, any genetic modifications at all. A classic, and rather common, example is that of a change in the host plant of an insect that happens to lay its eggs on a different plant species than usual, a simple instance of a behavioral change that can be heritable (and of evolutionary significance) without having a genetic change as its basis; this phenomenon is thought to be a significant factor in the patterns of speciation in butterflies (see Nice et al. 2002; Jaenike 1990) and some varieties of *Drosophila* (see Etges and Ahrens 2001). Similar cases involving birds with parasitic brood behaviors have also been reported (see Sorenson, Sefc, and Payne 2003). More robust examples, involving heritable changes in phenotypes, are easily found. Cell structure, for example, is inherited not primarily through genetic pathways, but through physical membrane-membrane interactions, and changes in these interactions can result in alterations in the cell structure that are inherited. Indeed, such changes have been implicated in speciation events (see Moss 2003, 96) and are thought to be central to the rise and diversification of symbiogenetic life (such as the role of chloroplasts in plants) (see Cavalier-Smith 2000). Similarly, epigenetic mechanisms involving heritable variations in chromatin condensation and in methylation patterns (parental imprinting) are increasingly recognized as sources of heritable variation that are not based on genetic variation as it is ordinarily understood (see Sollars et al. 2003, with commentary in Pigliucci 2003b). While such heritable structural changes can, in time, come to be stabilized by genetic changes (see Moss 2003; Sollars et al. 2003; West-Eberhard 2003), it is unnecessary to require genetic changes before acknowledging that evolution has occurred.

Even in cases in which phenotypic variation is associated with a genetic difference, modifications of gene frequencies may fail to capture important evolutionary changes in the population. David Magnus (1998) imagines a simple case in which, due to a change in the environment, the fitnesses of three genotypes (*AA, aa, Aa*), which had previously been neutral, change such that while the homozygotes remain equally fit, the heterozygotic condition becomes lethal (fig. 3.1). Even if there is no change in allelic frequencies in the population after the change, the population is different in ways that are of evolutionary importance—for example, it seems more likely under these conditions that the population will split into two reproductively isolated groups, and hence that speciation might occur!

In general, we consider evolution to be any heritable changes in the frequency or distribution of any traits of the population in question, no matter by what mechanism(s) such traits are heritable.

(*continued*)

Box 3.2 (*continued*)

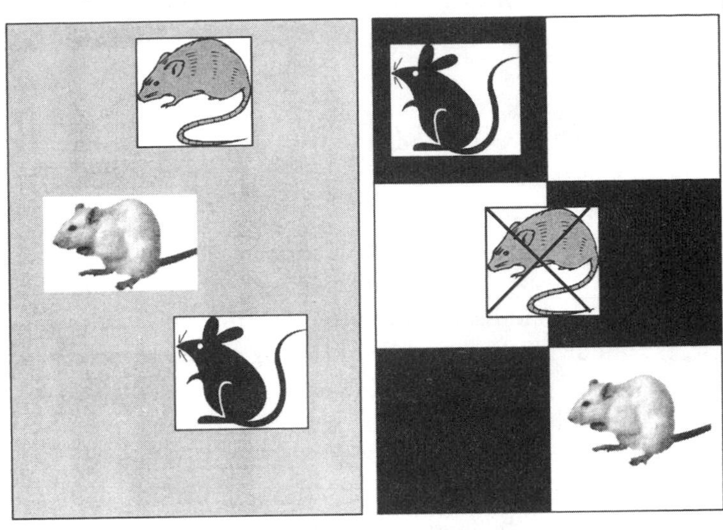

Figure 3.1. David Magnus (1998) recounts the following case, originally reported by Benson. (*Left*) Populations of the "pocket mouse" usually display extensive variation in color, and under "normal" field conditions, no particular color provides a selective advantage at all times. (*Right*) In the Tularosa Basin in New Mexico, however, the ground is either nearly black volcanic rock or nearly white sand; these unusual conditions are the result of relatively recent volcanic activity. Correspondingly, in these areas, either white pocket mice (in the sand dune areas) or black pocket mice (in the rocky zones) are found, but no "in-between" mice are present. Let us assume that color is determined by a single locus with two alleles (*A* and *a*), such that the original population is made up of three colors of mice corresponding to the *AA, Aa,* and *aa* genotypes. If the colors are selectively neutral and the alleles are in Hardy-Weinberg equilibrium (see chapter 10), the frequencies of the two alleles will each be 0.5 (and the frequencies of the three genotypes, respectively, 0.25/0.5/0.25). After the environment changes, the frequencies of the two alleles will still each be 0.5, but the distribution of the genotypes will be different (0.5/0/0.5), as will the likely evolutionary trajectory of the population. It is because of cases like this one that Magnus argues that defining evolution in terms of changes in allelic frequencies is inadequate.

levels of selection debate refers to informal selection, and informal selection can take place at several distinct levels, depending on what particular traits certain physical processes are interacting with in a potentially discriminate way. Those who argue that the levels of selection question has, in general, a pluralistic answer (e.g., those who broadly follow Lewontin 1970) are probably thinking of selection in this way. The general form of this argument against genic selectionism is that genes are "invisible" to many (but of course not all) of the processes involved in informal selection. The notion of visibility here is related to, but distinct from, the notion of discrimination as used above. Here, a trait is visible to selection via a particular physical process if that process can interact with that trait in a way that could affect the informal fitness of the entity sporting the trait.[5] Note that the issue of visibility is about whether the trait in question could interact with the process in a discriminate way, whether it in fact does so or not; therefore, not every visible trait actually affects informal fitness, and visibility ends up being necessary, but not sufficient, for discrimination.

The difficulty, again, with moving between informal selection (and fitness) and formal selection (and fitness) is that the interactions of the many processes and traits involved with the former are, in general, fiendishly complex. So, for example, while running speed might improve informal fitness with respect to predation, it might also be associated with, say, lower disease resistance, and hence might lower informal fitness with respect to disease (another physical process). How this plays out in any one case of an individual's reproductive success may not tell us much about how these traits will fare in the population as a whole, because the particular details of that individual's life may be idiosyncratic. These kinds of facts will therefore tell us even less about how these traits will perform, on average, in populations of this kind. When one considers not just two related traits, but all the traits and their interactions, the problem is likely to be intractable (attempts to measure selection on correlated individual traits notwithstanding; see chapter 2).

However, to continue the example above, just because running speed may be negatively associated with disease resistance, that does not imply that running speed is not selected for (in the informal sense) with respect to predation. Since the process of predation "sees" running speed, informal (individual) selection can operate on it. That is, when we observe organisms with variations in running speed, and consider the physical process of predation, we recognize that that process is discriminate with respect to that trait. This is true even if, at the level of predictive fitness, there is no difference between subpopulations selected on the basis of that trait because of the negative correlation of that trait with another discriminate process (or because, for example, the variations in that trait happen not to be heritable in a particular population).

So, assuming that—for example—Lewontin's explanation of the prevalence of the *t* allele in natural populations of house mice is correct (see

5. This notion of visibility shares with Brandon's use of Salmon's terminology of screening off (see Brandon 1982 and, e.g., Sober 1992 and McClamrock 1995 for discussion) an attention to proximal rather than remote causes. It differs, however, from the idea of screening off in that the causes that characterize visibility are physical interactions between particular processes and traits, and so the notion of visibility does not take into account the features of the causal chain necessary for the technical property of screening off to apply (for example, that the probabilities of particular links in the causal chain not be equal to one—see Sober 1992, 142).

Futuyma 1998 and citations therein for discussion), informal selection is operating at several different levels simultaneously. The segregation distorter called the *t* allele (actually several linked loci) operates at, or very close to, the genic level, and sperm with the *t* allele are selected for; there is, in other words, a physical process that is discriminate with respect to the *t* allele. However, male mice with two copies of the *t* allele are sterile, and here informal selection operates at the level of the physical traits involved in the production of viable sperm; that is, there is another physical process that is discriminate with respect to *t* allele homozygotes. Finally, groups with a higher than usual prevalence of the *t* allele may go extinct at a higher rate, whereas groups with a lower than usual prevalence may found more new groups; here informal selection works at the level of group traits, and there is a discriminate physical process that works on the founding of new groups.[6] When these three kinds of informal selection are combined, the net result in natural populations is that there is no difference in the growth rates of populations divided by *any* of these traits—the mean ratio of the growth rates of subpopulations selected on the basis of the *t* allele (absence, heterozygosity, homozygosity), or on the basis of the individual phenotypic features of the mice, or on the basis of related features of groups of mice, will be one. Hence, there will be no difference in formal fitness for any of these traits, and no selection at the formal level for any of these traits.

One possible complication in the story we have outlined so far emerges from the observation that different levels of selection and different traits may not make separable contributions to the organisms' fitness—that is, there may be no way to separate out in the neat way suggested above the contributions made by having or lacking the *t* allele, belonging to a particular group, and so on. Robert Wilson refers to this problem as "entwinement" and argues that different levels of selection often "do not make isolatable, distinct contributions to the ultimate evolutionary currency, fitness" (2003, 544). This is surely correct, but it is much more of a problem for moving between the informal (individual) level and the ensemble (formal) level than it is for thinking about informal fitness in the

6. This last is the most controversial claim about this case in several ways. First, empirically, this may not be the best explanation for the difference between the expected ratio of *t* alleles and the number in natural populations (for a summary, see Futuyma 1998). Second, even if this is what occurs, many would argue that it does not count as a true case of group selection, since the relevant features of the groups are determined wholly by the features of the individual organisms. However, those who support calling this a case of group selection might point out that what selection (in the informal sense) "sees" in this case are features of the group, and that the extinction or founding of groups are primarily the results of group features.

sense developed in chapter 1. Informal fitness is in no sense part of any "ultimate evolutionary currency," and the kind of informal selection suggested in the example above does not involve having to figure out how the various causal factors interact. The latter problem is, as we argued in chapter 1, likely to be intractable in many cases, which is why one cannot move from the causal physical processes that take place at the level of individuals (of whatever kind) to predictions about the average outcomes in similar populations.

In Dawkins's early arguments for the primacy of the gene, he insisted that the causal root of all heritable fitness differences was genetic. Indeed, for the early Dawkins, genes (not organisms) were the only things that could properly be said, in an evolutionary sense, to be fitter or less fit in an environment (box 3.3 addresses the tricky question of what to count as a "gene," whose answer depends on the context). The only variation that could be of evolutionary significance was, therefore, in traits that are genetic in the sense that variations in those traits can be mapped neatly onto variations in genes. If in fact the relationship between genetic and phenotypic differences were always (or even predominately) so straightforward, selection at the individual level might well be reducible to genic selection, in some sense. But even here, we would suggest that the target of selection is the phenotypic trait; that the phenotypic trait is caused by the presence of a particular gene does not change what the physical process interacts with. A gene that spreads throughout a population because it is associated with a particular trait that confers a major fitness advantage may, we suggest in chapter 6, be properly called a gene *for* that trait, but that does not imply that the gene is itself a target of selection. This is therefore not a case of genic selection, if we are thinking of selection as a process (thinking of selection in the individual, informal sense).

In any event, for the most part, the relationship between genes and phenotypes is nothing like that straightforward; the ubiquity of epistasis and pleiotropy, combined with the complex systems of epigenetic inheritance that research is uncovering (see, e.g., Sollars et al. 2003; Moss 2003; Oyama, Griffiths, and Gray 2001; Jablonka and Lamb 2005), suffice to make any view that takes the gene as causally central at the level of informal selection and fitness untenable. Indeed, even relatively straightforward complications of the genotype-phenotype mapping function can result in problems for the gene-centric view. As Sober and Lewontin (1982) noted, when there is heterozygote advantage, as in the famous case of the relationship between certain forms of the hemoglobin-coding gene, resistance to malaria, and anemia, it does not make sense to calculate the fitness of one or the other allele at the locus, nor to think of selection (for,

Box 3.3 What Are Genes?

There are at least two different questions we might be asking when we ask what "genes" are. The first is normative: What definition of "gene" *ought* we to use in biology? The second is descriptive: How do working biologists *actually* use the term "gene" in their practice? Neither is trivial to answer. The second question is currently the subject of a research project aimed at analyzing the correlations between the different definitions of "gene" used by practicing biologists and the different branches of biology they work in, when, where, and by whom they were trained, where they work, and so on (see Stotz and Griffiths 2004). How we answer the first question depends at least in part on what we think reasonable answers to the second one include, and on how we view the history of the concept.

In Dawkins's (1976) original articulation of genic selection, genes were something like the evolutionarily stable entities that were the causes of phenotypic differences; Dawkins claimed that these entities were nucleotide sequences. This position turned out to be untenable on empirical grounds, as nucleotide sequences are neither stable in the way Dawkins suggested, nor are they the causes of phenotypic differences, except under very particular conditions. Dawkins (1982) later acknowledged the latter point, but argued that all evolutionarily significant events could be rearticulated in terms of the fitness of particular alleles (a position made clearer by Sterelny and Kitcher 1988); however, the definition of "gene" in Dawkins's later articulation remained vague at best.

Lenny Moss's conception of the gene is suggestive in this context. Moss (2003) argues that genetic research has been shaped by the conflation of two very specific gene concepts, which he terms "Gene-P" and "Gene-D." The Gene-P concept emerges (roughly) out of Mendel's elements, and genes in this sense are simply those things associated with statistically significant differences in phenotypic outcomes, whatever they may be and however they may work. In this sense, knowing what version of a Gene-P an organism has gives one some information about what the likely phenotypic outcome will be; knowing, for example, what combination of alleles associated with pea color a pea plant has helps one to predict what color the plant's peas will be (*ceteris paribus*).

The Gene-D concept, on the other hand, refers to particular DNA nucleotide sequences that play a role in organismal development. Here, the specific nucleotide sequence is one of a number of resources that are necessary for development; while variation in the availability or type of this resource can make a difference developmentally, so too can variation in many other resources. In the Gene-D sense, one cannot properly speak of the gene as carrying information about phenotypic organismal features. Moss suggests that the discovery that DNA is intimately connected with the formation of the myriad proteins involved in organismal development (and cellular function more generally) permitted the conflation of the two concepts. Genes, then, became both nucleotide sequences and predictors of phenotypic differences, although there is no one entity to which both concepts refer, and any attempt to make one definition refer to both will eventually flounder on the incoherency of the conflation.

While we agree that there is massive confusion in the literature between, for example, genes as something like markers associated with particular differences in phenotypes in specific environments and genes as DNA sequences involved in development more generally, we are less convinced by Moss's particular historical account. For example, we think it

(*continued*)

Box 3.3 (*continued*)

likely that talk of "genes for" phenotypes has been, since its earliest days, wrapped up in the functional and hence evolutionary significance of the genes in question. So we tend to read H. Morgan's (and other early Drosophilists') claims to the effect that "the future of genetics [lies] in its connections with development and evolution" (from Kohler 1994, 177) as implying that, despite the technical limitations of the day, those early researchers hoped that the genes they were discovering would at least provide inroads into developmental and evolutionary biology. And later, the work with the *PERIOD* gene in *Drosophila* (carried out by Benzer, Hall, and others; see Kohler 1994 and citations therein) seems to us to have been wrapped up from the beginning in searches for biological meaning (and evolutionary significance), not just for a gene-phenotype relationship that happened to hold in a particular population of fruit flies in a particular laboratory environment (Weiner 1999). Elsewhere, we suggest that, if done well, it is possible for this kind of "gene for" talk to be neither strictly Gene-P nor Gene-D, and yet to avoid most of the pitfalls of conflation that Moss points toward (see chapter 6 and Kaplan and Pigliucci 2001).

Similar arguments can be raised against the other articulations of the gene that are currently popular (e.g., Waters 1994). For our purposes, no single conception of the gene is adequate. Rather, it is important to recognize that the concept of the gene that it makes the most sense to use depends critically on the level of analysis one is focused on. At the ensemble (formal) level of predictive fitness and natural selection, genes are essentially markers—it doesn't matter what (if anything) they actually do, as they are just a way of keeping track of the trait populations under observation. At the individual level of informal fitness and selection, genes must be functional units—binding sites, templates, and so forth—because it is only functional units that physical processes can interact with in discriminate ways. The details will, as usual, depend on what sort of research project one is engaged in.

say, malaria resistance) acting on a particular allele, because these fitness values depend on which other allele is present in any given diploid organism as well as on the external environment (presence or absence of malaria).

Later, Dawkins (1982) backed off the extreme gene-centric view in which the only things selection saw and acted on were genes, opting instead for a kind of pluralism. However, he maintained that any population changes of evolutionary significance could be rearticulated as changes in gene frequencies, and that any fitness differences could be reformulated as differences in the fitness values of particular alleles, given a particular selective regime.

Before we move on to a discussion of the so-called bookkeeping argument, it is worth noting that it is irrelevant if one is focused on the individual level. This is because a process is discriminate with respect to a particular trait if it interacts with that trait in a way that discriminates between trait variants. This means that, by definition, every discriminate process influences informal fitness. Physical processes are discriminate (or not) with respect to certain traits because of the causal interactions they are involved with; it is not enough for particular kinds of outcomes to be statistically correlated with the processes involved. However, at the ensemble (statistical) level, one cannot distinguish between cause and correlation, and it is to that level that we turn our attention now.

ENSEMBLES OF POPULATIONS, PREDICTIVE FITNESS, AND THE UNITS OF SELECTION

As Walsh, Lewens, and Ariew (2002) point out, when we are thinking at the ensemble level, there are no physical processes involved. Physical processes affect particular organisms that escape predators or not, reproduce or not, and so forth, but ensembles of populations are inherently statistical in nature, and hence are not the kinds of things that interact with particular physical processes. That is why the ensemble level can be used to understand predictive fitness and formal selection, and why predictive fitness and formal selection can be reliably determined only by attention to the ensemble level. As noted in chapter 1, predictive fitness is simply the mean ratio of the growth rates (or some other measure of relative success) of the subpopulations picked out by the trait of interest; where the mean ratio is one, there is no fitness difference, and where it is different from one, there is a fitness difference. Predictive fitness *supervenes*[7] on the physical properties of the population involved: it is, of course, because of those physical properties that populations have the predictive fitness they do, but many different sets of physical properties can map onto the same differences in predictive fitness. Because the physical processes involved are, for reasons of complexity, epistemologically uninformative about the statistical facts of the mean and distribution of growth ratios, there is no "right" answer to the question of what units one should use to keep track of those statistical facts.

However, if the rearticulation of Dawkins that Sterelny and Kitcher (1988) propose is correct, then for any ensemble in which the ratio of the growth rates of the subpopulations (picked out by any traits of interest) is not equal to one, there will be a way of dividing the population into mul-

7. Property A *supervenes* on property B if and only if changes in B do not necessitate changes in A, but changes in A necessitate changes in B. So, for example, the amount of money in a coin dish supervenes on the actual coins in the dish. One can change the coins in the dish without changing the amount of money in the dish (by replacing, say, a nickel with five pennies), but one cannot change the amount of money in the dish without changing the coins in the dish.

tiple trait populations on the basis of genetic markers that captures that difference in predictive fitness. And of course, insofar as evolution is defined as changes in gene frequencies in populations, there will always be a way of picking out the trait populations on the basis of genetic traits such that evolutionary change is identified. Whenever there is a change in the mean allelic frequencies, there is natural selection that can be found by dividing up the population on the basis of genetic traits. In these cases, dividing up the population on the basis of other kinds of traits (individual phenotypic traits, or even group-level traits, for example) might serve to identify the same shift in the mean growth ratios of the trait populations so identified, but it might not. If this line of reasoning were correct, then the genic level would be the "maximally adequate" level to use to represent evolutionary change (Sterelny and Kitcher 1988, 358), although in specific cases, traits at other levels might be adequate as well, and might, for example, make some aspects of the situation more perspicuous.

On the other hand, if, for example, one takes the insights of developmental systems theory seriously (see Oyama, Griffiths, and Gray 2001 for an overview), the position that evolution just *is* changes in gene frequencies will probably seem untenable (box 3.4; see also box 3.2) (this view, incidentally, was captured early on by neo-Darwinian synthesist Ernst Mayr with his famous characterization of population genetics as "bean bag genetics" [Mayr 1980]). Any instance in which variation in some trait interacts with a discriminate physical process such that there is an informal fitness difference, and in which that variation is reliably reproduced, by whatever mechanism (i.e., not necessarily genetic), will be available for discriminate physical processes to interact with, and hence available for informal natural selection (note that taking these insights seriously is also what makes the "replicator/interactor" distinction untenable; see box 3.5).[8] At the formal level, this implies that in these cases, there may be a difference in predictive fitness between trait populations identified by heritable differences in traits where those heritable differences do not have a genetic component and would therefore be missed by an analysis that divides up the population into trait populations on the basis of genetic differences (see box 3.2 and fig. 3.1 for examples).

What is clear, however, is that sometimes (but not always), dividing a population up on the basis of multiple different traits can yield (roughly)

8. Wagner and Laubichler (2001), also unhappy with the simplistic distinction between replicators and interactors, have begun developing a "context-dependent" mathematical approach that focuses on whatever units of inheritance—not necessarily genes—may be the focus of attention. Their aim is to provide a more general treatment of the mathematics of evolution that is not constrained by the classic quantitative genetic framework.

Box 3.4 What Is Developmental Systems Theory?

Developmental systems theory (DST) refers to a perspective on evolution and organismal development (see Oyama, Griffiths, and Gray 2001 for review). Supporters of DST reject the view that genes guide or control development in any interestingly robust way, and reject the notion that genes can be usefully understood to be any sort of repository of information (similar conclusions, though by different means, have been arrived at by Schlichting and Pigliucci 1998, West-Eberhard 2003, and Jablonka and Lamb 2005).

Oyama, Griffiths, and Gray (2001, 2, table 1.1; see also Robert, Hall, and Olson 2001) list six major "themes" in DST:

1. Joint determination by multiple causes: Every trait is produced by the interaction of many developmental resources. The gene-environment dichotomy is only one of many ways to divide up these interactors.
2. Context sensitivity and contingency: The significance of any one cause is contingent on the state of the rest of the system.
3. Extended inheritance: An organism inherits a wide range of resources that interact to construct that organism's life cycle.
4. Development as construction: Neither traits nor representations of traits are transmitted to offspring. Instead, traits are made—reconstructed—in development.
5. Distributed control: No one type of interactor controls development.
6. Evolution as construction: Evolution is not a matter of organisms or populations being molded by their environments, but of organism-environment systems changing over time.

Note that these claims are in part empirical and in part represent a conceptual shift. DST proponents ask that the empirical evidence they provide for the inadequacy of, for example, the view that genes control development be taken seriously as evidence that the standard view of genes as providing information about development, and everything else being essentially raw material, is insufficient.

the same subpopulations and reveal (roughly) the same mean differences in growth rates between the trait populations so identified. In these cases, the way one divides up the population will be a matter of convenience. Such pluralism about the levels of selection is a pluralism of multiple adequate representations, not of multiple causal factors operating simultaneously.[9] Indeed, because the predictive level makes no reference to causal factors at all, it is obvious that the different ways of generating the same results demonstrate the equivalence of particular kinds of mathematical representations rather than causes. Now, if it were always the case that, for

9. For example, the ability to keep track of certain kinds of group dynamics using only individual fitness parameters is elegantly defended in Kerr and Godfrey-Smith's "Individualist and Multi-level Perspectives on Selection in Structured Populations" (2002); in their response to commentators, however, they note that the mathematical equivalencies they develop do not address the causal structures of the changes in the populations in question (2002, 549).

Box 3.5 Replicators, Interactors, and All That

Following Dawkins (1976), there has been an extensive literature on the different conceptual (and perhaps physical) roles played in biology by *replicators* (the things that replicate) and *interactors* (the things that interact with the environment) (see Sterelny and Griffiths 1999 and citations therein). For Dawkins, genes were the only replicators–the only things that were reliably replicated and maintained in reproduction through evolutionary time. Bodies were the interactors ("vehicles" in his terminology)–the things that interacted with the outside world–but had significance only insofar as the traits that interacted were linked to the genes that could be replicated.

While the conceptual distinction between those things inherited across generations and the phenotypes that interact with the world has a long history (see, e.g., Mayr's [1982] discussion of Weismann's work in the nineteenth century), we regard any sharp distinction between replicators and interactors as having outlived its usefulness. Replication itself demands, in general, the organism's ability to interact. Genomes do not "self-replicate"; rather, their replication requires the coordinated actions of a cell, and the replication of a cell requires, at the very least, a compete cell situated in and interacting with the right kind of environment. There is, therefore, no way to distinguish in general between the things that are replicated, the things that do the replicating, and the things that interact.

example, any differences in the fitness of groups within a population could be discovered by dividing the population up into individuals that differed with respect to the particular traits those individuals possessed, the resulting level-of-selection pluralism would be what Robert Wilson calls (tongue firmly in cheek) "*Neckerphilia*": performing calculations at different levels of selection would be like viewing a Necker cube in different ways—changing views would not change the information available (Wilson 2003, 535).

However, there are good reasons to reject the claim that model pluralism of this sort is generally true. While in many cases there are likely to be a number of ways in which to divide up the population of interest and still reveal (approximately) the same predictive fitness differences, there is, we think, no kind of trait (genetic trait, individual phenotypic character trait, group phenotypic character trait, etc.) that can be used to do so in a way that will always reveal any fitness difference that there is to find; a fortiori, there is no level (gene, individual, group) that can capture all the changes in a population that are of evolutionary interest. Rather, different situations will call for dividing the population up on the basis of distinct kinds of traits if differences in predictive fitness are to be discoverable. This is a different kind of pluralism at the formal level: to capture all the evolutionary changes in a population that are of biological interest, multiple different ways of dividing up the population will generally be required. The way in which one chooses to divide up the population can not only change one's perspective, but can also result in a shift in what kinds of changes one finds.

It is worth stressing here that, as a matter of practical reality, the information necessary to accurately determine the predictive fitness of a particular trait is very rarely available. The reason for this conclusion is straightforward: the predictive fitness of a trait is given by the mean ratio of the growth rates of subpopulations identified on the basis of that trait; it is, however, very rare that enough populations that are sufficiently similar can be observed to generate an accurate approximation of the mean ratio. Indeed, in natural cases, usually only one population can be observed, and unless the standard deviation from the mean ratio of the growth rates is known to be very small on independent grounds, very little information about the mean can be gleaned from a single trial (a specific empirical example will be discussed in chapter 4, in the context of the study of the evolution of genetic variance-covariance matrices in *Drosophila*). In most extant natural populations, the standard deviation around the mean predictive fitness (i.e., the mean ratio of the growth rates of subpopulations) is likely to be nontrivial, so what happens to the subpopulations in any individual extant population will not accurately reflect the predictive fitness of those subpopulations, as this, again, is a fact about the mean ratio in ensembles of populations.[10] The only way to discover the predictive fitness of traits is with replicates of the population of interest; in laboratory populations, these can be literal replicates (see, e.g., Rose, Nusbaum, and Chippindale 1996 and citations therein), whereas in natural populations, the best one can do is to approximate replicates by looking at changes in a single population over time (though here, of course, changing environmental conditions make this an approach of dubious effectiveness).

If predictive fitness, as Matthen and Ariew (2002) sometimes suggest, is mainly used by and of interest to mathematical population geneticists, the conclusion that we usually do not have access to the relevant information about predictive fitness is, perhaps, not terribly surprising. John Dupré's often harsh criticisms of population genetics, for example, hinge on the claim that, for the most part, population genetics as a field has had

10. The distinction here is between being confident that there is a difference in predictive fitness and knowing, quantitatively, just what that difference is. In the first case, confidence might be generated from a combination of observed changes in the population and field studies that point toward particular differences in informal fitness; the second case, however, requires accurate determination of the mean value, and hence generally requires replicates or substantial additional information.

little success (or even apparent interest) in interacting with the real world to predict or even explain what occurs in natural populations (see, e.g., Dupré 1993, 131–42). Dupré blames this failure in part on population genetics as a field directing attention toward elements that are at best not obviously causally significant to the processes at work. But, if we take Matthen and Ariew's interpretation of population genetics seriously, that criticism misses the mark; predictive fitness makes no mention, nor can it make mention, of the physical processes that operate at the level of individual interactions. Predictive fitness is a statistical fact about ensembles of populations, and whatever causal processes are at work at the individual level are not a part of those statistical facts. It is only if population geneticists believe themselves to be discovering underlying causal structures that they are open to the attack Dupré suggests; while some surely do, the interpretation suggested here reveals why they should not.

It is for this reason that a focus on predictive fitness and formal selection can be said to miss key aspects of the evolutionary dynamics of populations. In the case of, for example, heterosis (i.e., heterozygote advantage), a population at equilibrium will show no mean formal fitness differences between the genotypes. In the formal sense, there are no fitness differences. To recall an oft-cited textbook example, in an area where malaria is endemic, the *HbS* "mutant" allele will be in a stable equilibrium with the *Hb* "wild type" allele, due to the heterozygote advantage of the *HbS/Hb* combination (partial malaria resistance). At the individual (and informal) level, the phenotype associated with the heterozygotic condition has a higher informal fitness vis-à-vis malaria resistance, and malaria is the process that results in informal selection for that phenotype. Similarly, the homozygotic condition *HbS/HbS* is less fit with respect to, say, the processes involved in oxygen exchange and blood flow, and those processes select against the homozygote. Hence, the homozygote *Hb/Hb* is fitter than *HbS/HbS* vis-à-vis gas exchange, but less fit than *HbS/Hb* vis-à-vis malaria. However, because of the relationship of the phenotype to the developmental substrate that produces it, the population as a whole will be at equilibrium (see Matthen and Ariew 2002, 71–77, for a discussion of the notion of a substrate in this context, and for details about how the specification of a substrate within a particular evolutionary context can account for so-called constraints).

Contrast this case with a superficially similar situation in which the genetic variants under consideration are really neutral—for example, where they are noncoding sequences. Here, again, there are no differences in formal fitness between the variants, and the mean expected outcome may be the same as the expected outcome in the case of heterosis just discussed (see fig. 3.1 for a related case). But the two cases are importantly different.

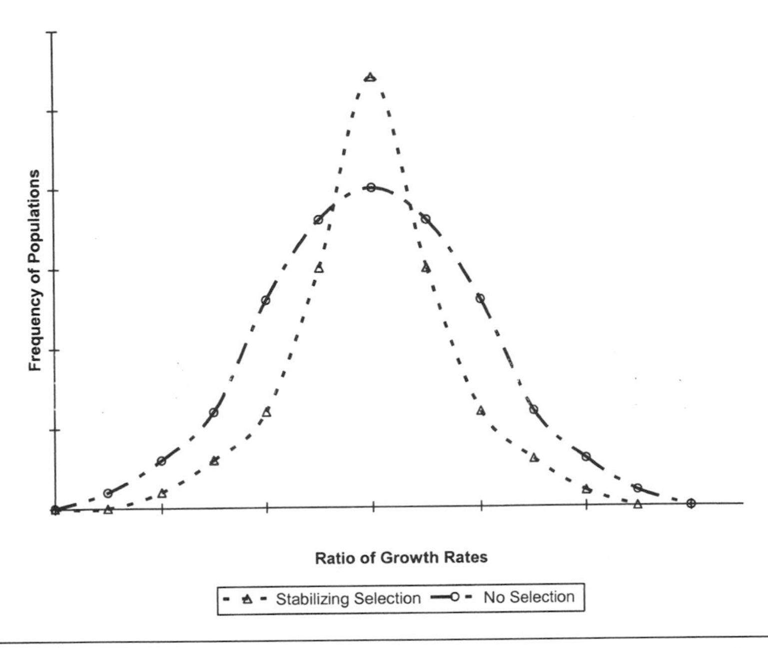

Figure 3.2. Hypothetical distribution of ratios of the growth rates of the trait populations of two populations. While there is no formal difference in fitness between the variants of the trait used to divide up either of these populations, in one case there is stabilizing selection (as evidenced by the reduced variance) and in the other there is *no* selection. Note that this comparison is proper only when the populations are of the same breeding size and basic structure.

In the case of heterosis, we can think of the case of the equilibrium distribution of genotypes as a case of stabilizing selection, and here we would expect the variance in growth rates around the mean predictive fitness difference in the ensemble of populations to be smaller than in the case of a set of truly neutral variations, given populations of the same size; while the fitness is the same, selection is different in the two cases (fig. 3.2). For heterosis, the formal (ensemble) level of fitness misses some of the evolutionary dynamics of the population: physical processes differentially impinging on the different phenotypes, followed by the re-creation of the original gene and phenotype frequencies due to the particular process of reproduction in play, followed by another round of informal selection, and so on. While nothing is happening at the predictive (formal) level, the informal fitness variations in these traits may be of great interest if one is attempting to understand the evolutionary significance of the traits in question. In the case of variation that is neutral at the informal level (neutral for every physical process that might matter), there is, of course, nothing to miss.

This problem can perhaps be seen even more clearly in distinguishing

Box 3.6 The Hawk-Dove Game versus Heterozygote Advantage

The "Hawk-Dove" evolutionary game is a classic case in which the evolutionarily stable strategy (ESS) is a stable equilibrium of strategies, rather than a single strategy. In the Hawk-Dove game, we imagine that players must adopt one of two strategies for resolving conflicts over resources. Hawks refuse to back down, and they will fight for the resource in question until one is seriously injured. Doves will back down before a fight breaks out. When a Hawk interacts with a Dove, the Hawk gets the resource. When a Dove interacts with a Dove, each has a 50 percent chance of getting the resource, and so, on average, earns half the resources. When a Hawk interacts with a Hawk, each has a 50 percent chance of getting the resource, and (at least) a 50 percent chance of being seriously injured. If being seriously injured is costly compared with the value of the resources, there will be a stable equilibrium of Hawks and Doves in the population. A population composed entirely of Doves can be invaded by Hawks; similarly, a population composed entirely of Hawks can be invaded by Doves. A stable equilibrium of strategies will therefore be maintained by frequency-dependent selection in the population.

Imagine further that the Hawk and Dove strategies are associated with the following genotypes: *hh* = Hawk, *hd* = Hawk, *dd* = Dove (the Hawk allele is dominant). Under these conditions, there will be a stable frequency of *h* and *d* alleles maintained in the population.

While the maintenance of particular proportions of alleles in the Hawk-Dove case is similar to the maintenance of particular proportions of alleles in the case of heterosis (both show no difference in formal fitness coupled with stabilizing selection at the formal level), in one case there is frequency-dependent selection at the informal level, and in the other case there is not.

cases of "true" frequency-dependent selection from heterosis. In the above example of heterozygote advantage, the fitness of each allele depends on the frequency of the other allele in the population; however, at the informal level, there is no physical process that is frequency-dependent in this way. The fitness of each of the three *phenotypes* is independent of the frequency of the other phenotypes, and again, it is only a quirk of the developmental and inheritance systems that produce the phenotypes that results in there being stabilizing selection for a particular proportion of each allele. Contrast this case with, for example, the "Hawk-Dove" evolutionary game, in which the strategies (the phenotypes) themselves are subject to frequency-dependent selection through the action of the same physical process that maintains the population at the equilibrium frequencies (box 3.6). Again, the lesson here is that very different processes at the informal level can result in the same patterns at the formal level.

The intuition, then, that there is something sterile about genic selectionism rearticulated as a bookkeeping device emerges from at least two different considerations. First, the units of formal fitness and selection are causally impotent with respect to formal selection and predictive fitness—formal selection and predictive fitness do not make mention of causal processes, nor can they. Second, in natural populations, it is usually impossible in practice to discover the predictive fitness of a trait, and impossible in principle to observe formal selection (which is, after all, a statistical distribution, and therefore cannot be observed by looking at any particular instance); this intuition implies that the units of selection chosen when looking at the formal models of evolutionary theory can have little to do with the processes and events one focuses on when one is looking at natural populations (where informal selection and informal fitness are appropriate). However, as long as there are targets of (informal) selection available for consideration as well (where these are the traits that the discriminate processes in question actually interact with), one ought be satisfied with any kind of units of (formal) selection that reveal the evolutionary processes of interest in the particular case, whether they are causally relevant traits or not.

CONCLUSIONS AND OTHER ISSUES

Making a sharp distinction between the individual level and the statistical level in evolutionary biology has, we suggest, profound implications for many areas of biology and philosophy of biology. Matthen and Ariew (2002) focused on the implications this distinction has for thinking about the causal structure of evolutionary theory, a theme we took up in chapters 1 and 2. In this chapter, we have paid particular attention to its consequences for the levels of selection debate. Traditional arguments regarding the invisibility of genes to many kinds of selection processes, or the ability of facts at the phenotypic level to screen off the genotype from selection, are irrelevant when applied to the statistical (the formal) level. Claims about genic selection's ability to represent evolutionary change are likewise irrelevant to the individual (the informal) level. No single conception of selection fits both levels, and so our inability to reconcile the different intuitions regarding the levels of selection is rooted in a philosophically important distinction (table 3.1).

This conclusion suggests that other topics of discussion in biology and the philosophy of biology would be affected by the distinction we have been examining. For example, if these two different ways of thinking about fitness and natural selection are correct, the key information regarding (predictive) fitness and (formal) selection is the mean of the ratios of the growth rates of the subpopulations identified and their distribution around the mean ratio (as much as this may often be epistemologically

Table 3.1. Summary of the characteristics of the three levels of analysis at which to think of selection and fitness used throughout the book

Level	Processes	Entities	Fitness	Selection	Drift
Individual (i) level	Physical processes that influence survival and reproduction (food gathering, predation, attracting mates, mitosis, etc) of the entities in question, which can be either discriminate or indiscriminate with respect to particular traits, and are usually stochastic in nature	The individual entities that the processes affect the survival and/or reproduction of (genes/proteins, organisms, groups)	**Informal fitness:** the propensity of an entity with some trait to survive and reproduce, given the way that trait interacts with the particular kinds of physical processes under consideration	The **discriminate processes** that influence the differential ability of organisms with particular traits to survive and reproduce may be said to "select" on the basis of the traits they are discriminate with respect to	Not well defined
Population level	Processes are the same as in i-level, just summed over all the organisms that make up that particular population	The actual subpopulations (trait populations) picked out by the traits the processes at the i-level affect	Not well defined. What happens to a subpopulation happens because of the particular, unique set of individual interactions at the i-level. No sense of "fitness" is given by the actual outcome of these interactions.	Not well defined. What happens to a subpopulation happens because of the particular, unique set of individual interactions at the i-level. Natural selection is inherently statistical, and not given by any particular actual outcome of these interactions.	The difference between the actual ratio of growth rates of the subpopulations and the ratio of growth rates given by predictive fitness (as an outcome measure)
Ensemble (e) level	Not well defined. Processes are features of individual physical interactions or series of interactions, *not* of statistical trends.	Collections (ensembles) of subpopulations picked out by traits of interest (ensembles of trait populations)	**Predictive (formal) fitness:** the difference in the mean ratio of the growth rates of the subpopulations under consideration from one (or some similar measure of the relative success of the relevant trait populations)	The **statistical distribution of outcomes** around the mean ratio of growth rates of the subpopulations	Not well defined; although the scatter around the mean ratio of growth rates of the subpopulations can perhaps be thought of as drift.

unobtainable). If the distribution of the ratios of growth rates in a population undergoing formal selection is radically different from the expected distribution of a population undergoing a nonselective "random walk" (see chapter 10), such that it is relatively infrequent that possible outcomes of the two cases are similar, then most of the actual changes that we observe in particular populations of this sort will probably be attributable to formal selection; on the other hand, if the distribution of the ratios of growth rates in a population experiencing formal selection overlaps substantially with the distribution of the ratios of growth rates of a population that is not experiencing selection, then we will be unable to identify formal selection in many cases (although we may be able to identify informal selection and, through careful study of the population, make some reasoned guesses about formal fitness differences). For example, in a case of directional selection in which the formal fitness difference is relatively large compared with the standard deviation of the distribution of the ratios of growth rates, most of the actual changes that we observe in particular populations will reflect the difference in formal fitness of the trait variants (the actual changes will probably reflect the correct *direction* of the differences in fitness, although those changes may not accurately reflect the *magnitude* of the differences). Conversely, where the standard deviation of the distribution of the ratios of growth rates is large compared with the distance of the mean ratio of growth rates from one, most of the actual changes that we observe in particular populations of this sort will be due to the stochastic nature of the physical processes involved, rather than to formal selection. As suggested in chapter 1, one way of rearticulating the difference between "neutralists" and "selectionists" (chapter 10) would therefore be by focusing on the empirical question of which kind of situation more often obtains, and under what conditions we can expect each kind of situation to obtain (see fig. 1.6).

Again, we do not mean to suggest that the particular way in which we have articulated the distinction between the informal and the formal versions of selection and fitness is the only possible way to make this distinction; rather, we think that it has so far been too easy for both practicing biologists and philosophers of biology not to make any such distinction at all, and that various problems and confusion in the literature can be traced to such failure. Simply being aware that the distinction is important, and that one must be clear about the particular sense in which one is using these concepts, could go a long way toward clearing up the conceptual confusion and pointing toward the real empirical problems. That, we think, would certainly constitute a worthwhile contribution of philosophy of science to evolutionary biology.

4 Studying Constraints through G-Matrices

The Holy Grail of Evolutionary Quantitative Genetics

CHARACTERS MAY EVOLVE IN PATTERNS THAT ARE UNPREDICTABLE FROM G.
—DAVID HOULE, "GENETIC COVARIANCE OF FITNESS CORRELATES: WHAT GENETIC CORRELATIONS ARE MADE OF AND WHY IT MATTERS"

In chapter 2, we argued that Lande-Arnold-style regression analyses of selection present sufficient problems to make them useful for a much more limited range of applications than commonly assumed. In this chapter, we expand that critique to the use of the genetic variance-covariance matrix (**G**), one of the central concepts in such analyses. Specifically, we argue that for both conceptual and empirical reasons, studies of **G** cannot be used to directly infer so-called constraints on natural selection. Again, the search for a general solution to the difficult problem of identifying causal structures given observed correlations has led to the substitution of statistical modeling for the more difficult, but much more valuable, job of teasing apart the consequences of the different possible underlying causes (a processes that begins with, but does not stop at, statistical analyses). Our critique of **G** as a measure of "constraints" sets the stage for chapter 5, where we examine the language of constraints on natural selection more generally.

THE EMERGENCE OF A NEW FIELD?

In the previous chapters, we used a particular analysis of the conceptual underpinnings of natural selection to critique the use of the results of Lande-Arnold-style regression analyses as a tool for making inferences about the causal pathways of natural selection. In chapter 3, we analyzed what kinds of things can be selected, and we discussed the relationship between the kinds of things that can be selected in biological evolution and the problem of measuring (and, to a lesser extent, detecting) selection in natural populations. In this chapter, we turn to another use of Lande-Arnold-style regression analyses; namely, to discover and measure "constraints" on evolution of traits in natural populations. This is done through a particular way of interpreting the so-called genetic variance-covariance matrices (**G**) (the multivariate equivalent of heritability; see chapter 2) used in multiple regression analyses, an approach that has been greeted with enthusiasm by many evolutionary biologists. For example, Steppan, Phillips, and Houle (2002) have declared the study of **G** to be "a new emerging field" that will "provide one of the most promising frameworks with which to unify the fields of macroevolution and microevolution." Similarly, Shaw et al. (1995) have claimed that "one motivation for estimating **G**-matrices is that they will reveal the most likely paths of evolution," and Roff (2000) has called the multivariate breeder's equation (see below and chapter 2) that features **G** "a central equation of evolutionary quantitative genetics." Moreover, Jones, Arnold, and Bürger (2003, 1747) write that the multivariate breeder's equation can be used to "reconstruct the history of selection or to predict the future trajectory of the phenotype as a consequence of selection" and can provide "a connection between microevolutionary processes and macroevolutionary patterns." They claim further that the **G**-matrix "can be used to assess the population's capacity to respond to selection . . . and test genetic drift as a null model" (2004, 1639). Nor are these isolated instances; the biological literature on **G** has been copious for the last decade or so, and the pace is, if anything, quickening, with major reviews on empirical and statistical issues surrounding this topic being published in top journals (e.g., Turelli 1988; Cowley and Atchley 1992; Shaw et al. 1995; Pigliucci and Schlichting 1997; Phillips and Arnold 1999; Roff 2000; Johnson and Porter 2001; Steppan, Phillips, and Houle 2002).

Given how prominent the study of **G** and the related multivariate equations has been in evolutionary biology, it is odd that very little philosophy of biology has yet been focused on comparative (i.e., phylogenetically informed) quantitative genetics and the related concept of the genetic variance-covariance matrix. A philosophical analysis and critique of the uses and interpretations of comparative quantitative genetics and **G** is long overdue and of vital importance, not only because the conceptual underpinnings of these ideas are fundamental to modern biological theory, but also because **G** is an especially problematic concept in many ways. It is, we will argue, difficult to study empirically, and of very limited value for conceptual reasons. In this chapter, we present the beginnings of such a critique, focusing on four related points. We first attempt to explicate what evolutionary biologists who study and make use of **G** mean by the concept and why they think that **G** can be used to discover and understand constraints on evolution. Next, we explore some of the practical problems faced in studying **G**; these problems point toward some of the conceptual limitations of **G** (our third point). Finally, we address what we consider to be the most important issue of all; namely, the ways in which concepts such as **G**, if they represent the final outcome, rather than the very beginning, of a research program, can obscure more fundamental issues in evolutionary quantitative biology.

In short, we argue here that **G**, and the related multivariate analyses, cannot be properly interpreted as measures of constraints on evolution, nor do they provide us with reasonable long-term predictions of the likely pathways of evolution in natural populations. Though the statistical techniques for working with **G** are well established, the scope of its application is as limited as that of the Lande-Arnold approach discussed in chapter 2, and for similar reasons. Those statistical techniques are certainly measuring something of value to biologists, but the relationship between what the techniques are measuring and most biologists' interpretations of such measurements is rather more distant than is generally thought.

BACKGROUND: WHAT EXACTLY IS G?

G is a crucial component of the multivariate extension of the classic breeder's equation that Fisher (1930) used as the centerpiece of his analysis of phenotypic responses to selection in both natural and artificial systems. Recall from chapter 2 (eq. 2.1) that the univariate (i.e., applied to one trait at a time) breeder's equation is

$$R = h^2 S,$$

where R is the response of a trait to selection applied with strength S, and h^2 is the heritability of that trait. As noted earlier, there are a number of problems with the idea of heritability. As these problems translate to the multivariate level of **G**, it is worth briefly reiterating them here: (1) Heritability is a local measure that can, and often does, change with changes in a population's gene frequencies and the environments it encounters. (2) Heritabilities do not reveal the causal pathways in development, nor, a fortiori, do they reveal the extent to which traits are responsive to differences in genes or to differences in environments. (3) Therefore, heritabilities do not provide a useful measure of the long-term ability of traits in a population to respond to selection. All of these limitations emerge from the simple fact that heritabilities measure a covariation (which is a statistical attribute) between phenotypes and genotypes, and therefore are not necessarily an indication of particular causal pathways (or, to put it in terms of our framework in the prelude, many different causal pathways project the same "statistical shadow," which is in turn quantified by the heritability estimate).

We also noted in chapter 2 that Lande and Arnold (1983) were the first to propose that a multivariate version of the breeder's equation could be used to study the more interesting problem of multi-trait evolution. One version of the multivariate breeder's equation (see box 2.2) is

$$\Delta \mathbf{z} = \mathbf{G}\boldsymbol{\beta},$$

where $\boldsymbol{\Delta z}$ is a vector of phenotypic responses to selection applied with a trait-specific intensity, summarized by the vector $\boldsymbol{\beta}$ of selection gradients, and mediated by the genetic variance-covariance matrix, **G**. As explained in box 2.2, there are other versions of the multivariate equation, but the key concept is **G**, which is thought of as a matrix relating the different phenotypic traits (fig. 4.1), where the diagonal elements are genetic variances (i.e., unstandardized heritabilities) and the off-diagonal elements are pairwise genetic covariances (which can be positive or negative, and assume values from zero to infinity). If the pairwise genetic covariance between two phenotypic traits is zero, there is no correlation between the variation of the two traits that is attributable to genetic variation; this is generally interpreted to mean that the traits can evolve independently of each other.

Recall also from chapter 2 that Lande and Arnold (1983) proposed the use of the multivariate version of the breeder's equation to approach two long-standing problems in evolutionary biology: predicting long-term multivariate responses to selection and estimating past selection gradients. Their idea was that, given **G** and a multivariate selection regime, one could read off the likely response to selection from calculations of $\boldsymbol{\Delta z}$, the phenotypic vector. Conversely, they reasoned, one could invert the equation to derive estimates of past selection gradients $\boldsymbol{\beta}$ from knowledge of current phenotypic values in the population and, again, the genetic variance-covariance matrix.

Another conceptually important interpretation of **G** was then articulated by Cheverud (1984, 1988), who pointed out that the set of variance-covariances can be interpreted as a measure of genetic constraints on future evolution. In this sense, in the same way that the diagonal elements of **G** measure the short-term readiness of a character to respond to selection (just as h^2 does in the univariate breeder's equation), the off-diagonal elements measure how much the evolution of a given trait is slowed down (or accelerated) by the coevolution of another one. If two traits have a genetic covariance of around zero, then by this line of reasoning they are able to evolve independently of each other. If, on the other hand, they share a positive covariance, then selection on one automatically drags the other in the same direction (in proportion to the amount of covariance);

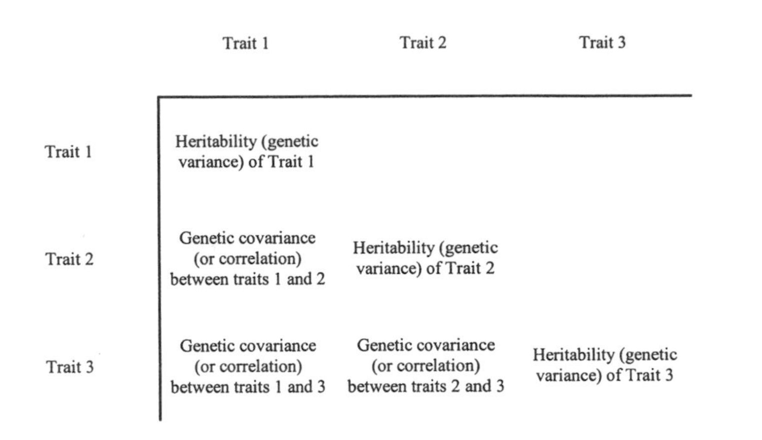

Figure 4.1. The basic idea behind the concept of a genetic variance-covariance (**G**) matrix.

whereas if the traits are linked by a negative covariance, selection to increase the value of one results in a correlated reduction of the value of the other. The higher the absolute value of the covariance, the more tightly linked to each other the evolutionary fates of the two traits would be.[1] In the same way that Lande and Arnold suggested that the equation could be inverted to detect selection in the past, Cheverud's interpretation suggests that one can invert the equation in order to detect the constraints that influenced the outcomes of past selection.

MANY PROBLEMS WITH G, MUCH TROUBLE USING G

The study of **G** and its related equations is often thought to provide insight into long-standing important problems in evolutionary theory, so it is no wonder that it has been actively embraced by practitioners of the field. But the enthusiasm surrounding the use of **G** ought not to blind us to the concept's limitations, both practical and, especially, conceptual. Indeed, if these problems are taken as seriously as they deserve to be, many of the promises of **G** turn out to be rather overstated. In this section, we first explore some of what we regard as the basic difficulties with **G**, including its likely long-term instability, the related problem of the locality of **G**, the attendant difficulty of specifying the traits and environments

1. Of course, things are not really that simple, even under the best of circumstances, because any two traits are then influenced by their joint covariance with any third trait; the trio is then influenced by its relationship to any fourth character, and so on. However, mathematical analyses and conceptual discussions of **G** rarely, if ever, step beyond the bivariate setting. We thank Lev Ginzburg (SUNY Stony Brook) for pointing this out to us.

that should be used in estimating **G**, and the problems involved in choosing a statistical method to use in estimating **G** from among the available competitors. Next, we discuss some more fundamental conceptual difficulties, concluding that the techniques that employ **G** are unable, in principle, to distinguish between cases of drift and selection, and unable—again in principle—to distinguish different underlying causal pathways more generally. This conclusion ought not to come as a complete surprise; since "drift" is not a process at all (see chapter 1) and multiple regression analyses run into conceptual problems when used as measures of selection (see chapter 2), the problem of distinguishing between selection and drift is in general not solvable by any statistical method. Further, the conceptual underpinnings **G** shares with heritability result in the two sharing certain limitations as well; as it is well established that heritability cannot be properly interpreted to reveal causal pathways, the similar conclusion regarding **G** follows logically. Finally, we discuss some of the difficulties that arise when one attempts to study **G** in practice, and we show how these difficulties make methods for generating **G** difficult to apply to many natural populations.

Some Basic Problems with **G**

The first difficulty we would like to point toward is that **G**, like h^2, is a local measure (see chapter 2); that is, any measure of **G** applies only to the particular population (with the particular genes and allelic frequencies actually extant), in the particular environment, in which **G** is actually estimated. As with h^2, if the available genes or the gene frequencies in the population change, so too might **G**; similarly, in a different range of environments (or if the population becomes differently distributed among the environments encountered), **G** might well change. Given these observations, if one wants to make use of **G** in simulations of long-term evolution (e.g., Via 1987), one must assume that the matrix stays constant (or at least proportional to the ancestral state) over the period one is investigating (generally thousands of generations). But, as several authors (e.g., Turelli 1988; Pigliucci and Schlichting 1997) have pointed out, such constancy is highly unlikely because—as mentioned above—evolution per se changes gene frequencies, and therefore **G** itself; nor is it unreasonable to suppose that the environment encountered by the population may change as well. This is not just a matter of the external environment changing in autonomous ways with respect to the population, though that surely happens often as well, but even more so of the population as it evolves finding itself experiencing a different environment *because* of its evolution (see, e.g., Lewontin 1983, Odling-Smee, Laland, and Feldman 2003). While it is reasonable to assume near-constancy of **G** for short-term applications (e.g., crop or animal breeding, artificial selection experiments, or perhaps even evolution in wild populations over a very few generations), it is quite clear that it is not reasonable to maintain the same assumption over tens of thousands of generations—that is, over evolutionary time scales.

The question of the stability of **G** over these periods is, of course, empirical, and it has generated a significant number of studies aimed at settling it. While a few authors have found constancy in the elements of **G** (e.g., Brodie 1993 for antipredator traits in two populations of garter snakes; Arnold and Phillips 1999, also in garter snakes), several others have demonstrated evolutionary divergence between populations (Waldmann and Andersson 2000 in species of the plant genus *Scabiosa;* Phillips, Whitlock, and Fowler 2001 in *Drosophila melanogaster*) or species (Paulsen 1996 in buckeye butterflies; Steppan 1997b in leaf-eared mice; Waldmann and Andersson 2000 between species of *Scabiosa*). Computer simulations undertaken by Jones, Arnold, and Bürger (2003) suggest that while under some conditions, a **G**-matrix of particular traits can be quite stable over evolutionarily significant periods, under other conditions it can vary significantly for those same periods; this means, they note, that "the empirical question regarding **G**-matrix stability is not necessarily a general question" (1747). Since changing the environment can change, for example, the kinds of selection a population is undergoing, and since the kind of selection a population is undergoing is one of the factors that Jones, Arnold, and Bürger (2003) found to influence the stability of **G**, there are good reasons to suppose that, as in the case of heritability, even if a particular **G**-matrix remains constant over one shift in environments, this does not provide evidence that it will remain constant given other environmental variations.

Questions regarding how likely it is that a **G**-matrix will be stable over time, and the difficulty of the empirical task of testing the extent to which **G** actually remains stable or changes in particular cases, lead directly into another more or less straightforward concern; namely, how one ought to go about empirically estimating **G**—how, in other words, one can actually generate the variance-covariance matrix for the traits of interest. Even when limited to measuring **G** for a particular population and environment at a particular time, this task is not trivial. While the difficulty of solving particular empirical problems is not generally considered to be a philosophical issue, in this case, the empirical difficulties are tied up with various conceptual problems, and exploring the former provides an excellent entry into the latter.

The first difficulty faced in studying **G** is finding a nonarbitrary way of choosing the traits to be used to construct the variance-covariance matrices. Asking how a *generic* **G**-matrix evolves is meaningless, as there is no such thing as a generic **G**; empirically, we will always be dealing with specific matrices measured on a small subset of all possible or even all biologically meaningful traits. Most often, in fact, we will be dealing with matrices comprising just a few traits. An obvious implication is that the stability of **G** over time depends critically on the traits actually used to construct the matrix, as well as on the phenotypic plasticity displayed by those traits in ecologically relevant environments; this has been shown both empirically and through computer simulations (see Jones, Arnold, and Bürger 2003, 2004). These facts should be obvious, but they rarely (if ever) emerge in the biological literature on **G**.

To make this problem concrete, notice that some pairwise combinations of traits tend to show stable covariances not only within, but also across species, or even across higher taxonomic/phylogenetic levels of analysis; in these cases, **G** seems quite stable. For example, most of the species of the plant family Brassicaceae have flowers characterized by two sets of stamens (the male sexual organs), in which four stamens are longer than the remaining two. While the reason for this is far from being clear (Karoly and Conner 2000; Conner 2002), the sizes of the two sets of stamens show covariation at the very high taxonomic level of an entire family of plants. On the other hand, it is well known that many traits can be uncoupled not merely within a population or species, but even within an individual organism! For example, many semiaquatic plants produce two distinct kinds of leaves in response to the particular environment in which they find themselves (below or above water; see Wells and Pigliucci 2000 for a review); parameters describing the shape of these two sets of leaves are completely uncoupled by the phenotypic plasticity that allows the plant to produce the two kinds of structures in an environment-dependent fashion. A **G**-matrix constructed using leaf shape as one of the traits would, therefore, be unstable not just over evolutionarily meaningful time scales, but even within an individual organism's developmental life cycle.

The upshot is that quantitative geneticists who focus on stamens in Brassicaceae reach completely different conclusions about the stability of genetic variance-covariance matrices from their colleagues who study leaf evolution in semiaquatic plants. Indeed, research by Waldmann and Andersson (2000) in *Scabiosa columbaria* and *S. canescens* populations found, among other patterns, that "the magnitude of (co)variances was more variable among characters than among populations"; in other words, the results of a given study of **G** depend more strongly on which traits the investigators choose to focus on than on the species being studied! Again, Jones, Arnold, and Bürger (2003) arrived at similar results through computer simulations, noting that their simulations implied that "some kinds of characters are more likely than others to have stable **G**-matrices" (1757). Similarly, no one would suggest, for example, that the stability of the heritability of a trait is independent of the trait chosen! Interestingly, however, much of the literature about **G** is written in a way that implies that the evolution of **G** is a single kind of thing, and that it makes sense to think of **G** itself, independently of the particular traits used to calculate it.

The same locality issue, of course, applies to the environments of the populations whose traits we are measuring in order to estimate **G**. Before the relatively recent resurgence of interest in the concept of phenotypic plasticity—the genotype-specific property of producing distinct phenotypes in different environments (e.g., Bradshaw 1965; Schlichting 1986; Sultan 1987; Scheiner 1993; Pigliucci 2001; West-Eberhard 2003)—it was customary for quantitative geneticists to ignore environmental effects and assume (hope?) that their results would hold regardless of what environments were used in their experiments. There were always very good theoretical reasons not to do so (Lewontin 1974), but there is now overwhelming empirical evidence that the parameters important in statistical genetics can be highly sensitive (plastic) to environmental conditions. We noted above some studies concerning the special case of heritability, but the literature is also becoming clear in the more general case of components of **G**-matrices. For example, Bégin and Roff (2001) studied genetic (co)variances in two species of crickets, each reared in two environments, concluding that "the expression of the genetic architecture can vary with the environment" and admonishing that "future studies should compare **G** matrices across several environments."

Laudable as the recommendation that **G**-matrices be compared across multiple environments might be, actually doing so will not prove easy for a variety of reasons (and, in fact, it is rarely done). First, quantitative genetic experiments, by their nature, already tend to be very large. Because one is studying (often subtle) differences among many traits, and because the statistical power of even the best crossing designs is fairly limited (see Mitchell-Olds and Shaw 1987), large samples are needed if statistically significant results are to be obtained. In order to study genetic (co)variances in multiple environments, the size of the experiment must be increased by at least as many times as the number of distinct environments one wishes to consider; the situation is even worse if more than one kind of environmental factor (e.g., temperature and quantity of light) is to be explored. Expanding the already large-scale experiments designed to estimate **G** to include multiple environments presents very serious logistical (and funding) difficulties.

Even if such difficulties can be tackled in particular cases, it is generally not at all clear what set of environments should be chosen for such experiments. It goes without saying that one cannot test all possible environments, nor even all those environments a species is likely to have encountered in its recent evolutionary history (of which we are often ignorant anyway). If one wishes to carry out research in the field, the logistical challenges quickly become daunting (e.g., Mitchell-Olds and Bergelson 1990); if one opts for controlled conditions in the laboratory, growth chamber, or greenhouse, then a whole different set of problems is raised by the fact that these conditions usually represent "novel" environments for the population in question, and these new environments may alter the genetic parameters of interest in unpredictable ways (e.g., Weigensberg and Roff 1996; Sgrò and Partridge 2000; Hoffmann et al. 2001). All of this does not imply that one cannot do meaningful research in this area, but only that researchers have to deal explicitly with a variety of problems that make evolutionary quantitative research extremely subtle and challenging.

There is yet another practical—and far from trivial—matter to be dealt with: there is currently a very complex, and sometimes acrimonious, debate over which statistical methods one ought to use to determine whether genetic (co)variances are indeed statistically different between the groups being examined, be these populations, species, or whatever. Despite the widespread interest in and alleged importance of studying variation in **G,** the biological community has not been able to agree on what analytical methods can be expected to work in what circumstances. Several approaches have been proposed and evaluated, from an element-to-element comparison (Roff and Mousseau 1999) to matrix correlations (Cowley and Atchley 1992), from maximum likelihood approaches (Shaw 1992) to common principal components analysis (Steppan 1997a; Phillips and Arnold 1999) to multivariate analyses of variance (Roff 2002), with review papers comparing different methods (Roff and Mousseau 1999; Steppan, Phillips, and Houle 2002). We will not enter into the technical details of this debate here, fascinating though they are. Rather, we simply note that the major difficulties of choosing a method of analysis seem to be related to the need for very large sample sizes to estimate genetic variances and covariances, and to the fact that even these estimates come with large standard errors. These features may make it artificially difficult to reject the most common null hypothesis—that the matrices are equal. Furthermore, especially for large matrices representative of many phenotypic traits, the problem of collinearity (high correlation between two or more variables) makes the matrices themselves close to singular and the resulting statistical analyses unstable (see chapter 2 for discussion). In summary, a researcher interested in this field is faced with a bewildering array of possible statistical tools, with no clear guidance on which to use or avoid. While a variety of statistical methods for approaching a given problem is a common and even welcome feature of modern quantitative science, it is also desirable to be able to settle on a subset of techniques that the community agrees will tend to yield similar and reliable results. This is not, at the moment, the case for studies of **G**.

More Fundamental Conceptual Issues Concerning **G**

The problems sketched above make studies of the genetic bases of multivariate trait evolution empirically difficult, and they imply that the results of such studies will be of limited generality. In this section, we focus on more conceptual difficulties with the way in which genetic variance-covariance matrices are used in evolutionary biology, and we argue that **G** cannot deliver all the goods that many researchers seem to expect of it. Studies of **G**, we contend, cannot be used to distinguish between the effects of selection and the effects of drift, nor are they a reasonable way to approach the study of the underlying architecture of trade-offs and constraints, nor can they help to reliably predict the direction or magnitude of long-term evolutionary change (although they do work approximately in the short term). As these are the areas in which the study of **G** is supposed to be most helpful, its failure to adequately address the main questions posed by researchers should, we think, give some pause to practitioners in the field.

As noted above, a significant part of the interest in studying **G** reflects an old problem in quantitative evolutionary theory: how to distinguish between the effects of selection and drift in the evolution of natural populations. This is part of what Shaw et al. (1995) were getting at when they wrote that "one motivation for estimating **G**-matrices is that they will reveal the most likely paths of evolution." But, as we noted in chapter 1, there is no reason to think that this problem is generally tractable, and good reasons to think that it may not be (see also chapter 2). The difficulty, as always, is that evolutionary biology is by its very nature a historical science, and hence one in which the problem of many processes resulting in similar patterns is particularly acute (Shipley 2000; see the prelude).

The hope was that the study of **G** would skirt this problem, in that at least some kinds of selection were expected to generate patterns of changes in (co)variance matrices different from those typically generated by drift. The prediction (see Roff 2000 for review) was that drift would affect all elements of a (co)variance matrix in the same (i.e., indiscriminate) way, thereby causing proportional changes in the matrix over time. Selection, on the other hand, was expected to act on subsets of characters in different ways, thereby altering the very structure of **G**. Therefore, by comparing (co)variance matrices over time, or between different populations, one could detect the presence of selection and drift and differentiate between the two. Unfortunately, in a survey of available studies, Roff (2000) concluded that in most of them, "the null hypothesis that most of the variation can be attributed to drift rather than selection cannot be rejected." This means that, in many cases, the expected effect of drift on a (co)variance matrix and the predicted effect of selection on the same (co)variance matrix cannot be distinguished with a reasonable degree of confidence. Roff (2000) concluded his review with the hope that "further development of statistical tests" would help "distinguish these two forces" (see chapter 1 for critique of this language).

But in fact the problem is not statistical in nature, and neither larger samples nor more sophisticated math will help. Work by Phillips, Whitlock, and Fowler (2001) shows that, except in very special situations, **G**-matrices will not distinguish selection from drift, because in any *particular* case the patterns that emerge as changes in the (co)variance matrix are consistent with both the action of selection and that of neutral drift. Phillips et al. demonstrated this elegantly by establishing several populations of *Drosophila melanogaster* and subjecting them to conditions likely to produce genetic drift in the form of severe bottlenecks in a selectively benign (quasi-neutral) laboratory environment. They then compared the **G**-matrices of the resulting descendant populations. When these authors considered the *average* **G** across populations (a construct that is statistically useful, but biologically undefined), it did indeed follow Roff's expectations: the populations that had gone through bottlenecks (and the attendant inbreeding) showed an average **G** proportional to that of the outbred controls, and the degree of proportionality was roughly correlated with the degree of inbreeding, as predicted by the theory *for a set of replicated populations.* If these results had held for the individual populations that underwent drift, rather than for the statistical construct created by averaging them, this would have been good news indeed. But when Phillips and co-workers examined the **G**s of individual populations that had been subjected to the bottlenecks and attendant inbreeding, they found that several were not proportional to those of their outbred controls: that is, these matrices *appeared* to have been produced by selection, not drift. Again, while, on average, the shape of the matrices did not deviate from the one predicted for the action of drift alone, in individual cases, it did.[2]

Again, it is interesting to note that Jones, Arnold, and Bürger (2004) arrived at roughly the same conclusion through their computer simulations. They found that while the average **G**-matrix from the simulation run could be used to accurately estimate the selection gradient, estimates made using a *particular* **G**-matrix were generally not as reliable (1649). Further, repeated sampling for a small number of generations (five, in their simulations) did not improve performance (because their **G**-matrices were highly correlated over such short periods) (1649).

The problem can be summarized in this fashion: *if* evolutionary biologists had access to replicated historical events (as in the artificially constructed case of Phillips et al.), *then* they could use changes in the **G**-matrices to distinguish between selection and drift. But usually what biologists have access to is a series of populations that evolved naturally, generally from a set of ancestors whose phenotypic and genetic features are largely unknown. Under these conditions, it is entirely possible that what looks like the result of selection (when a regression analysis is performed) is in fact just the result of variation around the mean (at the ensemble level) in a population that is not experiencing selection (at the formal level)—that is, that the outcome one is observing is "drift" (see chapter 1, esp. fig. 1.6). As we will see below, the problem emerges at the other end as well: not only can what looks like the result of selection really be an example of drift, but what looks like an example of drift can be hiding selection (see chapter 1, esp. figs. 1.3 and 1.6). The process-pattern problem bites back with a vengeance! The situation is not hopeless, because it is often possible to gather other kinds of evidence that point toward a particular outcome reflecting formal selection and a formal fitness difference, but that evidence will not come from purely quantitative analyses of the changes undergone in that particular population.

Another conceptually important issue is the possibility of using **G** to predict phenotypic evolution, which is predicated on **G** revealing constraints on evolutionary change. The idea is that trade-offs between traits to which an organism can allocate available resources (for example, be-

2. This is another example demonstrating that selection is a statistical phenomenon that can be detected at the ensemble level, but cannot be reliably detected by quantitative approaches at the level of individual populations (see chapter 1). It is also important to keep in mind that drift is not, and should not be thought of as, a "mechanism" of some sort; rather, drift is best thought of as a description of stochastic outcomes expected in populations of a given size.

tween survival and reproduction) should manifest themselves as (negative) genetic covariances between the traits in question. If this were true, then studies of **G**-matrices could reveal features of the underlying trade-offs that influence the direction of phenotypic evolution[3]—a major goal of evolutionary biology. Unfortunately, studies by Houle (1991) and Gromko (1995) have dealt what should have been devastating blows to such uses of **G** to infer constraints; but despite their key articles being published in *Evolution,* the premier journal in the field, these authors' arguments have scarcely made a dent in the literature.

Houle (1991) began by asking, quite sensibly, what genetic correlations reveal about underlying genetic and developmental structures, and then attempted to explain why the answer matters (see also Pigliucci 1996b in the slightly different context of models of phenotypic plasticity). In particular, he was interested in testing the possibility of using estimates of genetic covariances between two traits as indicators of underlying trade-offs. To do this, Houle (1991) constructed a relatively simple model of a trade-off (fig. 4.2), in which two traits z_1 and z_2 are allocated resources by two sets of genes: the set {**R**} is made up of regulatory genes that partition the available resources during development to z_1 or z_2, while the set {**A**} contributes to the acquisition of the resources to be allocated. Houle argued that the expectation of a negative correlation between z_1 and z_2 is based on the fact that {**R**} partitions the resources; hence materials that go to z_1 cannot go to z_2, and vice versa. But in reality, the observable genetic covariance between z_1 and z_2 depends on the relative contribution to the phenotypic covariance of both sets of genes, {**A**} and {**R**}, as well as on the features of the developmental environment. For example, if there are many more loci of type **A** than of type **R** (as is likely in biologically realistic systems), the effect of the former on trait covariance might overwhelm that of the latter, so that a researcher might actually observe a *positive* covariance, even though there is in fact a trade-off between z_1 and z_2!

Worse, even if a trade-off can be detected in principle, in practice it may go unnoticed. For example, under stressful environmental conditions, the contribution of type **R** genes may be more visible phenotypically, because there will be relatively fewer resources to be gathered by the type **A** genes (however many of them there are). But in the course of lab-

3. These trade-offs are what are often meant by "constraints" on natural selection, since selection on traits that are involved in trade-offs would be "constrained" by the other traits involved (see chapter 1, fig. 1.1). The ability to measure these trade-offs would therefore be one way of addressing the debates surrounding the relative importance of constraints and adaptation in understanding the biological world. See chapter 5 for a critique of this interpretation of "constraints" and of the view that selection and constraints can be usefully opposed.

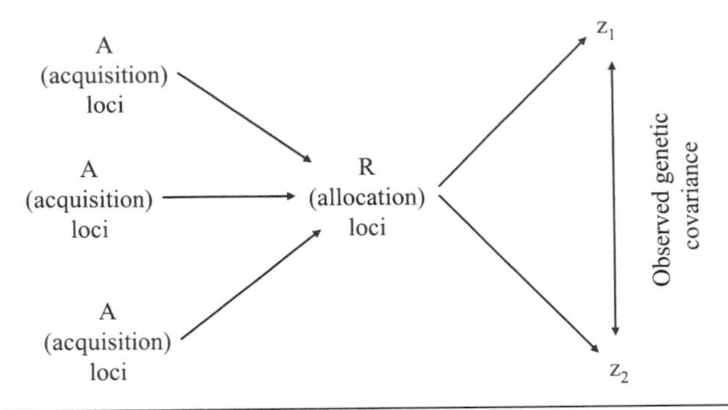

Figure 4.2. A rendition of Houle's (1991) model of the relationship between trade-offs and genetic covariances.

oratory experiments, it may be more difficult to detect trade-offs at all, because such experiments tend to be carried out under favorable growing conditions, given that researchers do not wish to impose conditions that may lead to high mortality among the experimental subjects and correspondingly low sample sizes (and statistical power). The difficulty of detecting trade-offs in favorable environments is sometimes referred to as the "silver spoon" effect (Sultan 1995), and it is related to the problem that led Rausher (1992) to propose that one should attempt to eliminate environmental effects statistically when carrying out studies of genetic covariation (but see our discussion in chapter 2).

Houle's observations imply that even when suitable studies are logistically feasible, the goal of deriving information about the genetic architecture of observable phenotypic traits from studies of (co)variance matrices is problematic at best, because genetic covariations do not necessarily reveal underlying constraints. Gromko (1995) revealed that the opposite is true as well: real underlying constraints are not necessarily detected by measurements of genetic covariation. That is, even where there is no observable (i.e., statistical) genetic covariation between two traits, those traits may in fact severely constrain each other's phenotypic evolution!

Gromko (1995) was interested in pleiotropy, the genetic phenomenon by which the same gene (or set of genes) can affect multiple characters, some of which may not necessarily be related in any logical fashion (i.e., developmental or selective) to each other. Genetic covariances are often interpreted as the result (mechanistically) of one of two phenomena: linkage (i.e., two genes are physically close to each other on the same chromosome, so their effects are statistically difficult to decouple because recombination is rare) or pleiotropy. Most authors think that only the latter is interesting biologically because—at least in outbreeding organisms—

linkage is eventually eliminated by continuous recombination.[4] If linkage is indeed biologically insignificant, genetic covariances become important indicators of pleiotropy, which is itself an entry point into the developmental genetics of organisms that are not yet amenable to detailed molecular studies.

The goal of Gromko's work was to investigate the correlated responses to selection of characters tied to each other pleiotropically. He ran a series of simulations in which virtual populations were characterized by different combinations of pleiotropic effects. One of the results was that many different patterns of pleiotropy produced the same genetic correlations—in other words, the same **G**-matrices could be produced by many different genetic architectures. To put it as Gromko (1995) did, therefore, "the genetic correlation does not uniquely determine a set of pleiotropic effects." Further, the sign and magnitude of the genetic (co)variances could not be used to predict the way in which the pleiotropic effects would influence actual evolutionary outcomes. Some sets of pleiotropic relationships allowed correlated responses to selection that, if observed in nature or the laboratory, would have led to the conclusion that there was little or no pleiotropy connecting the traits in question. Indeed, in some cases, the actual correlated responses to selection were in the opposite direction of that predicted on the basis of the observable genetic covariance. Again, as Gromko stated: "Whereas it has been previously established that genetic correlations are not necessarily constraints [by Houle's work discussed above], the alternative is also true: correlated response can be strictly constrained despite a genetic correlation of zero."

In short, genetic (co)variance matrices do not reflect the underlying genetic or developmental structures in any straightforward fashion, and therefore cannot be used to infer genetic architectures directly, nor to predict long-term evolutionary change. Where the genetic (co)variance matrix implies that the traits in question are independent (where there is no covariation), the traits can in fact be causally linked; conversely, where the genetic (co)variance matrix implies a strong correlation between two characters, they can in fact be causally independent, or even correlated in the opposite direction of that implied by the genetic covariance. These findings impose strict limits on the usefulness of estimating **G**, in the

4. In practice, this can take hundreds or thousands of generations, depending on the actual level of outbreeding and the population size. The dismissal of linkage as a relevant evolutionary phenomenon, therefore, may have more to do with the bias induced by the fact that we and most other mammals are highly outbred than with the existence of any real biological generality. Moreover, the literature on this point puts an undue emphasis on "equilibrium" situations; however, evolving populations cannot reasonably be considered to be in some sort of long-term stable state.

same way (and, again, for similar reasons) that the general usefulness of h^2 has serious limitations.

More Issues Concerning **G**

There are two additional conceptual problems with the way in which biologists think of and empirically study **G** that we wish to address. One of these problems concerns the idea of "sampling" genetic variance-covariances in natural populations, and the other concerns the question of whether one can say that **G**s evolve at all. A corollary of these problems results in yet another reason why the concept of **G** turns out to be of much more limited application than is widely acknowledged. Let us examine these points in turn.

Quantitative geneticists studying **G** empirically often create an artificial set of offspring derived from carefully designed crosses among individuals sampled at random from a natural population (Falconer and Mackay 1996).[5] The most effective breeding design for estimating components of **G** is the half-sib approach, in which one compares the phenotypes of half siblings (individuals that share only one of two parents). This design is better than alternative ones (such as the full-sib design or the comparison of clones) because it allows a finer partitioning of the phenotypic variance. For example, while comparing clones means that it is possible to estimate only a gross genetic variance (as distinct from environmental, interaction, and error variances), a half-sib design permits the experimenter to further distinguish additive and nonadditive genetic variances, and even maternal effects.[6]

There is a major problem with this approach: One can reasonably wonder whether, in carrying out such experiments, one is really studying the genetic (co)variance matrix of the population in question. While we argued above that there is no such thing as a "general" **G**-matrix for a population (since it depends on the traits chosen, etc.), in this case we are questioning whether the **G**-matrix estimated from controlled crosses should be thought of as representative of the matrix of the natural population from which the crosses were derived. The researchers here are esti-

5. There are other methods, such as those proposed by Lynch (1999), for studying quantitative genetic parameters in natural populations, but these have rarely been used because they come with their own set of formidable sampling and estimation problems. Given the scarce use of such methods to date, we will not treat them in any detail here.

6. In practice, though, this can be achieved only by the use of very large sample sizes, which often do not provide much power for the complex statistical analyses necessary to make sense of this breeding design. Furthermore, these "variance components" often turn out to be of difficult or dubious biological interpretation.

mating genetic parameters in an artificially created set of genotypes, one that is not actually found in nature and has little chance of ever being there, even if the population is mostly outbreeding, because it is vanishingly unlikely that the individuals in the population in question would ever cross in precisely (or even approximately) the same pattern as required by statistical tests and laboratory experiments. This problem seems to be completely ignored by practicing biologists, who instead focus on the relative merits of various breeding designs. But breeding designs are experimental devices that make sense, as their name clearly implies, when the goal is to *breed* certain characteristics in a population. The problem that evolutionary quantitative geneticists are trying to solve is very different: it is to *estimate* a quantity that is allegedly characteristic of natural populations. But this cannot be what they actually do. This problem of the resemblance, if any, between estimates of **G** obtained with breeding designs and whatever **G** would be derived from directly sampling a natural population has never, to the best of our knowledge, been addressed.

One way to understand the problem is this: Imagine that we wish to know the average height of individuals (a statistic) in a population of plants. What we do is sample individuals from the population, measure their heights, and then calculate the mean and some measure of dispersion (the variance, for example). We do *not* pick some individuals, cross them in specific combinations, and then measure the heights of their progeny. The resemblance between the results obtained with the first and the second methods is a matter of empirical determination, but the further the breeding design is from the actual mating pattern in nature, the less reliable the experimental approach will be.

The second question we wish to raise in this section is whether **G** itself in some sense evolves. Genetic variance-covariance matrices are often interpreted as the result of past evolutionary "forces," such as selection and drift. But this cannot mean that selection acts directly on the components of variance; variance components are statistical constructs that summarize individual attributes at the population level, and are therefore not the sort of thing that discriminate physical processes can interact with. The physical processes that make up informal selection (see chapter 1) act on particular entities, usually the individual organisms involved. While some physical processes are discriminate with respect to groups, and hence result in group selection, group selection has never—to our knowledge—been invoked in discussions of **G**, nor would invoking it really address the problem. We are forced to conclude that what those biologists who refer to the evolution of **G** mean is that **G** provides a summary indicator of the more complex and multifarious processes that occur at the level of individuals. But in this case we fall back to the kind of problem for the use of **G** raised earlier when we examined the combined works of Houle (1991) and Gromko (1995): what the physical processes that result in selection "see" are the different combinations of phenotypes present in a population, and indirectly, what these processes select are whatever genetic and developmental pathways lead to one phenotypic combination or another. But since there are many different combinations of genotypes and phenotypes that yield the same statistical estimate of **G**, how exactly is one to use **G** to understand individual-level forces acting on multivariate phenotypes? Changes in **G** can be the result of many different kinds of processes, and the raw statistical summation of the resulting outcomes cannot be used to disentangle the processes that caused them (although it can be used to suggest causal hypotheses to be tested by independent means; see the prelude).

This discussion brings us to a final interesting corollary: In what sense can one measure **G** in non-outbreeding organisms? For the many species of organisms that are more or less inbred (indeed, outside of mammals and some other groups of vertebrates, this is by no means the exception), the effective population size may be as small as a dozen, or sometimes only two or three, individuals. But all methods devised so far to estimate **G** require large sample sizes, often on the order of 90 or more families of genetically related individuals. That is more than the actual number of genotypes in many natural populations of inbreeders (and of some outbreeders as well). What then? Does this mean that **G** does not exist in such populations? Or, if it does, is it possible to sample it only by creating a highly artificial outbred population from the few naturally available (inbred) genotypes? If the latter is the case, **G** is beginning to sound like something that may materialize with great effort and expense in the biologist's laboratory, but that is not likely to be informative about the natural populations that are the actual objects of study.

WHAT NOW? GETTING PAST G

If the very concept of **G** is so fraught with problems (see table 4.1 for a summary), and if those problems are made even worse by countless analytical and empirical difficulties, why do evolutionary quantitative geneticists use **G** as the centerpiece of their conceptual framework (rather than, more reasonably, as a simple entry point into their research program)? Furthermore, if we have made a convincing case so far, are there alternative approaches to addressing the questions that **G** is supposed to, but cannot, answer?

Table 4.1. A summary of our critique of the use of G-matrices in evolutionary quantitative genetics

Category of problem	Problem	Why/What consequences	Selected references
Empirical difficulty	Components of **G** change with the environment (phenotypic plasticity).	Since it is difficult to study **G** in nature, all controlled-condition studies need to be considered carefully.	Paulson, Roberts, and Staley 1973; Mazer and Schick 1991; Ebert, Yampolsky, and Stearns 1993; Bennington and McGraw 1996; Sgrò and Hoffmann 1998; Bégin and Roff 2001
Empirical difficulty	**G**, at least for certain traits, is unstable over medium to long evolutionary time scales.	**G** cannot be used to predict long-term evolution, or to infer past selection patterns.	Shaw et al. 1995; Paulsen 1996; Steppan 1997a, 1997b; Roff and Mousseau 1999; Waldmann and Andersson 2000
Theoretical tractability/ empirical difficulty	It is not possible to distinguish between effects of drift and selection on **G** unless artificially replicated populations are available.	This is a particular version of the processes ↔ patterns many-to-many relationship. However, it is considered a crucial reason to study **G**.	Roff 2000; Phillips, Whitlock, and Fowler 2001
Theoretical tractability	No theory is available to tell us how evolution changes **G**.	It is too complex a dynamic problem, currently amenable only to empirical investigation.	Turelli 1988; Pigliucci and Schlichting 1997
Theoretical tractability	There are no agreed-upon statistical methods that work satisfactorily when comparing **G**s or their components among groups.	Available techniques test uninformative null hypotheses, have very low power, or often yield unstable estimates.	Cowley and Atchley 1992; Shaw 1992; Steppan 1997a; Phillips and Arnold 1999; Roff and Mousseau 1999; Steppan, Phillips, and Houle 2002; Roff 2002
Empirical difficulty/ conceptual impasse	Breeding designs used to estimate components of **G** may have little to do with patterns of mating in nature, and therefore with how **G** is actually influencing a population's evolution.	Breeding experiments are designed to maximize statistical power, but may end up *constructing* rather than estimating **G**.	This chapter
Conceptual impasse	The choice of traits making up any particular **G** to be studied tends to be arbitrary or idiosyncratic.	One cannot search for general answers about **G**.	Waldmann and Andersson 2000; this chapter
Conceptual impasse	**G** gives no indication of causal pathways.	G cannot be used to study genetic architecture because different architectures result in the same observable **G**.	Houle 1991; Gromko 1995; Pigliucci 1996b

(*continued*)

Table 4.1 ***(continued)***

Category of problem	Problem	Why/What consequences	Selected references
Conceptual impasse	**G** cannot be interpreted as a measure of "constraints" on further evolution.	There are patterns of underlying genetic trade-offs that do not show up as genetic correlations, and vice versa: sometimes measurable genetic correlations do not reflect underlying trade-offs.	Houle 1991; Gromko 1995
Conceptual impasse	For selection to act on **G**, it would have to be at the level of groups, since **G** is a population-level property.	Since few, if any, authors have proposed a group selection scenario for the evolution of **G**, what do we mean exactly when we say that **G**s evolve?	This chapter

The answers to these two questions are likely to be tightly linked. Quantitative geneticists may be continuing to use **G** for the same reason that many of them keep using the concept of heritability: on the one hand, many practitioners (though by no means all) rarely pause to seriously consider the limitations inherent in these tools; on the other hand, it is not clear what alternatives, if any, are available at the moment. Finally, several (we would argue too few) researchers use statistics for what they are meant to be: convenient summaries of complex data sets, useful for generating testable causal hypotheses. As we pointed out in the prelude, one of the roles that philosophy of biology can play is the negative one of providing thorough criticism of the currently used tools (Bacon's famous *pars destruens*). The hope, then, is that if these difficulties are clearly presented, they will be taken seriously, and the biological community may become more interested in investigating what alternatives are available (Bacon's *pars construens*).

A cursory perusal of the literature reveals that even staunch supporters of the current methods of evolutionary quantitative genetics feel uncomfortable about the situation they now find themselves in. Above, we noted Roff's (2000) call for entirely new analytical tools to distinguish the effects of drift from those of selection on **G** (and pointed out that this is probably not possible). Even earlier, in 1989, Barton and Turelli, two of the most prominent theoreticians in the field, published a rather skeptical article titled "Evolutionary Quantitative Genetics: How Little Do We Know?" The answer they proposed—that in fact, we know very little—has not changed much in the intervening fifteen years. Shaw et al. (1995)

stated that "short-term predictions based on sound estimates of genetic parameters are likely to be *qualitatively* informative [our emphasis]. In contrast, both empirical results . . . and theoretical considerations suggest that quantitative predictions for long-term selection and retrospective analyses of selection should be interpreted with caution." This is rather an understatement of the problem, but represents a significant statement nonetheless.

The success of plant and animal breeding reveals the adequacy of quantitative genetics theory for making short-term predictions. But, as Shaw and colleagues note, even within that restricted (and evolutionarily rather insignificant) time frame, the predictions are mostly qualitative; fortunately, for plant and animal breeding, that is all that is ever necessary. But over realistic evolutionary time spans, even qualitative predictions are probably not applicable. We do not have (and probably cannot have) a theory of how **G** changes during long-term evolution, though we know that it will change. To say that "quantitative predictions for long-term selection and retrospective analyses of selection should be interpreted with caution" is a good beginning, but we have no guidelines for how cautious we should be in any particular case, nor for what a "cautious interpretation" would involve in these cases.

Furthermore, while the recent flare of research focusing on the identification of quantitative trait loci (QTL) in natural or artificial populations represents a welcome empirical advance that has made it possible to pinpoint large genomic regions that are statistically correlated with naturally occurring phenotypic variation (Lander and Botstein 1989; Mitchell-Olds 1995; Kearsey and Farquhar 1998; Phillips 1999; Mezey, Cheverud, and Wagner 2000; Asins 2002), it has not done as much as was hoped for to advance our understanding of the causal structure underlying developmental variation. This, again, is because, despite being carried out with the use of molecular markers, QTL studies still use an essentially correlational/statistical method, founded on the same conceptual framework as classic quantitative genetics. This results in their being characterized by the same difficulties as more traditional methods, as well as by additional conceptual and empirical limitations that are intrinsic to the QTL approach itself.[7] QTL studies, therefore, can provide interesting additional information on the particular genetic architecture of a given set of traits

7. These limitations include the bias toward uncovering genomic regions with large phenotypic effects; the dependence on specific crosses that have to be limited in number for logistical reasons; and the strict limits on resolution, which depends on statistical power and hence—again—on logistical limits imposed on the size of experiments; not to mention the complexity of interpretation of statistical hypothesis testing when many multiple nonindependent tests have been carried out simultaneously.

in a particular population and set of environments, but no more. In other words, they suffer from the same problem of locality already pointed out by Lewontin back in 1974, and are therefore likely to be of little general theoretical help.[8]

And this, perhaps, is the crux of the matter: too many evolutionary biologists (though, again, by no means all) insist on pursuing an approach resembling that of "hard" sciences like physics (Pigliucci 2002), looking for general results that do not depend on specific local conditions. This amounts to a refusal to embrace the essentially historical nature of evolutionary biology (Cleland 2001, 2002; Kaplan 2002; Trout 2002). The very language used by most quantitative geneticists betrays this attitude when, for example, they refer to *the* evolution of "**G**," as if genetic (co)variance matrices were the kinds of things about which one could make general statements. If one takes seriously the results of theoreticians such as Houle (1991) and Gromko (1995), or the extensive series of empirical results, showing such great variability of quantitative genetic parameters in time and space, that we examined briefly above, then asking whether an unspecified **G** changes or remains constant over evolutionary time does not seem to be very sensible.

This is not to say that there are no general statements one can make about evolutionary biology, selection, or quantitative genetics. Bell's (1997) book on selection, for example, succeeds in listing several of them. But these generalizations are, by necessity, limited in number and paint a very broad picture; they can do little to explain the actual specific situation of any real natural population. Again we embrace Dupré's (1993; see the prelude) point that the historical nature and complexity of living organisms means that reductionist explanations of a general nature can often tell us *why* we observe certain patterns (e.g., we can form a solid conceptual account of selection in natural populations), but are of little use when it comes to predicting exactly *what* we will observe in any particular instance (e.g., we are in no position to say what selection intensities will be found in any particular population at any particular time).

This, then, is the conundrum of quantitative genetics: it provides a useful general framework for thinking about evolution in natural popula-

8. Indeed, the general conclusions of QTL studies are becoming rather simple to predict: yes, one will find QTL associated with a given trait; yes, some of them will "explain" (i.e., statistically account for) a relatively large amount of phenotypic variance; yes, some of them will have (unpredictable) environment-specific effects; and yes, some of the same QTL will be found in different crosses within the same species, and some will not. Most of the meat of these studies is to be found in very local information about which genomic regions are associated with what in a particular population. Like politics, much of evolutionary biology seems to be local.

tions and for summarizing the actual status of any particular one of them. Such a summary can then be used to generate causal hypotheses, which in turn can be tested by the same methods that we referred to at the end of chapter 2 in the context of studying selection. But if the goal is to directly infer anything about the genetic architecture of a population or trait, or the causal story of how that structure got to be the way it is, the statistical approaches of quantitative genetics cannot succeed alone. The only way to tackle these questions is through painstaking biological detective work—work that may include QTL studies and estimates of selection coefficients, but will also have to involve manipulative experiments, structural equation modeling, attempts to unravel the molecular bases of phenotypic plasticity, and a host of other approaches. There is no shortcut and no first approximation—just a lot of hard work. The evolutionary biologist is more akin to a Sherlock Holmes than to a theoretical physicist: while generalizations about the behavior of criminals are indeed useful, they will not lead us to find the culprit in any particular case. But of course, that dependence on local conditions is the beauty and essence of both good detective novels and good evolutionary biology.

5 A Quarter Century of Spandrels

Adaptations, Adaptationisms, and Constraints in Biology

THINGS CANNOT BE OTHER THAN THEY ARE. . . . EVERYTHING IS MADE FOR THE BEST PURPOSE. OUR NOSES WERE MADE TO CARRY SPECTACLES, SO WE HAVE SPECTACLES.
—DR. PANGLOSS, IN VOLTAIRE'S *CANDIDE*

Some twenty-five years ago, Gould and Lewontin published "The Spandrels of San Marco and the Panglossian Paradigm: A Critique of the Adaptationist Programme" (1979). In it, they criticized a position they termed "adaptationism" and argued that too many evolutionary biologists were not taking seriously enough the constraints on natural selection while indulging in the tradition of untestable "just-so stories." Since its publication, debate has raged between those arguing that Gould and Lewontin's focus on constraints incorrectly de-emphasizes good adaptive thinking in biology, and those arguing that the tendencies toward sloppy adaptive thinking identified by Gould and Lewontin continue largely unaltered. But what was this debate really about? The compelling metaphor presented by Gould and Lewontin hid many of the ambiguities in the claims of both sides in the debate. Uncovering the difficulties involved in making either position precise is, we argue, a good first step toward a more pluralistic approach to testing hypotheses about the origin, spread, and maintenance of phenotypic traits, as well as the developmental pathways that reliably (re)produce them.

TWENTY-FIVE YEARS OF SPANDRELS

Gould and Lewontin's controversial paper, "The Spandrels of San Marco and the Panglossian Paradigm: A Critique of the Adaptationist Programme," was published in 1979. The core of the piece was an attack against what they described as "the adaptationist program": an attempt to explain the existence and particular forms of nearly any phenotypic trait as the result of natural selection. The paper's exotic title derived from the example with which the piece began. The authors noted that the tapered spaces ("spandrels") between the archways supporting the domed roof of the basilica of San Marco in Venice were beautifully decorated with mosaic images, in a way that made admirable use of the triangular space. They claimed that while this space was put to great artistic use, it was an architectural by-product of employing arches to support a domed roof, and was not designed for that artistic use. The lesson they thought should be drawn from this example was that when faced with a "trait" that is being put to good use, one ought not to jump to the conclusion that that particular use is the reason the trait is there. Gould and Lewontin argued that a sloppy form of evolutionary thinking did just that, and that its practitioners went so far as assuming that (nearly) any and every trait had to have a good use that explained its presence, even if none was obvious. This contention generated endless debate, and even produced a measurable change in attitude in the post-1979 published literature. In this chapter, we examine both the original charge and the responses it elicited as an entry into the complex questions of adaptations, constraints, and the current thinking regarding biological explanations for complex phenotypic characters. Our view of the controversy is informed by our discussion of the nature of selection and constraints (the latter in the form of the **G**-matrix) that we have developed in the book so far.

Gould and Lewontin provided a detailed criticism of the assumptions that they thought the adaptationist program depended on (box 5.1) and laid out an alternative approach that they felt better accounted for the complexities of the evolutionary process. Specifically, they argued that, rather than viewing organisms as collections of more or less optimized individual traits, it would be more valuable to acknowledge that organisms are and must remain complex, integrated wholes, and hence that there are limits to adaptive explanations of individual traits, as well as other kinds of explanations available. They wished that these alternative explanations would be explicitly acknowledged (box 5.2), and hoped that biologists would find ways in which predictions made on the basis of alternative models could be rigorously tested. They argued that this would entail recognizing the legitimacy of alternative approaches to understanding evolution, in which constraints on optimization through natural selection would play a more central role. Of course, Gould and Lewontin freely admitted that adaptation by natural selection has been and is a powerful causal factor in shaping the phenotypic traits of organisms. But they hoped to shift the focus away from simple adaptive hypotheses and toward more sophisticated analyses that take account of the developmental processes that make the formation of certain traits possible. In addition, they wished to emphasize two related questions regarding adaptive hypotheses: first, what kind of evidence is necessary to support the hypothesis that a trait

Box 5.1 Adaptationism: The Object of Gould and Lewontin's Attack

According to Gould and Lewontin, the "adaptationist program" is characterized by the telling of stories involving natural selection that account for the presence and particular forms of traits by reference to their hypothesized adaptive significance. Gould and Lewontin charge adaptationist biologists with making the following assumptions and utilizing the following methods in their work:

Ontological assumptions:

1. Organisms can be usefully considered as assemblages of traits, the adaptive nature of each of which can be examined independently of the others.
2. Constraints on the power of natural selection can be assumed to be minor, so adaptation by natural selection is the proper null hypothesis (see chapter 10) for particular traits, either behavioral or physical.

Methodological practices:

1. Consistency between the observed trait and an explanatory story told in terms of natural selection is sufficient for the preliminary acceptance of the hypothesis that the trait is adaptive and evolved in just that way.
2. The failure of one adaptive story leads immediately to the search for another story of the same sort.
3. Any failure of particular traits to be optimal is accounted for by evoking "trade-offs" with other adaptive traits.

Because of these features, adaptive stories are very easy to create and very hard to falsify–hallmarks, Gould and Lewontin argued, of poor scientific hypotheses. Indeed, they claimed that it is inordinately difficult or impossible to ever reject the hypothesis that a trait is adaptive (in some way or other) within the context of the adaptationist program, as new stories can continually be cooked up.

is an adaptation formed by natural selection? and second, what evidence could point toward some other cause of the current state of the trait in question?

Over the past twenty-five years, the "Spandrels" paper has been much cited and criticized. Criticisms of it have ranged from the potentially biologically significant—for example, that Gould and Lewontin overstated the role of constraints in evolution (see, e.g., Rose and Lauder 1996)—to the seemingly bizarre—that the architectural feature they referred to is really called a "pendentive" and not a spandrel (see, e.g., Houston 1990; Dennett 1995). Nonetheless, the influence of the Spandrels paper has been so obvious that even now, it is commonplace to pay at least lip service to the kinds of difficulties involved in demonstrating that a trait is adaptive that Gould and Lewontin pointed out (see Rose and Lauder 1996 and citations therein).

Box 5.2 Gould and Lewontin's Alternatives to Adaptationism

Gould and Lewontin suggested that at least the following alternative hypotheses be considered in attempts to understand the etiology of particular traits:

No adaptation and no selection. The trait in question may be the result of, for example, genetic drift.

Indirect selection. The trait was not the subject of selection; its features are the result of its association with another trait (which may or may not itself have been the subject of natural selection).

Selection without adaptation. A trait may increase in frequency due to natural selection, but not be "adaptive" as the term is generally understood. Lewontin's example involves a resource-limited species and a genetic mutation that doubles fecundity: this mutation does not increase the population's mean fitness (as it is usually understood), it only alters the population dynamics.

Adaptation without selection. The trait may be adaptive (in the sense of being advantageous to the organism), but not itself the product of selection explicitly for that particular form of the trait. Some types of phenotypic plasticity may be a special case, as is behavioral flexibility. Although these abilities may themselves be selected for, the traits that are the result of the plasticity/flexibility are not necessarily the result of selection. Or the adaptive nature of the trait might be an accidental by-product; hermit crabs, for example, are often found in well-camouflaged shells, but that is because hermit crabs make use of shells they find locally, which tend to be well-camouflaged in the local environment (having been selected for camouflage in their original species).

Adaptation and selection, but no basis for distinguishing between adaptations. While the trait may be adaptive and may have been selected for, there may be no way of distinguishing between different forms of a trait on the basis of their adaptive significance (the problem of multiple "adaptive peaks"; see chapter 8).

Adaptation and selection, but the particular adaptation represents a secondary use of a trait already present for other (generally historical) reasons. This is the case Gould and Vrba (1982) later referred to as "exaptation."

But to what extent have empirical and conceptual advances over the past quarter century helped to ameliorate the difficulties pointed out by the Spandrels paper? What claims did the paper actually make, and to what extent has research since then supported (or contradicted) those claims? And why is it that the paper had, and still continues to have, such an impact? In this chapter, we argue (1) that the advances of the past twenty-five years have helped to ameliorate some of the difficulties that Gould and Lewontin drew attention to; (2) that empirical evidence has pointed toward complexities that neither supporters of Gould and Lewontin nor their critics have properly addressed; and (3) that the paper's enduring impact is due in large part to the compelling nature of its central metaphor, as well as to the politically and emotionally charged issues the authors were involved with at the time. We suggest further (4) that much of the continued debate surrounding adaptationism is misplaced—it is time both to give up the rather confused metaphor of the Spandrels paper and to move beyond the kind of simple-minded adaptationist thinking Gould and Lewontin criticized. Perhaps most important of all, (5) we need to clarify the often sloppy language and thinking that surrounds "constraints" on evolutionary change (a point briefly made by Antonovics and van Tienderen 1991).

WHENCE SPANDRELS?

Gould and Lewontin's attack on the adaptationist program was motivated primarily by two related phenomena. One was the rise of sociobiology (especially human sociobiology), attributable in large part to E. O. Wilson's 1975 book by the same name; the other was the popularization of the narrowly gene-centric approach favored by Dawkins and defended explicitly in his 1976 book *The Selfish Gene.* The features that Gould and Lewontin identify as defining the theory and practice of adaptationism (see box 5.1) emerge directly from their critiques of the styles of reasoning used in those arenas.

As Lewontin has always been quick to point out, human sociobiology has primarily been concerned with telling stories about human behavior that make such behavior out to be adaptive (see, e.g., Lewontin 1979, 1992). Lewontin argued that the move from telling those stories to those stories being accepted, given the assumptions of the adaptationist program, could be made much too quickly, as all that seems necessary is the consistency of some observed behavior with a story that accounts for it in terms of its supposed selective advantage at some time in the past. Arguments such as Dawkins's, meant to demonstrate that the focus of selection is individual genes made visible to selection through their direct relationship with particular phenotypic features, were then an integral part of the adaptationist program. These assumptions, when applied to human behavior, yield conclusions with dramatic political and social implications (see Lewontin 1992). The attack on adaptationism expressed in the Spandrels paper, and which Lewontin and Gould each pursued in many other works, would probably have been far less aggressive if all that was at stake was the adaptive significance of the variation in the color of snail shells.

The idea that selection (in the informal sense; see chapter 1) primarily acts more or less directly on genes is, as we noted in chapter 3, effectively dead (with a few exceptions arising from special circumstances). Recent advances in molecular developmental genetics have led to the recognition of epigenesis as a central, and largely still poorly understood, player in mediating the genotype-phenotype mapping function (see Alberch 1991; Wray 1994; Müller and Newman 2003; Jablonka and Lamb 2005). Even some sociobiologists, including Wilson himself, have backed down from talk that seemed to imply the primacy of genes in order to embrace the rather vague, but certainly more realistic, talk of "epigenetic rules" (Wilson 1998). But the difficulty of understanding the evolutionary history and import of human behavioral phenotypes still remains, as do the problematic social and political implications of work purporting to do so. Part of the problem is that studying the evolution of adaptive traits in humans is fiendishly difficult, which has left a void that is continuously filled by sloppy storytelling (see chapter 7 for discussion).

Aside from a move away from simplistic assumptions about the genotype-phenotype mapping function, theoretical evolutionary biologists have, in the past decade, been focusing on the fact that the environment varies extensively at every time scale, from the course of an individual organism's development to evolutionary time. While models of environmental heterogeneity had been proposed before the Spandrels paper (see, e.g., Levene 1953), the effects of environmental heterogeneity (including especially stochastic elements) and the evolutionary strategies or outcomes that it can elicit have since been investigated with a wider range of theoretical tools, including quantitative genetic and optimization models (e.g., Gabriel and Lynch 1992; Yoshimura and Clark 1993; Zhivotovsky, Feldman, and Bergman 1996). The local heterogeneity of the environment imposes a limit on what kinds of adaptive evolution can take place, and the long-term heterogeneity of the environment constitutes a moving target for adaptive responses (see Pigliucci 2001, particularly chapters 9 and 10, for discussion). Further, the more complex the functions of a trait, and the more functions a trait is involved in, the more sophisticated the optimizing models necessary to describe those functions need to be, especially as multiple local fitness maxima become available (Niklas 1997); this makes the kinds of one-trait optimization analyses criticized in the Spandrels paper far less tempting and adds significant layers of complexity to the production of careful optimization studies. None of this, of course, means that populations of organisms do not evolve in response to selection; obviously, they do. However, these kinds of complexities are a part of what makes moving from informal to predictive fitness so difficult, and hence makes observations of particular discriminate physical processes interacting with particular traits less than fully reliable guides to what formal selection the population might be experiencing. Indeed, one way to articulate the change in view encompassed by the currently available theory is that it seems to point us away from the ideas of "survival of the fittest" and optimization, and toward a model of "survival of the barely sufficient"—analogous to the concept of "satisficing" in foraging theory (Ward 1992).

Some of these ideas were actually expressed before the publication of the Spandrels paper by François Jacob (1977), who elaborated a model of evolution as "tinkering" (*bricoleur,* in the original French). According to Jacob, natural selection is best viewed as something that works only with the materials that happen to be available, given the particularities of a specific time and a particular place (this is similar to Darwin's [1868] metaphor of the builder who must use stones that fall from a cliffside to build a wall). Natural selection, according to Jacob, is the result of the constraints faced by living things (1163); in this view, the idea of natural selection as being some kind of force fundamentally *opposed* to constraints is misleading (see our discussion of evolutionary "forces" in chapter 1). The difficulty addressed in both the Spandrels paper and Jacob's paper is that it is impossible to meaningfully separate the action or result of natural selection from the complex physical embodiment of the organisms that make up the population in question; one way of understanding the charge of adaptationism is that too much writing on the adaptedness of particular traits has done just that (see Jacob 1977, 1164; Gould and Lewontin 1979, 558).

ADAPTATIONS AND CONSTRAINTS: TROUBLED CONCEPTS?

In our view, the real value of the debates that followed the Spandrels paper lies not in their all-too-often acrimonious claims and counterclaims, but rather in the way they pointed toward problems with the major concepts employed—that is, adaptation, adaptations, and constraints. In fact, the discussion resulted in at least some potentially useful ways of rethinking these concepts. In this section, we briefly review some of the literature that has emerged on the ideas central to the Spandrels paper and the discussions that followed it. We examine how clarifications of these concepts, as well as changes in the ways they are employed, have influenced the practice of evolutionary biology. Finally, we identify places where we think more conceptual clarification is still necessary.

Adaptations and Adaptationisms

> Adaptation: A process of genetic change of a population, owing to natural selection, whereby the average state of a character becomes improved with reference to a specific function, or whereby a population is thought to have become better suited to some feature of its environment. Also, an adaptation: a

> feature that has become prevalent in a population because of a selective advantage owing to its provision of an improvement in some function. A complex concept. (Futuyma 1998)

Since the publication of the Spandrels paper, there has been a substantial literature surrounding the question of what constitutes an adaptation, what the process of adaptation is, and what the relationship between the process and the trait might be. It is generally agreed that, at the very least, to count as an adaptation, a trait must be maintained in the current population because of a selective advantage it confers on the individuals that have it. In order for selection to be measurable (see chapter 2), of course, there must be variation in the trait, for it is that variation that must be selected against if the trait in question is to be an adaptation. Further, as the Futuyma quote above makes clear, the trait must have spread in a given population *because* of the selective advantage that it had over other variants, although testing this proposition raises the complication of somehow having to quantify the action of past selection—something that is fiendishly difficult under most circumstances, as our discussion in chapter 2 makes clear.

Whether a trait—in order to be thought of as an adaptation—needs to be maintained in the current population because of the same selective advantage that explains its prevalence in the population, however, remains something of an open (semantic) question. Indeed, Gould and Vrba (1982) muddied the waters somewhat by introducing the term "exaptation" to refer to traits that arose and spread in a population for one reason, but are used by the organism for a different (adaptive) function. In Gould and Vrba's description, it is not clear that the current adaptive function of the trait in question makes it an adaptation—it may be the case that the trait is not currently being maintained by natural selection at all. Nor is it clear what relationship the new function of the trait has to have to the reason the trait became prevalent in the population to begin with. What is clear is that every adaptation must have been, at some time, an exaptation—as evolution is not forward looking, the early adaptive uses of a trait must have occurred before selection on those uses.

Griffiths (1992) followed up on these questions and introduced yet another term: "exadaptations." Griffiths argued that what separates adaptations from exaptations, and exaptations from exadaptations (1992, 119), is the history of the trait in the population. He suggested calling traits that have been "co-opted" for a new functional role, but are not being actively maintained by natural selection because of their association with that new functional role, *exaptations;* those traits that arose in one adaptive context, but are now being maintained by natural selection because of a different adaptive function they have become associated with, he calls *exadaptations* (1992, 119). In the latter case, the trait may also be actively maintained by natural selection because of its old functional associations; a trait can obviously do several different things, any or all of which may contribute positively to the organism's (informal) fitness.[1] Determining whether an exadapted trait is being maintained by both its new functional association and its previous one, or by only one of these, will most often require testing of counterfactual claims regarding what would happen if one or the other functional association were no longer relevant, and these claims will naturally be difficult to test.

What should be clear from this explosion of terms is that explaining the origin, spread, and maintenance of a trait are different tasks—they require different techniques, involve different assumptions, and answer different questions. In the case of the origin of a trait, the explanations will be, by necessity, primarily developmental and epigenetic (see Wolpert 2002; Müller and Newman 2003)—again, as evolution is not forward looking, the initial appearance of a trait cannot be due to natural selection acting on that trait.[2] The spread of a trait in a population, on the other hand, will often be explained in terms of selection; note that this explanation will involve selection in the informal sense of a trait interacting with a physical process in a discriminate way. Recall that formal selection is not causal, but merely describes the (mean) spread of a trait in populations of the kind in question (see chapter 1). Explaining the maintenance of a trait in a population, will, however, involve attention to both the selective advantage of the trait (given informal selection via discriminate physical processes) and the developmental mechanisms by which the trait is reliably (re)produced in the organisms in question (see Ariew and Lewontin 2004; Walsh 2003).

With such shifts in what "adaptation" (the process) and "adaptations" (the traits) referred to, it is hardly surprising that the definition of "adap-

1. A better way to put this, albeit a rather longer one, is to say that any particular trait can (and often will) interact with a number of different physical processes in discriminate ways; insofar as a particular variant of a trait has higher (informal) fitness with respect to a number of different (discriminate) physical processes, we can speak of its being (informally) selected by a variety of physical processes—that is, as "doing" a number of things that are associated with higher (informal) fitness (see chapter 6). Whether this will translate into selection at the formal level is, of course, an open question, and figuring out how (if at all) the particular processes identified at the informal level are related to the statistical facts that emerge at the formal level requires, as we have noted, significant ecological "detective" work (see chapters 1 and 2).

2. Though, as one reviewer of this book noted, selection, by altering some heritable features of the population, can result in different combinations of heritable features being brought together in a population, and hence can result in new phenotypes being developed; nevertheless, we think it is clear that explaining *how* those phenotypes develop remains primarily the task of developmental biology, and will not require attention to the local history that resulted in that population having members with those particular combinations of heritable features.

Box 5.3 Three (or More?) Kinds of Adaptationism

Peter Godfrey-Smith (2001) identifies three distinct interpretations of adaptationism:

1. **Empirical:** Natural selection is so powerful, and the constraints on it so few, that practically every trait will be optimally adapted for some function (or an optimal compromise for a set of functions).
2. **Explanatory:** Adaptation by natural selection is the answer to the most important and interesting questions in evolutionary biology, however ubiquitous (or not) it happens to be.
3. **Methodological:** The best approach to understanding the historical origin, spread, or maintenance of a trait is to first assume that it is an adaptation, then work through that assumption to alternative hypotheses.

Godfrey-Smith notes that while a supporter of empirical adaptationism would probably endorse both methodological and explanatory adaptationism, support for either of the latter does not necessarily imply support for empirical adaptationism; nor is support for explanatory adaptationism necessarily linked to support for methodological adaptationism, or vice versa.

Further, as Godfrey-Smith makes clear, these different versions of adaptationism would be bolstered (or not) by very different kinds of evidence, and in the case of explanatory and methodological adaptationism, even people agreeing on how to interpret all the available evidence might disagree about the extent to which these forms of adaptationism are in fact supported.

It is worth noting that what Gould and Lewontin called "adaptationism" (see box 5.1) is not exactly reflected in any of the three versions of adaptationism Godfrey-Smith describes.

tationism" was questioned as well. Gould and Lewontin's definition of adaptationism, summarized in box 5.1, makes it out to include several different assumptions about the world as well as several different kinds of methodological practices. It is not clear either from the Spandrels paper or from their other works, such as Lewontin's 1978 attack on adaptationism, just which aspects of the adaptationist position are most problematic, or how these different aspects relate to one another. Since then, Peter Godfrey-Smith (2001) has identified three different positions one might reasonably call "adaptationism," the assumptions behind those positions, and some ways in which they might relate to one another (box 5.3). Until one is clear about just what one means by adaptationism, what evidence one needs to accuse someone of being an "adaptationist" will remain vague; this rather diminishes the force of the accusation.

Constraints versus Natural Selection? The Necessity or Meaninglessness of Constraints

The Spandrels paper had the unfortunate consequence of suggesting that "constraints" could be usefully opposed to adaptation, and that research

Box 5.4 A Troubled Metaphor

In 1990, Houston pointed out that what Gould and Lewontin had called "spandrels" were more properly called "pendentives," spandrels being any space between arches, and pendentives being the particular tapered triangular structure between two arches that meet at right angles below a dome, like those used in the church of San Marco in Venice. Gould and Lewontin had used the beautifully decorated pendentives (not spandrels) as an example of an architectural feature that had to be there given the architectural requirement that a dome be set on arches, but which had been put to a good (secondary) use in displaying mosaic artwork. The lesson Gould and Lewontin wished the reader to draw from this metaphor was that even when a trait (be it in a building or an organism) is doing something that is obvious and useful, that function may have little to do with *why* the trait is there, or with *how* it came to be there.

However, Houston (1990) noted that pendentives are not in fact the only structural solution possible—there are other ways of mounting a dome on top of two arches. Therefore, viewing pendentives as necessary was to mistake the solution actually used for the only possible solution. Given this conclusion, Houston suggested that only a true optimality analysis taking into account actual design desiderata, as well as possible available alternatives, would be adequate to fully explain the presence and particular form of spandrels (in this case, the pendentives).

In 1995, Dennett took up this topic, but went further with the argument. He claimed that Gould and Lewontin had gotten their architectural history utterly wrong: what they had viewed as a secondary use (the artwork) of a preexisting structure (the pendentive) that was there for other reasons was, to a large extent, the reason the structure was there in the first place (1995, 273–74). The pendentive was chosen from among the range of possible spandrels because it was good for artwork; hence, it was an "adaptation" for artwork (at times, Dennett even seemed to suggest that the basic architectural design—domes mounted on arches—was chosen because it would permit the use of many pendentives, which would be great for displaying artwork). This was just what Gould and Lewontin had claimed was absurd. It was not much longer before Dennett's position, in turn, came under attack by Robert Mark (1996), a professor of architecture (!) at Princeton. Mark argued that the best historical evidence pointed toward pendentives being the only structural option seriously considered, and that their being well suited for mosaic artwork was not a consideration in their use. In any event, the architectural debate now makes a regular appearance in discussions of the Spandrels paper.

The architectural aspect of the discussion points toward an important ambiguity in Gould and Lewontin's attack on adaptationism. Houston argued that if Mark's claim that there were structural reasons for choosing pendentives is correct, then pendentives *are* adaptive (albeit for structural support and not for artwork), and the anti-adaptationist view is left "in ruins" (1997, 125). The problem is that what Gould and Lewontin meant by spandrels being an architectural by-product of the church's basic design was left rather vague. If they meant that there was no reason at all why pendentives were used, then Houston's objection is on target. If, however (as seems to us more likely), they merely wanted to point out that the artwork, while making excellent use of the pendentives, did not play a significant role in the choice of pendentives, then Houston's objection is off the mark. If the only

(*continued*)

Box 5.4 (*continued*)

reasonable way of mounting a dome on arches at that time was to use pendentives, but the basic architectural design of a dome mounted on arches was chosen because it would provide lots of pendentives for the display of artwork, it is anyone's guess whether they are adaptive or not in Gould and Lewontin's reading.

The vagueness of the spandrels metaphor, and the ambiguity of the point the reader was supposed to draw from that metaphor, are echoed by the difficulty of interpreting the paper's biological claims. Were Gould and Lewontin arguing that the explanations for the spread and maintenance of many or most traits in organisms did not involve *any* reference to adaptations shaped by natural selection? Or were they suggesting that those explanations were likely to be far more complex than the simplistic adaptationist storytelling they were criticizing would allow? These are not equivalent claims, and the Spandrels paper—and the debates that followed—too often wavered between them.

on constraints would reveal the ways in which natural selection was limited. Gould and Lewontin wrote, in fact, "that constraints restrict possible paths and modes of change so strongly that the constraints themselves become much the most interesting aspect of evolution" (1979, 594) (a position, incidentally, on which Gould [1980] partially reversed himself not much later, suggesting that constraints play "creative" roles in the evolutionary process).

Part of the problem is that what Gould and Lewontin meant by constraints was left vague in the Spandrels paper—it was never exactly clear where constraints emerged from, or what they were, or even what they were supposed to be constraining, either in the case of the central architectural example (box 5.4) or in actual biological cases. On the one hand, "constraints" seemed to refer to the limited ability of natural selection to alter correlations between characters (Gould and Lewontin 1979, 597); as we saw earlier (in chapter 4), statistical correlations between characters are often assumed to reflect constraints on natural selection. On the other hand, Gould and Lewontin's position could be interpreted as referring to the complexity of the developmental process and the limits it puts on what kinds of changes can take place in a piecemeal fashion. For example, it is obvious that the complex body plans of extant vertebrates are due not to multiple independent selective episodes, but rather to historical relationships, and it is unreasonable to suppose that traits not coherent with such body plans could arise in contemporary vertebrate populations. However, neither of these views, properly understood, truly presents constraints as something *opposed* to natural selection. Indeed, we would argue that, in an important sense, these are not constraints at all.

Our view here follows, broadly, positions developed by Matthen and

Box 5.5 Development and Developmental Systems Theory

One of the key questions one can ask about a trait is how that trait is reliably reproduced in the organisms in question. As we noted in chapter 3, the assumption that genes are the only entities for which there is heritable variation on which natural selection can "work" has been shown to be empirically false (for review, see Jablonka 2001; Moss 2003; Jablonka and Lamb 2005). But a similar point can be made regarding organismal development more generally. Developmental systems theory (DST: see Oyama 2000; Oyama, Griffiths, and Gray 2001; Moss 2003; for a broader philosophical discussion of developmental biology, including DST, see Robert 2004) suggests that too much of contemporary biology still thinks of genes as "guiding" development in a privileged way. We are urged by the advocates of DST to get out of the habit of thinking of genes as having this kind of special primacy—for example, as being some kind of storehouse of information that guides the development of the organism (see also box 3.4). These authors point out, quite properly in our view, that genes cannot coherently be assigned such a role in development—the process by which organisms develop is not one in which genes guide development and the other resources involved simply provide raw materials. Rather, the other materials available (the other developmental resources) codetermine the way in which particular nucleic resources (genes) will be used in further developmental processes. Neither can be assigned either temporal or causal priority: without the full complement of developmental resources (which include particular genes, proteins, subcellular membranes and organelles, and specific environmental resources), organisms cannot develop, and changes in any of these resources will be associated with changes in development. Furthermore, many of these developmental resources—not just genes—are reliably inherited, and in many cases, variations in these resources can be inherited as well.

Understanding how a trait came to be prevalent in a population requires not only that we be able to tell a selective story (in terms of vernacular fitness) about the advantages of that trait, but also that we understand how that trait came to be reliably (re)created in each generation. Understanding the reliable reproduction of that trait will naturally involve understanding the changes in some of the developmental resources available, or the ways in which those resources are used, during development. Which resources are used differently, or what different resources are used, is not something that can be determined without exploring the particular way in which the trait came to be reliably reproduced. We should therefore learn to think of the genotype as just one of many axes representing available developmental resources that can vary, including, of course, several aspects of the environment.

In chapter 6, we will suggest that these observations should have a profound impact on how we think of studies that aim to discover the functions of particular traits and the developmental resources used in their (re)production.

Ariew (2002), Walsh (2003), and Ariew and Lewontin (2004), which we introduced in chapter 1. Each of these authors points out that so-called constraints on natural selection are the result of the physical instantiation of the organisms involved, including the systems of inheritance available and the developmental systems that are reliably replicated through those inheritance systems (box 5.5). Of course, natural selection and development require *some* physical instantiation or other. Natural selection (either informally as a process, or formally as a statistical trend) can only take place, and hence can only be understood, given a particular physical "substrate"[3] (Matthen and Ariew's term)—that is, a particular set of systems of inheritance and reliably replicated developmental pathways.

Once we get out of the habit of thinking of natural selection as a force (see chapter 1) pushing populations toward some fitness peak (see chapter 8), it becomes clear that to talk about constraints on the action of natural selection is to make a categorical mistake. As we noted briefly in chapter 1, if one is concerned with informal selection on a particular trait, certain physical processes (death by predation, gamete incompatibility, etc.) either impinge on organisms or not, and those that do impinge do so in ways that are either discriminate or indiscriminate with respect to the trait in question. Nothing in the physical makeup of organisms can constrain selection understood in this way. While some processes are indiscriminate, or do not impinge on the organism at all, that is hardly a meaningful constraint on the "power of selection." At the statistical or formal level, populations either undergo a mean change in the frequency of particular trait variants or not, based on the mean relative success of the organisms with the trait variants of interest. Again, there is nothing here that constrains (formal) natural selection—formal selection is a statistical fact about the populations in question, not a process that can be subject to constraints. Finally, explaining apparent disconnects between the informal (or individual) and the formal levels involves not appeals to constraints, but rather to, for example, the particular developmental processes at work, or the other physical processes and traits that are significant in any given case (box 5.6). Neither of these constrain selection—on the contrary, they are what makes selection possible.

The goals of understanding how developmental processes arose, how and through what systems they are inherited, and how they become reliably replicated are all of fundamental importance, and nothing we have said so far should be interpreted as suggesting otherwise. One way to interpret Gould and Lewontin's push in the Spandrels paper is just this—that understanding how traits arose, spread, and are maintained in a population is something best done by careful attention to the particular inheritance systems and developmental pathways involved, not by a narrow focus on the functions the traits now serve. This is also a way of understanding Schlichting and Pigliucci's (1998, chapter 6) approach to constraints. Their suggestion, made in response to what they viewed as an

3. This concept is related to what Schlichting and Pigliucci refer to as the "infrastructure" necessary for evolution (1998, 74–75).

Box 5.6 Examining Constraints or Exploring the Substrate?

Antonovics and van Tienderen (1991) remarked that studies of constraints are hampered by the lack of a clear null hypothesis (see chapter 10 for a discussion of "null hypotheses" in the context of biology). What is supposed to be constraining what, and how are these constraints to be envisioned? What even counts as a constraint? For example, Gould (1989) posited constraints on the shell shape of the land snail *Cerion,* whereas a computer simulation of the shell morphospace has actually indicated that such space is essentially saturated, with no evidence of a limitation on form (Stone 1996); is this evidence that there are no constraints of the sort Gould suggested? As we noted in chapter 4, research on genetic variance-covariance matrices cannot be used to reveal constraints on long-term evolution, nor can these matrices be used to unambiguously detect the effects of selection.

One way to avoid the morass of constraints is to think instead about the particular question being asked. For example, if one is interested in how easy or difficult it is for some particular trait rather than another to evolve from some ancestral state, given a particular selective regime, transitional probability matrices can provide a reasonable entry into the research program. Transitional probability matrices specify what the probability of a given evolutionary change is compared with the probabilities of alternative evolutionary pathways for a particular population and environment. The theoretical biologist is therefore forced to explicitly formulate her hypotheses in statistical terms. One could then, in principle, measure actual probabilities (e.g., frequencies derived from a phylogenetically informed data set) and compare them with the predictions by means of any matrix comparison algorithm (see Manly 1985, especially chapter 6). Of course, these matrices are statistical in nature; if one wants to know *why* particular transitions are more frequent than others, more mechanistic approaches will be necessary. For example, the likelihood that some kinds of developmental pathways (rather than others) will arise can be explored by the empirical study of the likelihood of certain base-pair substitutions, or of crossing-over or transposition events, or of the duplication of particular kinds of pathways (see, e.g., Wilkins 2002 and citations therein).

For instance, it is impossible for an *Arabidopsis* plant to both flower early and have many leaves (Callahan and Pigliucci 2002)—physiologically, the resources necessary to do one preclude doing the other (see chapter 1, fig. 1.1c). It is also impossible for human populations to become homogeneous for the partial immunity to malaria provided by the *HbS* allele because the immunity is provided by heterozygotic superiority, which will always result in a genetic polymorphism, given Mendel's laws of inheritance. But these two impossibilities are of very different kinds, and to think of them both as constraints on the power of natural selection is to blur an important distinction.

almost comic proliferation of types of constraints and the confusing terminology that resulted (see Antonovics and van Tienderen 1991), was that all categories of constraints (with the possible exception of absolute constraints imposed by fundamental laws of physics) can be considered genetic or epigenetic in nature; that is, that the only constraints on evolution are the very systems that make natural selection (via development and the reliable inheritance of developmental systems) possible. One can read Walsh (2003) as following up this position: he claims that, insofar as we think of constraints as developmental processes and the mechanisms by which those processes are replicated, it is the constraints themselves that explain adaptive traits, and not natural selection (297). While we prefer to think of developmental processes and systems of inheritance as explaining the possibility and reliable replication of adaptations, rather than as the adaptations themselves, we think the basic point—that there is no understanding of constraints such that it makes sense to think of natural selection and constraints as opposing processes or forces—is very much on the right track.

DETERMINING THE QUESTION AND WORKING TOWARD ANSWERS

Given the confusion that surrounds adaptations and constraints, it would be difficult to articulate a "middle ground" between a focus on natural selection and a focus on constraints. Instead, what we must do is carefully specify the questions we are asking, then determine the best approach to answering them. Advances in the past twenty-five years have given us both the empirical and the conceptual tools necessary to meaningfully address more of the questions that we might ask. If, for example, we are interested in understanding trade-offs between certain characters (e.g., vegetative growth vs. seed output in a perennial plant), we might focus on (1) the potential costs and benefits of particular trade-off strategies, or on (2) how different strategies are implemented developmentally, or on (3) how these strategies arise and in what ways they are most likely to be modified (see, e.g., Gould 1980; Cheverud 1984; Maynard Smith et al. 1985). Two things should immediately be obvious: first, that none of these questions can be answered without reference to the developmental processes involved and the systems of inheritance that make such processes possible, and second, that each of these questions requires a different kind of empirical research project in order to be answered. The first question, for example, is one that can be addressed through optimization analyses, but only after substantial work has been done on the requirements of and the possible solutions available to the system (see, e.g., Marrow and Johnstone 1996; Seger and Stubblefield 1996; Niklas 1997). The third question, on the other hand, requires an understanding not only of the developmental biology that underlies the trade-offs in a given population, but also of the developmental systems used in the wider clade, as well as of the way in which developmental pathways can (and cannot) be co-opted to perform new functions (see, e.g., Wilkins 2002).

More generally, when faced with a trait of interest, we can ask (at least) the following questions:

1. How did the trait arise? By what developmental mechanisms did it originally appear? What developmental pathways were co-opted to produce it?
2. How does the trait develop? What are the developmental pathways by which it is produced now? How reliable are those pathways? What environmental perturbations are those pathways sensitive to?
3. By what mechanisms is the trait reproduced across generations? What systems of inheritance, in combination with what reliably reproduced developmental pathways, lead to this trait being generated?
4. How (and why) did the trait spread through this population? Why is it prevalent? Furthermore, what is the relevant contrast class with respect to this question? Is it the trait it supplanted? Or other traits that might arise given the available developmental processes?
5. Why (and how) is the trait being maintained in the current population? What are the other viable options? That is, what alternative options could arise (by mutation, developmental "errors," etc.), given the developmental systems of the organism in combination with the systems of inheritance involved, and how often do they arise? Why don't the other options that arise spread in the population?

While these questions are obviously not wholly independent of one another, each demands a different kind of answer, and answering each question demands a different research focus (and, obviously, different empirical data). Moving away from a focus on adaptations versus constraints permits one to see the ways in which particular research projects can and should be structured to answer specific questions, and permits us to see that part of the problem is that one ought not to expect any one research project to answer all, or even any large subset, of these questions.

The questions above are obviously not easy to answer; for many organisms, and many traits, we do not yet have powerful enough empirical or conceptual tools to gain satisfactory answers. Furthermore, some of these questions—those about the genesis and spread of traits—are historical in nature, and the information necessary to generate convincing evidence for particular scenarios may, in many cases, be forever lost to time. We may be able to show that some pathways are possible—that a trait might have emerged and spread in a particular way—but not to distinguish empirically between those possible pathways to pinpoint the one(s) that history actually followed. And yet, this is the kind of work that is necessary if we are to understand why some evolutionary trajectories are more common than others, or why the mean value of some trait in some population does not change under selection.

The fact that organisms are physical entities with complex developmental histories and particular systems of inheritance is not a constraint on the power of natural selection; rather, as we have seen, it is what makes natural selection possible. Of course, these physical features do *influence* the kinds of adaptations that are likely, as well as how quickly those adaptations will spread, and what their chances are of being maintained. But understanding how and why particular physical aspects of populations make some evolutionary trajectories more likely than others is not a matter of identifying constraints; instead, it is a question of understanding those complex developmental pathways that lead to particular traits, given certain developmental environments.

These are not counsels of despair. Gould and Lewontin's Spandrels piece pointed toward some of the fundamental difficulties in testing adaptive hypotheses. Work since then has gone some way toward ameliorating those difficulties, but it is obviously still harder to test adaptive hypotheses rigorously than it is to make them up. While we reject the language of constraints opposed to selection, we think one can interpret Gould and Lewontin as pointing toward the kinds of complex developmental pathways that we here stress as a sound alternative locus of empirical inquiry. Again, it is hard to rigorously investigate these pathways and the ways in which they influence the probability that evolution will take some trajectory rather than others—harder, certainly, than just motioning vaguely toward constraints or generating statistical correlations (though the latter will often provide a valuable starting point). But such work, while difficult, is rewarding us with a richer and more nuanced view of how phenotypic evolution actually works.

6 Functions and For-ness in Biology

BART (TO POLICE OFFICER LOU): WOW! CAN I SEE YOUR CLUB?
LOU: IT'S CALLED A BATON, SON.
BART: OH. WHAT'S IT FOR?
LOU: WE CLUB PEOPLE WITH IT.
—*THE SIMPSONS,* "SEPARATE VOCATIONS"

Functional ascriptions have traditionally been, and continue to be, an important part of biological thought. We argue here that the same kinds of reasoning that go into assigning functions for phenotypic traits should be applied to genes and other developmental resources as well. In this view, the notion of a gene (for example) being "for" a particular phenotypic trait or set of traits is understood to be about the relationship between the prevalence of the gene in the population in question, the functional/causal association between the gene and the trait(s) in question, and the selective history of the trait and gene in that population. The same reasoning will apply to other developmental resources (cellular membranes, niche modifications, etc.) that one might wish to identify as "for" particular associated functional traits. In all these cases, it will be necessary to gather statistical, biochemical, historical, and ecological information before properly claiming that a gene or other developmental resource is "for" a phenotypic trait. This will, by necessity, proscribe many contemporary uses of "gene for" talk. Our hope is that, rather than hampering practical use of "gene for" talk, our approach will stimulate much-needed research into the functional ecology and comparative evolutionary biology of gene action, and the relationship between genes and other developmental resources.

FUNCTIONS AND FOR-NESS IN BIOLOGY

In the previous chapters, we presented a number of reasons to be skeptical, or at least cautious, about some of the claims often made regarding the action of natural selection. In this chapter, we introduce and briefly defend a particular view of what it might mean to say that a biological trait has a certain biological "function," and develop a framework for deciding when particular developmental resources (including genes) can be said to be "for" the development of given biological traits. The problem of identifying the biological function of a trait and the problem of developing a view of what developmental resources are "for" vis-à-vis other functional traits are, we think, clearly linked. Indeed, we think that these questions should be answered in roughly the same way—by appeal to the recent evolutionary history of the features in question. This view has been well developed and extensively defended in the case of functional ascriptions to character traits; however, there has been comparatively little work done on the developmental side.

Whether attempting to determine the function of certain character traits or of various developmental resources, there is a very real danger of slipping into adaptationist thinking of one form or another (see boxes 5.1 and 5.3). Our hope is that this danger can be avoided, both by a careful focus on what it means to say that a trait (or a developmental resource) has some function, and by spelling out what we have to know about the relationship between an organism's development, its environment, and its evolutionary history if we are to speak properly about its traits (including developmental ones) having particular functions. So, for example, it is often claimed that a gene is "for" a particular trait; in our view, the only good sense that can be made of this locution is that the gene has a certain (biological) function. But what needs to be stressed is the difference between a gene being "for" (functionally) a particular trait and a gene merely being "associated with" a trait: the human *HbS* allele, we argue, is *for* (partial) malaria resistance and merely *associated* with mild anemia (in heterozyotes). Or, similarly, there is a difference between a trait having a function and its just *doing* a particular thing: the function of the heart is to pump blood, but while it also makes thumping sounds, the latter is just something it does, not its function. It seems obvious that we can (and do) know how to find genes associated with various phenotypic traits and how to find correlations between traits and things those traits do; further, we do know of a lot of genes and traits so associated. But what do we have to know to make the functional claims?

We argue here that the best sense that can be made of functional locutions in biology is that of the "modern history" view of functional talk that Godfrey-Smith developed for thinking about the biological functions of character traits (a.k.a. "direct proper functions": Godfrey-Smith 1994; see also Griffiths 1993). In this view as it was originally conceived, a trait has a certain function by virtue of its having been maintained by recent selection because of its ability to do that particular kind of thing. Similarly, we will argue, we can properly speak of a developmental resource (such as a given gene, or a particular cellular membrane) being "for" some trait (or some developmental process more generally) only when that developmental resource has been maintained by natural selection in the recent evolutionary history of the population by its causal association with the trait (or process) under consideration (a more precise definition will be developed below). Based on this interpretation, it will turn out that to say that, for example, a gene is for some (variation on) trait *x* is to say something not merely about what the gene happens to do in this particular case, but also about what biological meaning that association has.

While the sort of functional talk developed by Godfrey-Smith (and other authors, such as Millikan [1984] and Griffiths [1993]) has come under attack by, for example, Davies (1994, 1995, 2000) and Manning (1997), we argue that the etiological approach to functions that Godfrey-Smith and others have developed is still the best one for understanding the function of character traits in biology; while the critiques of various etiological proposals point toward some important complications, we suggest that, for many kinds of traits and for developmental resources, these complications ought to be embraced. In part for these reasons, it may well turn out that fewer traits have clear-cut functions than one might have thought—that the relationship between what a trait "does" and the ontogeny and phylogeny of that trait is often so complex and contingent as to resist straightforward interpretation in terms of recent selective history. Similarly, it may turn out that the functions served by particular developmental resources are often so multifarious and complex as to resist (easy) interpretation. But even if this is the case, finding out about, and gaining some entry into, these kinds of complexities will (and ought to!) remain a necessary part of evolutionary biology, and those places where functional ascriptions can be made may well turn out to be of particular interest.

Obviously, a comprehensive review of the relevant literature is far beyond the scope of this chapter—indeed, at this point, even book-length reviews of functional literature in biology are, by necessity, somewhat incomplete. Similarly, it will not be possible to respond to every objection that has been raised against the various etiological accounts of biological functions in the literature. Rather, what we hope to do here is sketch a way of thinking about a certain kind of functional talk in biology that will permit us to sensibly extend functional talk to developmental resources (such as genes and other sources of heritable variation) as well as traits, outline the difficulties facing scientists who wish to pursue research guided by this kind of functional talk, and distinguish this kind of functional talk from the too-often too-sloppy kinds of pseudo-functional ascriptions that get bandied around in the biological (and especially biomedical) literature.

CHARACTER TRAITS AND ETIOLOGICAL FUNCTIONS

As many authors have noted, one of the striking features of the biological world is the ease, and apparent appropriateness, with which functional ascriptions can be made. It seems entirely natural, and mostly unproblematic, to think that the function of the heart is to pump blood, the function of the eye is to see, the function of a sparrow's wings is to fly. This is strikingly different from other scientific endeavors—we may use features of the nonbiological natural world in particular ways, but that is not their function. Artifacts, whether produced by humans or by nonhuman animals, acquire their functions from the uses to which they are put by their users—chimpanzees may *use* twigs to collect termites, but a particular twig has that function only because of the way the chimpanzee actually uses it and the intention of the chimpanzee to so use it.

But how can features of the natural world acquire functions without their having been deliberately made to have those functions? It is now widely, if not quite universally, accepted among biologists and philosophers of biology that the most plausible account of how biological character traits acquire functions is through their historical association with natural selection in evolving populations. The most general term for functions acquired in this way is *etiological* functions, the term *direct proper functions* being narrower and demanding a more particular association between the trait and some selective advantage. The basic idea behind etiological functions is simple; indeed, as far back as the 1960s, Lorenz wrote that "unless selection is at work, the question 'What for?' cannot receive an answer with any real [biological] meaning" (1963, 14). The difficulty has been to make precise just how selection has been at work, and what exactly it has worked on.

One response has been to claim something like the following:

> For a trait T (a token of an adaptive character trait) to have the function of doing some thing f in an organism O (an individual organism), traits of type T must have contributed to the fitness of O's (evolutionarily recent) ancestors by doing f and must have been selected against alternative traits $T^1 \ldots T^n$ in the ancestral populations in question because of T's ability to do f. (Adapted from Buller 1998, 507; see also Godfrey-Smith 1994, 359)

There are, naturally, many wrinkles, and formulating the definition in ways that attempt to account for all of them is clumsy in the extreme (compare the ways in which Godfrey-Smith [1994] and Griffiths [1993] deal with the issues discussed here). For example, it seems clear that the trait *T* in the above formulation must be a homologue of the *T*-type traits that were selected in the recent past, and that *T*-type traits must be reliably reproduced in populations of the organisms in question (see Godfrey-Smith 1994, 359). The sense of "selection" in the above formulation might also need to be made more precise, as a trait might have a function that increases the fitness not of *O*, but of *O*'s relatives; the function *f* might therefore not be, strictly speaking, a function that attaches to *O*, but rather to a population that includes *O* (see, e.g., Griffiths 1993, 416). A more general concern applies to the trait concept employed in the definition: if we think of *f* as a function of a trait in some system (whether it be the individual organism with the trait in question or a group that contains organisms with that trait), we might wish to be more precise about what constitutes a trait (see Wagner 2001, e.g., chapter 1). One possible approach is to think of "traits" as discrete functional units—the particular function performed does not define the trait, but that it performs some kind of function makes it a trait (see Wagner 2001, chapter 7). Unfortunately, using such a definition here would generate real worries about circularity, because we would need to identify the trait in question as functional before beginning our analysis.

It is not at all clear to us that all of these wrinkles can be avoided by sufficiently clever definitions; it may be that the biological world is complex enough that we will be unable to find necessary and sufficient conditions for some biological character trait to have a function (see chapter 9 on the strict limits of "necessary and sufficient" types of definitions in biology) or, for that matter, for something even to be a biological character trait! But this problem should not prevent us from making the obvious observation that the function of the heart is to pump blood, and that the thumping sound made by a working heart is not in fact part of its function. Between the extremes of the actions of traits for which functional ascriptions are clearly appropriate and those for which functional ascriptions are clearly inappropriate, there is probably a gray area, and any definition that permits the unambiguous assignment of functions and non-functions will probably produce results that at least some people will find counterintuitive.

For instance, Manning (1997) uses the example of segregation distortion (meiotic drive) to suggest that the direct proper function version of the etiological account is caught in a dilemma. Recall that segregation distorters are strands of DNA that have a higher probability of being passed on in sexual reproduction than do non-distorters at the same locus—that is, they have a better than 50 percent chance of being used during meiosis in the production of sperm or egg cells, or they increase the chances that those gametes that include them will be used (see box 3.1). Manning notes that proponents of the direct proper functions account recognize two ways of dealing with this problem. The first is to "bite the bullet" and acknowledge that the function of segregation distorters is to increase their own chances of being used in gametes; this solution is unattractive, however, because segregation distorters are not generally thought of as functional in biology (Manning 1997, 76; see also Godfrey-Smith 1994, 348). The other possibility is to argue that only traits that increase the fitness of the organism of which they are a part can have functions; while this might solve the problem of segregation distorters, it does so at the cost of introducing a very narrow conception of fitness into the definition (Manning 1997, 76; see also Godfrey-Smith 1994, 348) and excluding many plausible functional traits (morphological and behavioral characteristics of nonreproductive castes in eusocial insects, for example) from functional analysis. Godfrey-Smith attempts to avoid this problem by requiring that the putative functional trait "reside" in a biologically real system, and that it increase the fitness of that biologically real system in which it resides (Godfrey-Smith 1994, 349); this would permit the inclusion of many traits whose functions work through complex inclusive fitness and yet, Godfrey-Smith claims, eliminate nonfunctional traits such as segregation distorters. This definition still leaves rather vague what a "biologically real system" is, as well as what it is for a trait to "reside" in one; traits that are nested in complex ways within multiple biological systems might have a variety of functions in this view, and might or might not count as discrete multiple traits. In any event, it seems clear to us that attempts to find necessary and sufficient conditions for each term of contention will rapidly degenerate into a massive regress.

Perhaps worse than the problem with conflicting intuitions, attempts to disambiguate the various key concepts employed by the etiological theory may result in some aspects of the definition failing to cohere well. Indeed, several recent attacks on direct proper functions and etiological functions more generally have focused on the difficulty of generating a definition of etiological functions that, taken as a whole, manages to respect all of our biological intuitions, and yet in which every aspect is sufficiently clear to resist charges of vagueness. For example, Davies (1994, 1995) argues that there is no way of making precise the notion of the (apparently) functional trait in question being homologous with past versions, given what we know about the complex developmental processes by which traits are (re)produced and acquire their forms and (apparent) functions. In other words, since there is obviously no gene or small set of genes that produces a particular phenotypic trait, Davies argues that figuring out just what constitutes the reproduction of that trait across generations will be difficult, and that the task of identifying the trait via the mechanism of its reproduction seems likely to founder on either overspecificity of the developmental process or underspecificity of the trait in question (Davies 1994, 370–74).

This might sound like the sort of quibble that biologically informed common sense can easily avoid, but it points toward an interesting problem for the etiological view. Consider the case of traits that appear to be functionally similar in different organisms and which are produced via many, but not all, of the same developmental pathways. For example, Roth (1988) presents an interesting scenario in her discussion of "genetic piracy." According to Roth's definition, genetic piracy is the process by which "new genes, previously unassociated with the development of a particular structure, can be 'deputized' by evolution; that is, brought in to control a previously unrelated developmental process, so that entirely different suites of genes may be responsible for the appearance of the structure in different contexts" (Roth 1988, 7; see also Schlichting and Pigliucci 1998, 144–46; Ganfornina and Sanchez 1999). Roth cites de Beer on the eyeless mutant of *Drosophila,* in which it is possible to select for other genes that result in "a fly with restored eyes that still has the original mutation" (1988, 7). From this observation, Roth concludes that in such cases, "new genes are involved in the formation of eyes that previously had not been" (1988, 7). In this way, a different set of genes, and hence different developmental resources and pathways more generally, may become associated with the formation of a trait that appears to be "the same" functional trait. That is, it seems natural in this case to think of the trait in question, the new eye, as being homologous with the ancestral eye, despite the use of different genes (or more generally, different developmental resources) in its production.

It would be a mistake, however, to reject the direct proper functions interpretation because of issues like this. Rather, what these issues suggest is that the relationship between a particular character trait and the development of that trait is itself fluid, and that understanding how functional traits are reproduced, and indeed, what constitutes the reproduction of a trait, involves engaging in a number of different research projects using a variety of different tools. It is one thing to say that the function of eyes in *Drosophila* is to see, and that the developmental processes that make possible eyes that can see were selected in the recent past and are functionally associated with those eyes; *establishing* what those processes are is a rather larger and more complex task. If functional talk in biology is to be properly extended to the developmental resources and pathways that make the reliable reproduction of adaptive traits possible, these complexities will need to be addressed. It is to this task that we now turn.

ONTOGENY, DEVELOPMENTAL RESOURCES, AND TRAITS

To what extent ought we to think of the developmental resources associated with the formation of adaptive phenotypic traits as being *for* those traits? This question, we think, lies at the heart of the controversy over "genes for" talk in biology (box 6.1). Following the same reasoning used in ascribing functions to particular adaptive traits, we wish to suggest that a particular developmental resource (such as a gene, a suite of genes, a methylation site, a cellular membrane, or an environmental resource) is "for" a particular adaptive trait when it has been maintained by recent selection in the population in question because of its functional association with that trait (box 6.2).

In some cases, it seems clear that a particular developmental resource is very likely "for" the production of a given adaptive trait. Consider, for example, the *FY-0* allele in humans. The *FY-0* allele is present in some 87 percent of the population in sub-Saharan Africa, but vanishingly rare in populations where malaria has not been a problem (see Cavalli-Sforza and Cavalli-Sforza 1995, 125). The *FY-0* allele is causally associated with the absence of the FY substance, and this absence "grants . . . a certain amount of protection against *Plasmodium vivax* [a particular malaria parasite] by making it hard for the parasite to multiply" (Cavalli-Sforza and Cavalli-Sforza 1995, 125). So it seems overwhelmingly likely that the *FY-0* allele is functionally associated with malaria resistance (itself an adaptive trait), and therefore that we can speak of *FY-0* as a gene for (a gene with the function of providing) partial malaria resistance.

Of course, no gene is ever sufficient for the production of any phenotypic trait. And while the *FY-0* allele is (generally speaking) necessary for partial malaria resistance via the absence of the FY substance, so too are many other genes and developmental resources. Our focus here is therefore on the particular *variations* in developmental pathways that can be coherently thought of as being "for" certain phenotypic traits, insofar as those variants were selected for their relationship to the production of the functional traits involved.

It is important to stress that this notion of for-ness can be applied to any heritable developmental resource. Cortical inheritance, for example, is mediated by membrane-membrane interactions rather than genetic

Box 6.1 Genes-For or Not Genes-For?

Historically, arguments about whether it makes sense to claim that some gene is "for" some (phenotypic) trait have centered on two extreme positions (see Kaplan and Pigliucci 2001 for discussion). Richard Dawkins is usually taken (fairly or unfairly) as the representative of the view that there are lots of genes for phenotypic traits, and that theorizing about the adaptive significance of traits is at least a good first step toward discovering that there are genes for it (e.g., 1976; 1982, especially chapter 2). Dawkins, in *The Extended Phenotype,* sometimes defends speaking of genes for a trait in a way that makes this locution roughly equivalent to the gene in question being statistically associated with the trait under study. He writes, for example, that the proper translation of the words of a geneticist who speaks of a gene for red eyes in *Drosophila* is something like this: "There is variation in eye color in the population; other things being equal, a fly with this gene is more likely to have red eyes than a fly without the gene" (1982, 21). Sober (1993, 186–87) attacks this use of the language of genes for traits by noting that "a gene *for* phenotype *x* presumably is a gene that *causes* phenotype *x*," and that simple correlations between genes and traits obviously do not imply causation; genes may be statistically associated with a trait for reasons other than developmental significance.

Sterelny and Kitcher (1988, 348–52) attempt a reconstruction of Dawkins's notion of a gene being for a trait that they hope will be more precise and will properly account for the complex interactions genes can have with their environments. The intuitive idea behind their proposal is that "we can speak of genes for *x* if substitutions on a chromosome would lead, in the relevant environments, to a difference in the *x*-ness of the phenotype" (1988, 348). They go on to claim that a particular locus, L, affects a particular trait, P, if there are allelic substitutions at L that will result in differences in P in "standard environments" (1988, 349). This substitution requirement is supposed to add a causal element to "genes for" talk by excluding mere associations without any causal import. While much work has been done in fleshing out the idea behind "standard environments" (1988, 350; see also Kitcher 2001, 404–7), the main problem with this version of "genes for" talk is that it suffers from the same basic defect as Dawkins's articulation: it does not do justice to the notion of something being for something else. It seems clear to us that "being-for-ness" is not just a matter of statistical association, nor even of causal mechanisms. This is a more general difficulty with accounts of biological functions that fail to take the selective significance of the functional ascription into account; indeed, Griffiths (1993) cites this problem explicitly as the motivation for developing direct proper functions (and rejecting Cummins's functions)* in biology. If we want to use the phrase "a gene for *x*," then we have to take seriously the implications of the language involved in that phrase; in our view, neither Dawkins nor Sterelny and Kitcher take the implications of this locution to their logical consequences.

At the other extreme, it is sometimes suggested that the attacks on "adaptationism"

* In Cummins-style functional accounts, entities have functions by virtue of the causal role they play as part of some larger system's capacity (what that larger system does or is able to do) (see Cummins 2002 for review). A modification of Cummins's approach, focused on biological entities, is developed and defended in Weber 2005.

(*continued*)

Box 6.1 (*continued*)

(see chapter 5) and genetic determinism most closely associated with Gould and Lewontin (e.g., 1979) imply that the language of genes for traits (at least if used at anything above the most straightforward biochemical level) is entirely misguided (Levins and Lewontin 1985; Lewontin 1992). Further, the criticisms made by Oyama and others of the supposed-centrality of the gene, and their proposed alternative ("developmental systems theory"; henceforth DST—see box 5.5), are sometimes thought to render "gene for" language obsolete (Oyama 1992, 2000; Gray 1992; Griffiths and Gray 1994, 1997). However, most of these criticisms are not, we think, best interpreted as suggesting that it is impossible to properly speak of a gene being for some trait or other, but rather as pointing out that this way of talking has been problematic, in large part because it has been interpreted to make genes out to have a special place in the developmental process, which is exactly what DST wants to deny (see, e.g., Griffiths and Gray 1994, 277; Odling-Smee 1988, 74).* Gray (1992, 199), for example, writes, "I wish to dislodge the gene from the privileged site it has occupied in our accounts of development and evolution." This dislodging task can be done, and done well, without having to argue that there is no possibility of finding genes that actually are for traits, a position we think is untenable.

* Genes, it is worth noting again, may often be necessary for the production of some phenotypic trait, but they are never sufficient. The same may be said of many other aspects of the developmental process. We can just as easily speak of particular variations in developmental pathways as being "for" certain phenotypic traits, insofar as those variants were selected for their relationship to the production of the functional traits involved, even where they are inherited primarily through epigenetic mechanisms (see Moss 2003). Nothing in our way of conceiving of "genes for" speaks against these observations.

systems, and it is known to be both reliable and associated with variations that can have clear phenotypic (and fitness) consequences. In ciliates, these consequences are known to include differences in cilia function, size, and cell number (see Landman 1991). If a particular change in the membrane structure (induced, say, through environmental trauma) results in a phenotype that permits the organism to utilize a new resource, and the organism outperforms similar organisms because of that new phenotype, then the new membrane structure will be spread and maintained in the population because of its causal association with that phenotype, and could be considered *for* that phenotype.

Of course, in cases in which variations in developmental resources result in variations in multiple traits on which informal selection can operate, identifying what trait the variation in question is "for" may be difficult or impossible. The problem is that, as always, merely finding that a particular change in phenotype is associated with increased reproductive success is not sufficient to identify what is actually being selected, or, for that matter, even to show that *selection,* in the formal sense, is occurring.

Box 6.2 Finding Developmental Resources "For" a Trait

In order to properly claim that a particular developmental resource is "for" a trait, we would, ideally, wish to know all of the following:

1. That variation in the resource is (causally) associated with variation in the trait
2. What the most important aspects of the biochemical and developmental pathways that generate the trait are, and how the resource is exploited by them
3. That the trait in question was the likely subject of natural selection in the recent evolutionary history of the species in question
4. That the maintenance of the developmental resource in question was actually the result of natural selection in the species' recent evolutionary history

The first requirement, that there actually be an association between the developmental resource in question and the phenotype, is the reason why it is sometimes suggested that a statistical approach is a good first step in finding developmental resources (such as genes) for traits. Note, however, that in natural populations, there might be no observable variation in the developmental resources associated with variation in a trait (when, for example, all variation in the trait is lethal), which means that lack of a statistical correlation by no means implies lack of a causal connection (see also the prelude and chapter 4).

Obviously, the second requirement is difficult from an experimental standpoint. There are not many organisms for which we know most of the relevant biochemical pathways involved in the reliable (re)production of a given trait. Indeed, we suspect that, in the case of the many kinds of developmental resources that are used in the production of many traits in complex ways, it will be difficult or impossible to talk sensibly of them as being for any trait in particular. Nevertheless, there are some resources and some traits that we have a rather good handle on, and so for some kinds of resources, we can expect at least limited success in meeting the second requirement.

The third requirement poses perhaps one of the most difficult challenges. We want to suggest by this requirement that we need more than a good just-so story about how the trait might have been useful; we need to actually understand something about how it might have been the object of (formal) natural selection. This understanding has to rely on a phylogenetically informed comparative analysis of the organisms in question—a necessity that originates from the core fact that evolutionary biology is a historical, not just an experimental, science (a good understanding of the ecology of the trait is, of course, also necessary).

Finally, we need to know how natural selection on a given trait could account for the prevalence (and particular form) of the corresponding developmental resource; again, this fourth requirement may be difficult to meet, especially when the resource is used in many different developmental pathways. In any event, it will often be empirically challenging to identify the pathways from selection on a trait to the maintenance of a resource, but it will sometimes be possible to at least sketch them.

It should be clear that it will rarely be possible to fully meet all of these requirements. In some cases, however, we may wish to suggest cautiously that a gene, for example, may be for some trait when not all of these requirements have been adequately met. For example, in the cases of purported genes for phenotypic traits in humans, it would be out of the question (ethically and practically) to perform many of the experimental manipulations that are at the heart of meeting the second requirement. Gene knockout experiments, gene substitution experiments, attempts to develop reaction norms for the relevant genotypes,

(*continued*)

Box 6.2 (*continued*)

environments, and traits in question, laboratory evolution experiments, and so forth, while part of the geneticist's and ecologist's standard toolkit for many organisms (see Schlichting and Pigliucci 1998, 16–20; Burian 1997; Culp 1997), are simply inconceivable in the case of humans (see chapter 7).

Figuring out what the targets of (informal) selection actually are in cases like these, and how they relate to formal differences in fitness, are tasks that might require access to multiple replicates of a population in order to distinguish the way in which formal selection operates (see chapter 1). Nevertheless, we wish to suggest that it is just this kind of information that is needed if we are to identify particular resources as being for particular phenotypes.

A related point is that it will generally be hard to identify resources as being for particular traits when the development of those traits is highly "canalized" and hence variation in particular resources is often not associated with variation in those traits. In the above example from Roth, it was noted that *Drosophila* with mutations normally associated with eyelessness could still form eyes, perhaps by utilizing other genetic resources not normally associated with the formation of eyes in fruit flies. Often, of course, variations in the genetic resources necessary for eye production that result in eyelessness will be actively selected against (in both the informal and formal senses of selection, due to the likely lethality of the resulting trait under natural conditions; see chapter 1). But of course, while some variations in the developmental resources (including genetic resources) that go into eye production are selected against, other variations are buffered by various other developmental systems. In these cases, it will be hard to identify just what systems are "for" the development of particular traits.

A fairly well-understood example of such a buffering system is that associated with the *hsp90* system in *Drosophila.* The *hsp90* allele codes for[1] a protein that is part of the complex heat-shock response present not only

1. We use the phrase "code for" here in the strict historical sense of the word developed by Godfrey-Smith (2000), which we feel is the most common use in the molecular genetics community. We mean by it only that the *hsp90* allele is the DNA sequence that (along with the rest of the molecular machinery used in protein synthesis) explains the presence of the HSP90 protein, the one it codes for (but fails to resemble—hence the "code" metaphor). We agree with Godfrey-Smith that this interpretation does justice to the way the term is actually used by careful molecular biologists (and follows early uses of the word). Neither the stronger interpretation of "coding" talk developed and attacked by Sarkar (1996), nor Kitcher's fully metaphoric understanding of it (Kitcher 2001), seem to us adequate to do so.

in fruit flies, but in virtually every organism in which it has been searched for (Komeda 1993; Krebs and Loeschcke 1994; Loeschcke and Krebs 1994; Maresca et al. 1988; Prandl, Kloske, and Schoffl 1995). Rutherford and Lindquist (1998) described flies in which the function of the HSP90 protein had been impaired, either by mutation or by chemicals (see also Queitsch et al. 2002 on the plant *Arabidopsis,* as well as Sollars et al. 2003). These organisms display an array of morphological changes, spanning virtually every aspect of *Drosophila*'s adult phenotype. These morphological changes are associated with a variety of heritable developmental resources, including both genes and various epigenetic inheritance systems (chromatin condensation systems, for example). Rutherford and Lindquist concluded that HSP90 buffers *Drosophila*'s developmental machinery against such variation, thereby allowing other loci to behave neutrally and to accumulate mutations; had the variation in these developmental resources not been buffered, it seems very likely that much of the morphological variation associated with it in unbuffered flies would have been selected against.

So, ought we to think of developmental resources associated with variation in *Drosophila* eyes, but usually buffered by systems like the HSP90 protein, as functionally associated with eyes? After all, selection has not maintained those resources in a pristine state because of their role in development (instead, it has produced systems that buffer the variation that happens to accumulate), and may not, in recent history, have maintained them against variation at all. Where the development of particular traits is the result of complex and continuous (but not necessarily unintelligible) interactions between various developmental resources and certain environments, many of which are also involved in the development of other traits, it will be difficult to pinpoint aspects of the developmental process as being "for" the trait in question.

A related case was described by Atchley and Hall (1991) in their theoretical work on epigenetic systems. They pointed out that tissue-tissue communication (as in the induction of the formation of eyes in vertebrates by the adjacent neural crest) is not a genetic effect in the strict sense, but rather a response to the internal environment of the embryo. Moreover, this epigenetic effect catalyzes further gene action, which is necessary for the production of proteins specific to the new structure being developed (e.g., the crystalline in the eye). Local (intracellular) genetic effects, internal environments (cell-cell or tissue-tissue interactions), and external environments therefore all combine to produce the phenotype and behavior of the organism. In cases of this kind, it is difficult to make good sense of the idea that any one of these resources is for specific attributes of the phenotypes and behaviors.

So, in these kinds of cases, it is likely that particular developmental resources are being maintained by natural selection not because of their role in forming any given phenotype, but rather because of any one of a number of multiple roles. From this perspective, one can certainly argue that *hsp90* is "for" buffering the fly's developmental machinery from heat shock; on the other hand, one might suggest that it was maintained by selection not for buffering in general, but only because of the particular adaptive traits whose development it buffered. Of course, this buffering system is involved in the development of nearly all of the fly's adaptive traits, and it seems like a stretch to argue that it could be for all of those traits. Here, attempting to understand the function of a developmental resource leads inexorably into an exploration of the complexities of the case; because these complexities are at the heart of developmental biology and phenotypic evolution, they ought to be embraced. Even when we are unable to say what *the* function of a developmental resource (or for that matter, a trait) is, attempting to figure out the function(s) of that resource can (and often will) result in valuable research.

FUNCTION OR HAPPENSTANCE? THE PROBLEM OF NONADAPTIVE PSEUDO-FUNCTIONS

In the above cases, particular developmental resources were used in the development of multiple adaptive traits, and it was assumed that in many of these cases, natural selection would maintain the resource because of its functional association with the various traits (box 6.3). But some features of some developmental resources or traits, while they play important evolutionary roles, seem not to be subject to selection themselves. In these cases, exploring the ways in which these resources or traits influence the evolution of a population can lead to a focus on the importance of nonfunctional roles in the evolutionary history of organisms.

A case that has received much attention follows from Rutherford and Lindquist's work on the HSP90 system (1998, discussed above). Recall that disabling the HSP90 system "releases" variation in developmental pathways that is normally "hidden" by the buffering system. That is, differences in genetic and epigenetic developmental resources (see Sollars et al. 2003 on *Hsp90* and epigenetic mechanisms; also Jablonka and Lamb 2005) that are normally hidden become visible as heritable phenotypic variation. Rutherford and Lindquist hypothesize that the HSP90 system results in the accumulation of a reservoir of genetic and epigenetic variation that, in times of rapid environmental change, is released and results in new phenotypes, some of which will be better adapted to the new environments than were any of the variants in the buffered population.

Box 6.3 Complexities in Development

There are a number of different kinds of complexities that can make identifying a developmental resource as being functionally associated with a particular trait either conceptually or empirically difficult. Here, because they are the best documented, we focus on the complexities of genetic resources; however, we anticipate that similar complexities will emerge in the study of other kinds of developmental resources (epigenetic and environmental) as well.

Pleiotropy. A gene may have measurable effects not on just one trait (however we define "trait"; see Gould and Lewontin 1979; Wagner 2001), but on several. In some cases, some of these pleiotropic effects will be either neutral or subject to negative (informal) selection, and hence will be merely associated with the gene in question (although they will be causally associated). Others will have been subject to (positive) selection; in these cases, some of the effects will be functional, whereas others will not.

Epistasis. Physiological epistasis (as opposed to the statistical concept with the same name, employed in quantitative genetics; see Cheverud and Routman 1995 for a discussion of the complex relationship between the two) is the result of actual physical interactions among gene products. If gene products interacted only in an additive manner—that is, if the metabolic or phenotypic results were always directly proportional to the action of each gene—epistasis would present few problems for our understanding of the functional role of genes in the production of phenotypes. The reality is that additivity is the exception rather than the rule. Sets of genes can interact in ways that are positive or antagonistic, as well as strikingly nonlinear, all of which can vary with the presence or absence of still other genes or other developmental resources. In these cases, uncovering the functional role of any particular gene without taking into account the various other developmental resources involved will obviously be exceedingly challenging.

Gene complexes. Related to the issues surrounding epistatic interactions of genes is the fact that many traits are associated not with single genes, but rather with complex suites of genes. While such coadapted gene complexes might be maintained by natural selection because of their causal link to particular adaptive traits, and hence might be candidates for functional ascriptions, almost nothing is known about the evolution of such complexes or the role they play in the development of phenotypes (at least in eukaryotes). In any event, such functional ascriptions will be further complicated by the fact that any individual gene in such a gene complex might participate in the development of numerous different traits, or indeed, in numerous different suites of genes.

Phenotypic plasticity. The same genotype often produces different phenotypes under different environmental conditions; such phenotypic plasticity can be the result of environmental triggers directing development down one or another of several alternative pathways (adaptive regulatory plasticity; see Pigliucci 2001, chapter 5). Particular genes may be involved in the generation of such adaptive "developmental conversions" and hence may be thought of as "plasticity genes" (see Schlichting and Pigliucci 1995). But ascribing specific functions to the genes involved in particular plastic responses, as well as those involved in the formation of the different traits themselves, will again raise conceptual and empirical challenges.

(*continued*)

Box 6.3 (*continued*)

The gene concept. Finally, it is worth noting that recent research points away from the model of a particular stretch of DNA that is responsible for a single stretch of mRNA, and hence for a single protein; rather, a given stretch of DNA is often involved in the generation of a number of different proteins (as well as in other regulatory tasks). A segment of DNA can be involved in the generation of different proteins via a number of different mechanisms, such as frameshifts, alternative reading frames, alternative splicing, cotranscriptional splicing, sense/combined antisense transcription, and various combinations of these mechanisms. The same stretch of DNA may therefore be used in the production of a number of different proteins, each of which is itself involved in the development of a number of different traits. Attempts to find *the* function of a gene (or indeed, to identify *a* gene) will therefore often reveal serious complexities, many of which may be of deep developmental and evolutionary significance.

Even if this is true, it is highly unlikely that *hsp90* is actually "for" accumulating a reservoir of genetic variation, or "for" the occurrence of alternative phenotypes under certain environmental conditions. In order for these to be functions of the *hsp90* gene, selection would have had to act not on the new phenotypes released by the breakdown of the HSP90 system, but rather on the breakdown of the HSP90 system itself; while such a use of HSP90 may have been adaptive (as Rutherford and Lindquist claim: 1998, 341) it is probably not an adaptation, as it seems very unlikely that selection has in fact acted in this way. Nevertheless, these additional roles played by *hsp90* and similar genes may be crucial for long-term phenotypic evolution.

Another example is given by Sterelny and Griffiths (1999, 209) in their analysis of the possibility of species-level (or higher-level) selection. Magnetotactic bacteria rely on magnetosomes to avoid oxygen-rich waters, which are toxic to them, by swimming downward. These bacteria respond differently to the direction of the magnetic pull depending on which hemisphere they are in. The system by which the polarity of the magnetite crystals in these bacteria is determined is nongenetic: they are reproduced directly from the parent's crystals. This system is also somewhat unreliable, such that a certain percentage (but a small one—perhaps 1 percent or less; see Bazylinski 1999) of the offspring of a particular genetic line respond to the direction of the magnetic pull randomly. From the standpoint of an individual bacterium with randomized magnetosomes, this reduces fitness, as half of its descendants will swim merrily to their deaths. However, for the species as a whole, this randomization prevents extinction during times when the Earth's magnetic field switches. Sterelny and Griffiths note that one might conjecture that all those species that developed more reliable ways of passing on magnetic directionality are now extinct, and so the unreliability of the system is an adaptation to the Earth's unreliable magnetic field.[2] Generally speaking, it seems reasonable to suggest that the system for passing on magnetic directionality is "for" the bacteria's ability to avoid oxygen-rich waters. Whether the unreliability of that system also has a function is a much harder question to answer—but it is *conceptually* possible that it does, and indeed, it is even possible to think of some ways of trying to find out. Even if it does not have a function, it might still be of critical evolutionary importance to the bacteria in question. Attempting to distinguish between "true" functions and handy happenstances will, therefore, be another endeavor in which a focus on functional ascriptions and "for-ness" may well result in valuable research projects, even when it turns out that the developmental resource in question does not have the function supposed.

LOOKING FOR VERSUS FINDING FUNCTIONS: THE VALUE OF FAILURE

Trying to decipher the function of a particular putatively adaptive trait forces the researcher to investigate the history of the population in question, the historical relationship between the trait and the developmental systems that reliably (re)produce it, and the selective regimes that maintain the trait and the developmental systems against particular kinds of variation. In the end, for many traits, it may be impossible to identify *the* function they carry out; traits that do many different things may be maintained by a variety of different functional associations and mechanisms (some of which may not be "functional" in any traditional sense). But although the search for the function of a trait may often end in an impasse, it is not thereby rendered pointless.

Similarly, it may often be challenging to point toward the function of any of the developmental resources necessary for the reliable (re)production of a (putatively) functional trait; any of a number of resources may be involved in the development of many different adaptive traits, and a resource may or may not be maintained by selection because of its association with some particular trait (box 6.4). Again, the search for a func-

2. Given the short generation times of bacteria and the very long periods between magnetic pole reversals, this particular example strikes us as massively implausible. The individual fitness advantages of a more efficient inheritance system (if it ever arose developmentally) would, it seems clear, take the more efficient system to fixation well before the next polarity reversal; there would, therefore, be no chance for selection at the species (or a higher) level to act. It is far more likely that the inefficiencies of the system are a happy (for some bacteria) accident. To the best of our knowledge, this topic has not been studied in any depth in any of the varieties of magnetotactic bacteria; however, a similar issue—whether the evolution of high mutation rates can be adaptive—has received extensive attention (see Sniegowski et al. 2000 for a brief review).

Box 6.4 An Illuminating Example: What is *PhyB* For?

The *phyB* locus in the weedy cruciferous plant *Arabidopsis thaliana* is involved in the production of one of a class of molecules named phytochromes. Phytochromes are light-harvesting molecules that function as photoreceptors in *A. thaliana* and in many other organisms, including all flowering plants, nonflowering plants, green algae, and cyanobacteria (Wu and Lagarias 1997). The specific functions and effects of these photoreceptors differ to some extent, but the PHYB gene product is sensitive to the ratio of red to far-red (R:FR) wavelengths of light. The molecule can exist in the cell in two forms, one capable of absorbing red light and thereby switching conformation to the alternative form; the other sensitive to far-red light and capable of shifting to the red-sensitive form. PHYB is present in *A. thaliana* in both forms, whose chemical equilibrium depends on the R:FR ratio (Quail 1997a, 1997b).

Phytochrome B has been associated with a number of phenotypic effects, generally grouped under the label of shade avoidance responses. Under natural conditions, the R:FR ratio is an indicator of impending competition: the surrounding vegetation absorbs the red component of the spectrum, which is photosynthetically active, but reflects the far-red, too weak to be energetically useful. If a plant can sense the level of competition by perceiving the R:FR ratio, it may be able to adopt alternative strategies, such as suppression of branching to concentrate resources on vertical growth of the main stem, or accelerated flowering to complete the life cycle before the quality of the environment deteriorates further (Smith 1982; Schmitt and Wulff 1993).

Indeed, physiological studies have confirmed that phytochrome B affects a variety of plant phenotypes, including seed germination (Shinomura et al. 1994), cell elongation (Reed et al. 1993), cotyledon expansion (Neff and Volkenburgh 1994), seedling appearance (McCormac et al. 1993; Casal 1995), and flowering time (Halliday, Koornneef, and Whitelam 1994), as well as leaf production and branching pattern (Pigliucci and Schmitt 1999). Furthermore, evolutionary biologists have amassed convincing evidence that the shade avoidance response is adaptive, not only in *A. thaliana* but in other plants as well (Dudley and Schmitt 1995; Schmitt 1997; although by no means in all plants: species that live under constant shade in the understory are not expected to, and in fact do not, show any appreciable shade avoidance syndrome; Bradshaw and Hardwick 1989). Finally, molecular genetic investigations have demonstrated that the mechanics of action of the *phyB* locus are complex. Its gene product interacts with a light-labile phytochrome termed PHYA (Reed et al. 1994; Shinomura et al. 1994), with at least one other phytochrome (PHYC: Halliday, Koornneef, and Whitelam 1997), and with a completely different class of photoreceptors sensitive to blue light and known as cryptochromes (Casal and Boccalandro 1995). Phytochromes A and B interact with blue receptors in a complex manner, both synergistically and antagonistically, depending on the environmental conditions and the characters studied (Yanovsky, Casal, and Whitelam 1995; Callahan, Wells, and Pigliucci 1999).

Of course, photoreceptors by themselves do not perform any biological function; they have to act in concert with other developmental resources (including other gene products) to do so. Here the literature is much more vague and incomplete, although we know that there are several "transduction genes" whose products carry the "information" about the environment from the phytochromes to other molecules (Ang and Deng 1994; Quail et al.

(*continued*)

Box 6.4 (*continued*)

1995; Lasceve et al. 1999). Eventually, the bioeffectiveness of light perception and transduction is mediated by one or more plant hormones; gibberellin is known to be involved in this process (Lopez-Juez et al. 1995; Yang et al. 1995; Chory and Li 1997; Pigliucci and Schmitt 2004), although other hormones (e.g., cytokinin) affecting some of the same traits are altered by light signals independently of the phytochromes (Su and Howell 1995).

So, what is the *phyB* locus "for"? The obvious answer is that its function is to gauge the R:FR ratio. But light perception per se cannot be the target of natural selection, unless the information so acquired is actually used in some biological function related to fitness. Plants, after all, are not just curious about their environment. And here is where the trouble begins. If we refer to the physiological studies associating mutations at the phytochrome B locus with their phenotypic effects, we are led to conclude that the gene is, at the very least, for the control of germination, cell elongation, leaf production, flowering time, apical dominance, branching pattern, *and* reproductive output. One problem with this sweeping generalization is that a lot of other genes and other developmental resources also seem to be "for" the same traits, showing only partial (and sometimes contrasting) overlap with the effects of *phyB*. And of course, *phyB* has these effects only under particular environmental conditions; it is a "plasticity gene" that has certain effects only under specific environmental conditions (Pigliucci 1996a). Furthermore, some of these traits—for example, leaf production—are altered by mutations at the phytochrome B locus not because the phytochrome has much to do directly with leaf production, but more likely because the general growth rate of these mutants is slower than that of the wild-type plants (this may also account for part of the difference in flowering time among genotypes). That is, some of the alterations associated with changes at the *phyB* locus are bound to be accidental by-products of the mutation, and not indicators of the wild-type function of the gene.

The *phyB* locus belongs to an ancient family of genes, certainly predating the origin and evolution of flowering plants (see Mathews, Lavin, and Sharrock 1995). These genes have diversified in their biological effects, which are now only partially overlapping. The phytochromes can be thought of as very ancient photoreceptors whose DNA sequence has been highly conserved for the past billion years at least (Quail 1997a, 1997b). Such a long phylogenetic history also strongly suggests that they must have been maintained by natural selection as part of photoreception complexes. But the effects of these genes have varied considerably during the evolution of cyanobacteria, algae, and plants; their products have always been more than just photoreceptors, but the additional functional components have constantly shifted through evolutionary time.

In the "modern history" view of functions, what constitutes the function of a gene will often be rather fluid and fuzzy. In this case, the *phyB* gene is involved in a particular kind of photoreception and—indirectly—in a series of environment-dependent changes in the phenotype and phenology of the plant. This flexibility does increase the fitness of the plant, and the gene qualifies as a plasticity gene. However, which bits of the response are adaptive and which are allometric by-products of altered growth rates is a challenging empirical question that remains to be answered. The function of the *phyB* gene—what it is for—is still somewhat unclear.

tional association between a particular developmental resource and a given trait may not succeed for any number of reasons; for example, variation in the resource in question may be associated with strong negative selection because of its association with some other trait that arises developmentally before the trait of interest. But not being able to demonstrate the possible functional association that one had been looking for is not necessarily a failure; rather, it can lead to the exploration of the complex ways in which developmental resources are used and the variety of different mechanisms by which they are maintained and reliably (re)produced.

Sometimes, the search for the function of a particular trait or developmental resource will yield relatively clear examples of functional associations; not every case generates the kinds of complexities that make such ascriptions seem strained. In these cases, we can properly speak of what the trait is for, or of what the developmental resource in question is for. But sometimes, the search for functional associations will reveal various complexities that make functional ascriptions difficult to support. How often such complexities will be revealed, and what kinds of complexities they will turn out to be, are of course empirical questions. Attempts to reveal particular functional associations, then, are guaranteed neither to succeed nor to fail; whether they succeed or fail, however, they will often generate insight into both developmental and evolutionary biology.

7 Testing Adaptive Hypotheses

Historical Evidence and Human Adaptations

I MUST SAY THAT THE BEST LESSON OUR READERS CAN LEARN IS TO GIVE UP THE CHILDISH NOTION THAT EVERYTHING THAT IS INTERESTING ABOUT NATURE CAN BE UNDERSTOOD. . . . IT MIGHT BE INTERESTING TO KNOW HOW COGNITION (WHATEVER THAT IS) AROSE AND SPREAD AND CHANGED, BUT WE CANNOT KNOW. TOUGH LUCK.
—R. C. LEWONTIN, "THE EVOLUTION OF COGNITION: QUESTIONS WE WILL NEVER ANSWER"

Because evolutionary biology is concerned with history, phylogenetic information is often necessary to distinguish between evolutionary scenarios. Recently, some prominent proponents of evolutionary psychology have acknowledged this, and have claimed that such evidence has in fact been brought to bear on adaptive hypotheses involving complex human psychological traits. We argue, however, that in practice, evolutionary psychology has failed to use phylogenetic information in a meaningful way, due in large part to the structure of the family Hominidae (the group consisting of the great apes, including humans). For many traits of interest, the closest extant relatives to the human species are too phenotypically different from humans for such methods to provide meaningful data. While phylogenetic information can be useful for testing adaptive hypotheses in humans, those hypotheses generally involve traits that are either (1) not widely shared within the species or quite widely shared in the larger clade, or (2) of a lower order of complexity than the kinds of traits evolutionary psychologists have so far been interested in. Unfortunately, this suggests a much less sexy, but more realistic, research program than proponents of evolutionary psychology would like to carry out.

ADAPTATIONS, EVIDENCE, AND HUMANS

The attempt to understand the adaptive significance of various human traits (especially behaviors) by focusing on their proposed evolutionary trajectory has a long (and troubled) history (see, e.g., Kevles 1985; Kitcher 1985; Dupré 2001, and citations therein for review). Currently, "evolutionary psychology" is the most prominent of the fields making claims about human psychological traits being adaptations (see Jones 1999 for a review of the field; see Buller 2005 for an extensive critique). In this chapter, we use the interpretations of natural selection, adaptation, and functions developed in the previous chapters in an attempt to get clear about what kinds of evidence ought to be appealed to when making claims about supposed adaptations, and the degree to which these kinds of evidence can be marshaled in support of purported human adaptations. Our conclusion, perhaps not surprisingly, is that the kind of evidence necessary for making well-supported adaptive claims is too often unavailable in the case of purported human adaptations (especially in the case of alleged psychological ones). On the positive side, we suggest that, despite the limitations imposed by the human organism, a variety of approaches can yield interesting results. Insofar as these efforts are focused on human *adaptations,* the results are likely to be rather narrower and less bold than those claimed by evolutionary psychology; insofar as they are focused instead on the contemporary *adaptive* value of traits (rather than on the historical question of adaptation), the results may be broader, but of less evolutionary interest (though perhaps still of significant biological interest, understood more broadly).

One of the results of Gould and Lewontin's attacks on adaptationism (see chapter 5) has been an increased focus on questions about the kinds of evidence necessary for an adaptive hypothesis (or a nonadaptive hypothesis, for that matter) to be considered well supported (see Griffiths 1996; Brandon and Rausher 1996; Rose and Lauder 1996, and citations therein). Obviously, the kinds of evidence that are accepted as legitimate will have serious implications for the success of evolutionary psychology in meeting its stated goal of providing strong empirical support for adaptive hypotheses concerning human psychological traits (or, more precisely, the underlying mechanisms that generate particular psychological traits and abilities in humans; see Lloyd 1999; Lewontin 1998). In general, it is widely agreed that there are a number of different kinds of evidence that can be brought to bear on hypotheses surrounding how and why a certain phenotypic trait arose, spread within a population, and has been maintained within that population. This evidence can come from laboratory and ecological field studies of the consequences of phenotypic manipulations of the trait, laboratory evolution experiments designed to test adaptive and nonadaptive hypotheses (such as those involving the robustness of proposed developmental pathways), optimization analyses, regression analyses of the trait against measures of fitness, and analyses taking into account the likely phylogeny of the populations or species involved. In

Box 7.1 Evolutionary Psychology Or . . . ?

"Evolutionary psychology" can name any one of a number of positions. At the most trivial, "evolutionary psychology" simply acknowledges that human psychological traits (including behaviors) are made possibly by the human brain, which itself is the product of evolution, and is almost certainly an adaptation. But in its most common current use, "evolutionary psychology" refers to a field with a particular set of assumptions and methodological practices (see Buller 2000 for an insightful review that attends carefully to the different senses in which the term is used). Evolutionary psychologists, in the latter sense, tend to affirm most or all of the following positions (see Cosmides, Tooby, and Barkow 1992):

1. Human psychological or behavioral tendencies are generated by special-purpose mental "modules."
2. These modules are "activated" in response to particular environmental "cues."
3. These modules evolved in response to particular "problems" the ancestors of humans faced during their (relatively recent) evolution.
4. Therefore, each mental "module" is an *adaptation* to a specific problem ancestral humans faced.

Evolutionary psychologists further argue that the best way to discover these "modules" is to theorize about what those specific problems were and how they might have been solved; one then looks for evidence that human minds actually implemented that solution, and if such evidence is found, interprets this as prima facie evidence that the adaptive hypothesis in question is correct (see Cosmides, Tooby, and Barkow 1992).

Other researchers accept that human psychological traits or behavioral tendencies are the result of natural selection on those traits or behaviors, but reject the particular mechanisms (such as the "massive modularity" of the mind) and methods proposed by evolutionary psychology's proponents. Some of these researchers—for example, those in "evolutionary developmental psychology"—suggest that attention to the way that human learning ability actually changes throughout development might better reveal the ways in which particular kinds of adaptive psychological traits or behaviors are reliably generated in the human case (see, e.g., Bjorklund and Pellegrini 2002).

While some of our criticisms are specific to the particular goals and methods of evolutionary psychology, for the most part our concern is with the kinds, and the quality, of evidence necessary in *any* field that makes claims about biological adaptations, including especially those that make claims about human adaptations.

general, more than one type of evidence will need to be gathered for an adaptive (or again, a nonadaptive) hypothesis to be reasonably tested against its competitors (see chapters 5 and 6).

In this chapter, we explore the extent to which the various techniques used to test adaptive hypotheses in nonhuman populations can be and have been successfully applied in the context of humans. While our focus is on those adaptive stories put forward by evolutionary psychology (box 7.1), much of what we say will apply equally well to other disciplines that have generated adaptive scenarios in the human case (e.g., medical genetics and physical anthropology). Comparisons of these domains will reveal that the focus of evolutionary psychology's main proponents on complex psychological traits and supposedly universal human adaptations (box 7.2) is problematic from the standpoint of putting forward testable hypotheses—as much as it is sound as a general idea. Studies focused on different—perhaps less "sexy"—traits, however, might prove to be more promising.

In the end, we conclude that while current techniques have been very successful at disentangling competing hypotheses in some categories of organisms and with respect to some kinds of traits, in other organisms (chiefly humans, unfortunately), we should not expect such successes. The historical nature of evolutionary biology not only makes certain categories of facts about populations easier to determine than others, but makes facts about some populations easier to determine than similar facts about other, otherwise comparable, populations. Humans, it turns out, despite being of deep interest to us for obvious reasons, are an especially difficult species to work with. This means that, regardless of our curiosity about ourselves, we may not be able to find the answers to all the questions that interest us. Indeed, even in some cases in which we have excellent grounds for believing there to be an answer, we may be unable to unequivocally determine it. Again, the reason for this is the nature of evolutionary biology as a partially historical science—when one is trying to uncover the historically relevant casual pathways that led to a particular set of results, it will sometimes happen that all the information currently available cannot distinguish between different plausible causal pathways (Cleland 2002). In these cases, we must resign ourselves to admitting that while it is reasonable to think that the pattern of interest was caused by some particular underlying causal structure or other, we cannot pinpoint one with sufficient confidence.

TESTING ADAPTIVE AND NONADAPTIVE HYPOTHESES

What techniques are available in the case of nonhuman organisms for testing adaptive and nonadaptive hypotheses about the origin, spread, and maintenance of traits in natural populations? In this section, a few such contemporary techniques (and their attendant conceptual machinery) will be briefly considered (see table 7.1 for a summary). A comprehensive review is, of course, well beyond the scope of this chapter, but we hope to give the reader a sense of what techniques have been and can be brought to bear on testing adaptive hypotheses (see Rose and Lauder 1996 and

Box 7.2 What Kinds of Claims Do Evolutionary Psychologists Make?

Evolutionary psychologists take their goal to be discovering the "evolved psychological mechanisms" that are adaptations to "the way of life of Pleistocene hunter-gatherers" (Cosmides, Tooby, and Barkow 1992, 5). Success would mean, they claim, uncovering a "universal human nature" that is biologically based and broadly independent of culture, although the expression of particular traits might well be influenced by cultural factors (Cosmides, Tooby, and Barkow 1992, 5). As noted above (see box 7.1), evolutionary psychologists are in broad agreement with the claim that the human mind is made up of many quasi-independent modules, each of which is an "evolved psychological mechanism" that is an "adaptation" to a problem faced by ancestral humans.

What kinds of "evolved psychological mechanisms" do evolutionary psychologists propose? Some examples include the following:

Potential mate recognition and evaluation modules. Evolutionary psychologists have proposed that human males and females each evolved several different kinds of mental modules used to recognize and evaluate potential mates. These include modules (in males) evolved to detect health and fertility (in females) and modules (in females) evolved to detect status and fitness (in males). It is proposed that females have evolved modules that estimate the likely costs and benefits of "extra-pair" copulations, and that males have evolved modules that estimate the likely costs and benefits of various strategies to discourage females they are in "in-pair" relationships with from "cheating" (see, e.g., Buss 1994, 2002 for review).

Social exchange modules. Evolutionary psychologists propose that the "problems" of making fair but profitable social exchanges and maintaining social hierarchies are solved by a set of modules that evolved to recognize social exchange situations and properly compute the required values of the exchanges. Similarly, modules are proposed for recognizing "in" versus "out" group relationships, detecting opportunities for, creating, and maintaining "coalitions," and detecting violators of agreements or social norms ("cheater detection modules") (see Cosmides and Tooby 1992 and Sell et al. 2003 for review).

Various hunting/foraging-specific modules. Evolutionary psychologists have proposed that human males and females have developed various modules that are keyed to solving the kinds of problems they faced in hunting or gathering. For example, it is argued that females were the primary "gatherers" and hence have evolved modules particular to spatial memory and reasoning, whereas males were the primary "hunters" and hence evolved modules particular to skills such as throwing accuracy (see, e.g., Kimura 1999; McBurney 1997, and citations therein).

Evidence for the existence of these "modules" and the psychological adaptations associated with them has been reviewed, and the claims made by evolutionary psychologists critiqued, in other places (again, see Buller 2005 for review). Here, we simply wish to remind readers that the claim that a particular trait is an adaptation implies a specific set of both biological and historical circumstances, and that defending that claim demands that these biological and historical circumstances be rigorously tested. Alas, as we shall show, sufficiently rigorous testing is sometimes not feasible.

Table 7.1. Some techniques for testing adaptive hypotheses

Technique	Evidence generated/hypotheses tested
Phenotypic manipulation, laboratory or field	Fitness consequences of the traits in question; causal mechanisms associated with traits and their fitness consequences
Transplant studies	Hypotheses regarding selective pressures; hypotheses regarding local adaptations
Laboratory evolution	Robustness of developmental pathways; strength of "constraints"
Optimization analyses	Qualitative assessments (for quantitative plausibility); sensitivity analyses; degree of path dependence
Phylogenetic analyses	History of trait, homology vs. homoplasy; historical relationship between trait and particular selective pressures
Regression analyses	Relationship between trait and environmental variables; relationship between trait and fitness measures

citations therein for a more extensive review of some of the more relevant literature).

Experimental Manipulations, Transplant Studies, and Fitness Consequences

If one has tentatively identified a trait one thinks is an adaptation for a particular function, a reasonable next step is to test whether or not modifications to the trait (including its absence) have the effects on fitness that the adaptive hypothesis being entertained would suggest. Techniques for carrying out such tests range from the straightforwardly mechanical (such as artificially shortening and lengthening the tails of birds) to sophisticated attempts to knock out key genes involved in the formation of the trait in question to subtle manipulations of embryonic forms to influence final phenotypic features. Field studies of fitness (such as transplant experiments), laboratory tests of fitness, and other methods can be used to check the consequences of these manipulations. While proper precautions must be observed, if the studies reveal a change in fitness in the direction suggested by the adaptive hypothesis, they will have provided at least some evidence that the adaptive hypothesis is on the right track.

Sinervo and Basolo (1996) discussed some of the many techniques for experimental manipulation of phenotypes and selective environments that have been used to test adaptive hypotheses. They laid out three general types of experimental techniques that can be used to study adaptation. First, the environment can be varied in a systematic way; this can be done either by raising the organisms in controlled laboratory settings or by moving them from one natural environment to another. The latter techniques are traditionally thought of as transplant studies, whereas the former generally are considered a form of research into phenotypic plasticity, but either can be used to obtain several kinds of information. The most obvious such information is relative fitness in the new environments. More importantly, one can check to see whether an environmentally induced change in one trait results in correlated changes in other traits, which is basically a way of studying the integration of plasticity responses (see Pigliucci 2001).

When environmental changes are maintained for a number of generations, these kinds of environmental manipulations segue into studies of laboratory evolution. Obviously, the extent to which laboratory evolution is useful for testing various kinds of hypotheses depends heavily on the organisms being studied—bacteria can be exposed to novel environments for tens of thousands of generations in the course of a relatively short experiment, whereas testing the evolution of even a rapidly cycling plant such as *Arabidopsis* for a few tens of generations would push the patience (and funding resources!) of most evolutionary ecologists. But depending on the organism in question and the perseverance of the researchers (or the length of their supporting grants), laboratory evolution experiments can be used to test the robustness of various developmental pathways, the likelihood of certain traits arising in response to particular environmental variables, and the most likely ways in which transitions between particular traits could take place. As we noted in chapter 5, studying the ways in which traits reliably coevolve and the different frequencies with which particular relationships between traits change can provide a useful entry into the study of one of the many meanings of "constraints." And it seems clear that if we wish to understand why traits have the functional associations they do, and how these were acquired in evolutionary time, we must recognize this kind of fluidity in the relationship between changes at the genetic and phenotypic levels (see chapter 6). Natural selection can obviously be involved in important ways in such changes, but so too can historical contingencies and evolutionary "accidents" involving chance mutations. Studying laboratory evolution, with attention to the ways in which the uses of particular developmental resources change with changes in the population, is one way of gaining insight into some of these issues (see chapter 6, especially the discussion of Roth 1988 and the idea of "genetic piracy").

The second type of experimental manipulation involves direct manipulation of particular phenotypes, which can take several forms. Adult phenotypes can be adjusted by the physical addition or removal of physical matter related to the trait (e.g., filing off birds' beaks, adding longer tail feathers, clipping leaves into different sizes or shapes) (see Sinervo and Basolo 1996 and citations therein). Organisms can also be manipulated phenotypically at the embryonic stage; such ontogenetic manipulations can sometimes result in significant and well-integrated phenotypic variations, the fitness consequences of which can then be studied more or less directly (see Sinervo and Basolo 1996, 161–62 and citations therein). In all these cases, care must be taken that the phenotype under consideration is being modified in a biologically meaningful way, and that other variables (e.g., the fitness consequences of being manipulated) are accounted for. However, well-designed studies can go a long way toward solving, or at least greatly ameliorating, these difficulties (e.g., proper systematic manipulation of controls; see Sinervo and Basolo 1996, 173–75 and citations therein).

Finally, Sinervo and Basolo (1996) suggested that one can, by manipulating a single physiological mechanism known to be or suspected of being associated with several traits, study the way those traits are correlated and maintained to get a feel for some of the trade-offs and genetic architecture (i.e., pleiotropic and epistatic effects) involved. Again, these authors stressed that such studies must be very carefully designed if they are to avoid spurious associations and not miss critical trade-offs (153–54 and citations therein).

These three types of techniques—experimentally manipulating the trait in question, the environment that the trait is thought to have evolved in response to, or the physiological mechanisms that a number of related traits are thought to be involved with—can provide evidence that the traits of interest are (or are not) adaptive in the way hypothesized.

Optimization Analyses

Arguments that a particular trait is "optimal" for the task it performs came under heavy attack in Gould and Lewontin's "Spandrels" paper (1979; see chapter 5). But Gould and Lewontin were not objecting to optimization analyses per se. Rather, they were objecting to the following methodological proposal: When faced with a trait that interests you, *assume* that the trait was shaped by natural selection to do some one thing or some set of things, and that it is now optimal for carrying out that task (in the case of multiple functions, the trait may not be optimal for each, but represents an optimal balance between the tasks; 1979, 585). Lewontin referred to this as the "backwards" mode of reasoning in optimality studies and rejected its legitimacy almost out of hand (Lewontin 1979, 6).

In the "forward" mode of reasoning, on the other hand, given some knowledge about the organism's general way of living and reproducing and its phylogenetic history, one considers potential "problems" the organism faces and attempts to calculate locally optimal solutions to those problems without knowing ahead of time how the organism in question actually solves them. Lewontin seemed less certain that such arguments are useless (1979, 6). In this mode, one starts not by considering what the trait happens to do now that may have been evolutionarily useful, but rather by asking what the trait should be like given the phylogenetic history of the organism in question, the evolutionarily significant (discriminate) processes organisms of that type probably faced in the relevant parts of their history, and the general developmental features of organisms of that type.[1] Once such a question is phrased in a specific enough way, optimization models can be usefully employed to test whether natural selection did in fact work on the trait being considered to shape it for the functions one is hypothesizing.

Of course, one must be careful to phrase the question properly; as Jacob (1977) pointed out before the publication of the Spandrels piece, evolution works only with what is available, "tinkering" with existing structures rather than starting anew. So it is senseless to ask what a trait should be like *simpliciter;* only given the actual historical circumstances thought to be associated with its having arisen and been maintained in a given population can such a question make sense.

However, even forward arguments become questionable if deviations from the result calculated as optimal are accounted for not by considering, for example, whether the problem has been misstated, but rather by assuming (as opposed to testing) other constraints on the optimality of the trait involving other necessary functions (Lewontin 1979, 11–12). The reason is that doing so just reduces the technique to a de facto version of the backward method, in which one assumes that the trait is optimal for some function, or an optimal balance between functions, before one starts. However, a forward optimality analysis that refuses to succumb to such ad hoc fixes may yet have some persuasive force.

Nor is such a method alien to Gould's thinking. The engineering analysis suggested by Gould himself in "Darwin's Untimely Burial" is just such a forward optimization argument, although he does not call it that (1976; see esp. 33). It is by such engineering or optimization analyses, performed ahead of time (sometimes literally so, as in laboratory evolution—see Rose, Nusbaum, and Chippindale 1996), that we can test adaptive hypotheses. But even when such analyses are performed after the fact, they can, as Gould recognized, help us to see the ways in which the results of natural selection can be creative and can produce organisms well designed

1. This is one way of articulating what is usually meant by evolutionary pressure—it is simply the physical processes that interact in discriminate ways with traits (at the informal level) such that the formal level reveals predictive fitness differences associated with those traits. As always, moving between the levels is difficult, but with careful ecological detective work, sometimes possible.

for their environment. Gould noted, for example, that "it got colder before the woolly mammoth evolved its shaggy coat" and that the coat represents "good design"—an appropriate response, in an engineering sense, to the increasing cold (1976, 33).

Even forward-looking optimization studies, however, are not a panacea. Part of the reason for this is that where a trait is associated with more than one function of evolutionary significance (as many are!), multiple solutions to the optimality function, even taking local features into account, become equally viable (Niklas 1994, 1997). When attempting an optimization study of such complex traits, then, determining ahead of time which solution will have been struck upon may be difficult. This problem echoes that of multiple adaptive peaks mentioned in the Spandrels paper—it may be possible to say that the trait has adaptive significance, but there may be no good reason, in terms of adaptive excellence, why that particular form of the trait resulted, rather than some other roughly equally fit form (only historical contingencies will separate these cases). Indeed, some researchers have argued that one of the most powerful uses of optimization analyses is precisely to reveal that a particular adaptive hypothesis is insufficient to account for the features of the trait in question (see, e.g., Seger and Stubblefield 1996, 117). This result is often taken to mean not that the adaptive claims in question are wholly false or uninteresting, but only that they are incomplete and in need of more research to determine with greater precision what, for example, the various constraints are, how the trait actually arose in the evolutionary history of the organism, and the like.

In any event, Seger and Stubblefield noted that formal optimization analyses are easier to perform on traits the closer those traits are to reasonable measures of fitness (1996, 115). For this reason, some of the best optimization analyses have been performed on traits such as egg laying in birds (Seger and Stubblefield 1996, 98–102). The more remote the trait in question is from reasonable fitness measures, the harder it is to calculate the effects of variations in the trait on fitness measures. In cases in which the trait is not reasonably close to fitness, Seger and Stubblefield recommended that formal optimization analyses be rejected in favor of comparative analyses that attempt to match particular (informal) selective regimes (i.e., particular discriminate processes, in our interpretation; see chapter 1) with specific kinds of variation across populations (1996, 115–17).

Comparative Phylogenetics and Changing Traits

One way to begin to answer questions about whether a particular trait is in fact an adaptation is to think about the history of the trait and of the populations involved. By finding out when a trait arose in evolutionary time and which related populations share or fail to share that trait, one can begin to find ways of formulating and testing increasingly precise adaptive hypotheses. If, for example, it can be determined that a trait arose during (or immediately after) the particular stressor (discriminate physical process) that it is hypothesized to be an adaptation for, that would seem to provide at least some evidence that it is an adaptation. If, on the other hand, the trait can be shown to have arisen at a time when the hypothesized stressor was not significant, this would seem to count against the adaptive hypothesis being considered. Such a test will usually involve comparing more or less closely related populations in which the stressors are known to have been present or absent and figuring out the evolutionary history of the trait compared with that of the populations and the environmental stressors. Further, where a particular discriminate process can be shown to impinge on certain populations more or less often, one can compare the extent to which the hypothesized adaptation to that stressor is present in the different populations. Work done comparing the risk of death from malaria with the frequency of the *HbS* allele in human populations is a simple example of this kind of approach (see Futuyma 1998, 384–85 and citations therein for review).

A related historical technique is the use of Lande and Arnold's multiple regression analyses for studying multivariate evolution; recall that Lande and Arnold (1983) claimed to have found a way in which retrospective analyses of observed phenotypic changes could be used to disentangle the effects of direct and indirect selection (see chapter 2). Lande and Arnold proposed that these methods could provide quantitative measurements of selection and "replace rhetorical claims for adaptation and selection" (1983, 1210). As we noted in chapter 2, using Lande and Arnold's methods to uncover evidence of past selection demands that one accept a number of doubtful assumptions. Even if one did so, it is worth noting that these analyses do not serve to show *why* particular traits were selected (in the formal sense), but merely *that* natural selection was probably involved in their spread and maintenance in some population (see Leroi, Rose, and Lauder 1994).

Putting It All Together: Testing Adaptive Hypotheses

As we noted above, it is usually the case that no one technique is sufficient to test a particular adaptive (or nonadaptive) hypothesis against reasonable competitors. Rather, a number of the techniques discussed above generally need to be employed before a particular hypothesis can be considered well supported. In this section, we use the extensive work done to test a particular example of an adaptive hypothesis to explore the ways in which a number of different lines of evidence can converge.

Males of the sword-tailed fish of the genus *Xiphophorus* have extensions of their tail fins called "swords." These swords have long been thought to be the result of sexual selection (Darwin 1871, chapter 12); that is, it has long been believed that the swords of the sword-tailed fish were adaptations to the female fish's preference for male fish with swords. Testing this hypothesis against reasonable competitors, however, has proved to be a long and difficult task. Sinervo and Basolo (1996) have provided a brief review of the more than twenty-five years of work done in this field.

The first question in this case involved the fitness consequences of the sword vis-à-vis female mate choice. If females do not preferentially mate with males with swords, the hypothesis that the sword is an adaptation to the females' preference is problematic; while it is still possible that the sword is an adaptation to a past female preference that has recently ceased, it is prima facie unlikely. In 1979, phenotypic manipulations revealed that females did in fact have a preference for males with swords, rather than for some other phenotypic effect merely correlated with swords. Specifically, researchers cut the swords off some male fish, then attached plastic swords of various lengths to their tails. The plastic swords were made from either colored (to resemble real swords) or clear plastic. The response of the female fish to the colored, but not to the clear, swords revealed that it was the presence of the sword, and not the effects of the manipulations or another covarying trait, that was causing the response (Sinervo and Basolo 1996, 174).

Note, however, that this test does not yet establish that the sword is an adaptation to females' preference for swords. If the sword preexisted the females' preference for swords, then, while sexual selection might have resulted in stabilizing selection for the sword's existence, the sword would not in fact be an adaptation to the preference. Initial approaches to answering this question focused on those species of *Xiphophorus* that make up the northern platyfish. Most of these species lack swords, but the addition of swords to males' tails revealed that the females, even of these species whose males normally lack swords, prefer males with swords (Sinervo and Basolo 1996, 174). However, rather than settling the question, this research revealed an additional complication, in that some attempts to resolve the phylogenetic history of the sword-tailed fish within the *Xiphophorus* genus implied that the sword might be primitive to the platyfish/swordtail clade—that is, that the sword-free platyfish may have lost their swords at some point after female preference for swords had been established (Sinervo and Basolo 1996, 174–75). Finally, the same kind of phenotypic manipulation was done on the closely related fish genus *Priapella* and some other, more distantly related genera, none of the species of which have swords. This work revealed that the *Priapella* females share *Xiphophorus*'s preference for males with swords, whereas the females of other, more distantly related genera do not (Sinervo and Basolo 1996, 175). The female preference for swords, then, is very likely a shared primitive trait of the *Priapella-Xiphophorus* clade, and the existence of swords in (some) species of *Xiphophorus* is, in fact, an adaptation to the females' (preexisting) preference for males with swords.

There are still a number of outstanding questions, such as why females of the *Priapella-Xiphophorus* clade share a preference for males with swords, whether the preference is for swords or for some more general construct (such as relative flashiness, however it is expressed), and why none of the males of the *Priapella* clade have developed tails, whereas so many males of the *Xiphophorus* clade have (are all the *Xiphophorus* swords developmentally or genetically homologous?). These are the kinds of questions that could be addressed via laboratory evolution and artificial selection experiments in model organisms that were easier to deal with. But in any event, the hypothesis that the sword of the male sword-tailed fish is an adaptation to the female's preference for males with swords is clearly well supported, and the supporting evidence certainly goes well beyond the kinds of "just so" stories Gould and Lewontin (1979) criticized in the Spandrels paper (see chapter 5). Testing adaptive hypotheses in a satisfactory fashion is indeed feasible, albeit somewhat more complicated than some would like.

THE TROUBLE WITH HUMANS: THE IMPOSSIBLE AND THE MERELY DIFFICULT

It is an unfortunate thing for programs focused on the evolutionary significance of particular human traits (especially psychological traits), but many of the techniques commonly used to test adaptive hypotheses in other organisms are of little use in testing purported human adaptations. Obviously, experimental manipulations designed to reveal the fitness consequences of modifications of putatively adaptive traits, in both laboratory and natural populations, fall into this category, as using these techniques on humans, except in the most limited of ways, would be ethically impossible. Partial exceptions are provided by "natural" experiments, in which phenotypic traits are "manipulated" accidentally (e.g., because of traumatic injury or genetic mutations), although in these cases it is hard to control for potentially causally relevant but confounding variables (e.g., variations in related traits). In the case of hypothesized human psychological adaptations, the phenotypic manipulations necessary to properly test adaptive hypotheses might well be beyond the current state of the art in neuropsychology, even if they were not ethically inconceivable. Similarly, laboratory evolution experiments designed to falsify alternative hypotheses (such as those involving so-called genetic constraints and allometry; see Schlichting and Pigliucci 1998, 178–88) or to test the repeatability of adaptive pathways (see Rose, Nusbaum, and Chippindale 1996 and citations therein) can be rejected as all but impossible in the human case.

Transplant experiments, wherein supposed adaptations to local conditions are tested by physically moving the organisms in question to other locales and observing the fitness consequences, can be done with humans in only a very limited way (e.g., following people who move on their own) and generally result in a self-selecting and probably atypical sample. Again, properly controlling for the effects of being moved is obviously difficult in the human case. While it is possible in principle, little research on human adaptations has been done using these techniques (a partial exception is provided by work on the *HbS* allele, malaria resistance, and migration; see below and, e.g., Das 1995).

Optimization analyses that take historical contingencies into account (see, e.g., Sober 1996; Orzack and Sober 1994; Lauder 1996; Seger and Stubblefield 1996, and citations therein) seem a more promising avenue for research into human adaptations because they primarily involve observation rather than manipulation. Similarly, techniques making use of broadly historical evidence about the genesis and spread of a trait should provide plausible approaches to testing adaptive hypotheses in humans. Indeed, this latter kind of evidence has been explicitly mentioned as relevant by prominent evolutionary psychologists: Miller (1998, 117), for example, has noted that "examining the distribution of traits across related species with known phylogenies" can be used to "discern when and where evolutionary innovations occurred" (but see box 7.3 on the topic of phylogenetic analyses and evolutionary psychology; see also Buss 1999, 54–64).

While these techniques are not quite as problematic as phenotypic manipulations of humans would be, there are still important questions about how successful one should expect such approaches to be at developing evidence that can be used to support or reject contemporary adaptive hypotheses in humans. In the case of some possible adaptations, these techniques have been applied with at least moderate success, and adaptive hypotheses have been well supported (see below for the example of the *HbS*

Box 7.3 Phylogenetic Evidence and Human Psychological Adaptations

Despite Geoffrey Miller's avowed desire to use comparative phylogenetic data to test particular adaptive hypotheses (1998, 117), Miller's (and other evolutionary psychologists') actual use of such data tends to be limited. Consider, for example, Miller's contention that artistic ability in human males is an adaptation shaped by sexual selection. Artistic ability, Miller suggests, displays traits (intelligence, creativity, physical coordination) that are key indicators of fitness, and thus artistic ability could be selected directly as a cue to fitness (see Miller 1998, 2000; Miller and Todd 1998). Miller claims that females came to desire artistic males because artistic ability was a cue to the fitness of those males, and over time female mate choice selected for artistic ability in males. Indeed, Miller claims that only the sexual selection hypothesis can explain why "males produce about an order of magnitude more art" than females (Miller 1998, 119), and that the sexual selection hypothesis is the only explanation of art that can avoid falling into "fallacious" theories about the "'self-expressive' functions of human art" being a side effect of "surplus" abilities (1998, 89). This hypothesis is of some interest, in part because it generated some significant coverage in the popular press (see *The Economist* 1999, 71).

One might quibble with Miller's evidence for artistic ability and interest in art being sexually dimorphic in humans, but in this case it seems rather besides the point. Consider what the sexual selection hypothesis demands. First, at least some females had to have a preexisting heritable preference for artistic males over less artistic males. In addition, there had to be heritable variation both in artistic ability and, perhaps more important, in the desire to produce art in the males of the ancestral population. Artistic males had to have, on average, higher fitness than nonartistic males. Finally, if there was variation among females in the preference for artistic males, then females with a preference for artistic males had to have, on average, higher fitness than females without such a preference. If these assumptions hold, then the sexual selection hypothesis could at least get off the ground. But the difficulty, put simply, is that there exists research that can be reasonably interpreted as suggesting strongly that *interest* in artistic achievement in primates clearly predates artistic *ability* (*pace* Miller's dismissal of previous research in this area; see Miller 2000). While this does not rule out the possibility that sexual selection played a role in shaping artistic ability in humans, it does rule out sexual selection on artistic ability as being an important part of the *explanation* for artistic expression arising in the hominin line.

While the classic text on the subject of artistic expression in nonhuman primates was written by Morris (1962), no one has summarized the conclusions of that early research more succinctly than Hutchinson (1965). Hutchinson notes that when certain great apes are given tools to paint or draw with, they will often engage in that activity with no "reward"—indeed, Hutchinson notes that for many of the animals tested, "any intrusion is resented more than if the animal had for instance been disturbed while eating" (1965, 103). He further notes the appearance that "providing the young ape with paints, brushes, and canvas gives it, for the first time in its life, something very important to do" (1965, 103). He claims that these results show "that the desire and capacity to engage in some sort of autotelic activity exists in animals that have diverged from the human line many millions of years ago and do not have the intellectual capacity to invent the mechanisms to provide the sort of satisfaction that is within their intellectual range" (1965, 103–4).

(*continued*)

Box 7.3 (*continued*)

Again, these findings do not rule out the possibility of sexual selection playing some role in the formation of artistic ability in humans. But the explanation for human males' (and for that matter, females') interest in art cannot be a preexisting female preference for artistic ability in males that caused males to develop artistic interests and skills. The interest in artistic expression predated the development of the skills necessary for that artistic expression. If this is correct, then the development of artistic expression would follow the development of the necessary skills, whether or not there was a particular female interest in artistic expression in males as a fitness cue.

Why human and some nonhuman primates have an interest in artistic expression is left unexplained, as is when and under what conditions such an interest may have arisen. But we think it is clear that we cannot appeal to sexual selection acting on human artistic ability during a time when "we had to amuse each other on the African savanna" (Miller's phrase, 1998, 116) to answer these questions, either.

allele and malaria resistance). However, despite the claims of some of evolutionary psychology's proponents, we argue here that such evidence is rarely available in the case of purported "universal" human psychological adaptations. The very limited information we can gather on the environments in which key aspects of human evolution took place make optimization techniques difficult to apply to the human case in practice. Furthermore, while in some cases phylogenetic information about the family Hominidae (the great apes, including humans) may provide evidence relevant to adaptive hypotheses in humans, nature and history have conspired to make the task of obtaining such information about our species (and, for that matter, our close relatives) much more difficult than it is in many other taxa. These difficulties, combined with those discussed above and those due to logistical and ethical limitations (see table 7.2 for a summary), continue to frustrate attempts at seriously testing many of the hypotheses of interest to evolutionary psychologists.

Human Adaptations: Discovery and Evidence

In the case of some physical adaptations in humans, there is general agreement regarding what needs to be explained and how well supported (or not) current explanations are (table 7.3). For example, while human bipedalism seems to most researchers to be a fact in need of explanation, there is a consensus that no current hypothesis (adaptive or otherwise) is so well supported by the evidence as to exclude all reasonable competitors (see Tattersall 1995 for discussion). The massive increase in brain size during human evolution is in a similar position—there is agreement that it stands in need of (probably adaptive) explanation, but no consensus

Table 7.2. The trouble with humans: Available techniques for testing adaptive hypotheses and their difficulties

Technique	Difficulties in the human case
Phenotypic manipulation, laboratory or field	Ethical constraints forbid most deliberate phenotypic manipulations; many confounding factors in accidental phenotypic manipulations (trauma, genetic disease, etc.)
Transplant studies	Ethical constraints forbid deliberate, controlled transplant studies; it is difficult to control for confounding factors in natural transplant experiments
Laboratory evolution	Ethical constraints make this impossible in the human case; even if it were possible, primates in general are poor model organisms for this sort of research
Optimization analyses	Possible in principle, but little is known about the relevant selective pressures in the case of putative human adaptations; little is known about the environments in which much of human evolution took place
Phylogenetic analyses	Possible in principle, but very sparsely populated clade; little is known about the environments in which key speciation events took place
Regression analyses	Possible, but there is little meaningful systematic variation at the genotypic or phenotypic level; little is known about relevant selective pressures

Table 7.3. A few widely shared human traits without well-supported adaptive (or other) explanations

Trait	Hypotheses considered
Bipedalism	Freeing the hands, thermoregulation, predator avoidance, increased visual range in tall grasses, some combination
Big brains	Machiavellian intelligence, sexual selection, social coordination, co-selection with tool use, general problem solving, improved ability to modify environment, some combination
Relatively sparse body hair/ relatively fine body hair	Thermoregulation, sexual selection, nonadaptive by-product (but of what?), hydrodynamics (!), some combination

on what the best explanation is. While many possible scenarios have been proposed, none are generally considered well supported, nor is there agreement on a plausible research program for solving this problem.

Famously, the high prevalence of alleles associated with sickle-cell anemia (the *HbS* alleles) in certain human populations is also a fact in need of explanation; however, in contrast to the cases mentioned above, there is agreement on what the correct explanation is (at least in outline; see also our discussion in chapter 6). It is beyond reasonable doubt that the relative frequency of the sickle-cell allele in some populations is an adaptation for partial malaria resistance; the alleles that cause sickle-cell anemia in homozygotic individuals are associated with resistance to malaria in heterozygotic ones. Evidence for this adaptive hypothesis comes from many sources. First, the alleles associated with sickle-cell anemia are found at high frequencies only in populations that have lived in areas with serious malaria problems (mosquito-infested areas). Second, work done on the history of populations vis-à-vis migrations indicates that the prevalence of *HbS* alleles and the prevalence of malaria show a clear relationship. Finally, reasonable estimates of the fitness of the *HbS* versus "normal" alleles in various environments correlate fairly well with their historical distributions (see Griffiths et al. 1996; Das 1995, and citations therein). There are still a number of questions to be answered in the sickle-cell case, and some fascinating work is being done on the extreme heterogeneity of the clinical manifestation of sickle-cell disease,[2] but the basic adaptive claim seems well supported. While there has been some research suggesting that other diseases caused by single recessive genes (such as the genes associated with cystic fibrosis, CF) may be the result of alleles with similar adaptive roles (in the case of CF, possibly resistance to typhoid; see Pier et al. 1998), none of these efforts have as yet gathered evidence strong enough to have been generally accepted. Similarly, a promising research avenue involves attempts to explain particular patterns of genetic variation through the testing of novel adaptive hypotheses, such as the work being done on the potential link between the apparently recent spread of the *CCR5-Δ32* "HIV resistance allele" in populations of European descent and outbreaks of bubonic plague in Europe (see Stephens et al. 1998).

Other phenotypic traits in humans for which adaptive explanations have been offered and seem reasonably well supported include skin color (fair skin for populations with diets poor in vitamin D in locales with low levels of sunlight) and some variations in body size and shape among populations in radically different climates (see, e.g., Cavalli-Sforza and Cavalli-Sforza 1995; Lewontin 1995; table 7.4). What is startling, though, is that so many human phenotypic features, even basic, universally shared ones (such as our large brains, bipedalism, relative hairlessness, etc.) that are generally agreed to have arisen after the lineage that gave rise to humans diverged from the lineage that gave rise to modern chimpanzees, have by and large not been explained by hypotheses (adaptive or otherwise) that have gained wide acceptance (see Tattersall 1995 and 2000 for reviews). This is certainly not the situation for many other species, for

2. For example, some published work points toward the existence of "co-mutations" inherited with the *HbS* allele that provide partial protection against some of the medical problems associated with sickle-cell disease, even in the homozygotic case (see Guasch et al. 1999 and citations therein). More work obviously needs to be done on this subject; it would be helpful to know, for example, when these (putatively) protective mutations arose compared with the *HbS* mutations, and what the relationship is between their predicted and observed pattern of spread in human populations.

Table 7.4. Less widely shared traits in humans, adaptive explanations currently in vogue, and the degree of support those adaptive explanations are currently said to have

Trait	Adaptive explanation	Degree/type of support
HbS allele	(Partial) malaria resistance in the heterozygotic case	Very well supported: quantitative fitness analyses, expected fitness patterns versus prevalence, analyses of migrations, analyses of phylogenetic histories
Light skin	Vitamin D production (for populations with diets deficient in vitamin D in areas with weak sunlight)	Fairly well supported: expected fitness patterns versus prevalence, analysis of fitness, analysis of phylogenetic histories
CCR5-Δ32 allele	Resistance to bubonic plague	Speculative: analysis of phylogenetic history, some quantitative work on expected fitness and spread
Alleles associated with CF in the homozygotic case	Resistance to typhoid in the heterozygotic case	Speculative: analysis of expected fitness versus prevalence, some phylogenetic work

which the literature is full of adequate adaptive explanations for typical phenotypic traits. Why, then, have acceptable explanations even for such striking human traits as our bipedal stance been so hard to confirm?

Phylogenetic Histories and Bad Luck

The basic difficulty of testing adaptive hypotheses for widely shared human traits stems from an unfortunate feature of our phylogenetic history. Our closest living relatives are the other great apes; however, as such relationships go, they are not really all that close to us. It is widely agreed that the most recent common ancestor that we share with any of the other great apes lived at least 6 million years ago (mya), and possibly rather earlier (see Tattersall 1995, 218; Goodman et al. 1998); the most recent common ancestor shared by all the great apes (including humans) lived at least 14 mya. Nor do the great apes themselves represent a particularly diverse range of species. Furthermore, only the two extant species of chimpanzees (chimpanzees and bonobos) share a relatively recent common ancestor with another extant species (Tattersall 1995, 218; Goodman et al. 1998), although some have suggested that DNA evidence implies that orangutans may be best thought of as representing two distinct species, having diverged perhaps 1.7 mya (see Gagneux et al. 1999); some similar arguments have been made about the two varieties of gorillas. There are, then, at most seven or so species in the family Hominidae,[3] which share a com-

3. According to Marks (2005), the systematics (and attendant nomenclature) of the groups containing humans, the other great apes, and the extinct ancestors of the great apes "is presently in its worst shape since the 1930s." For our purposes here, we are adopting the following conventions:

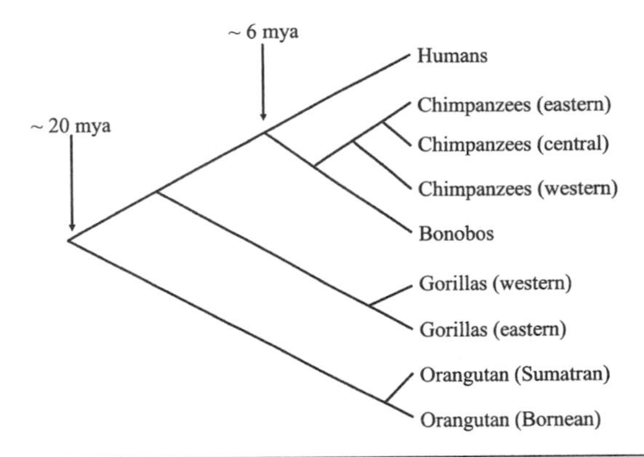

Figure 7.1. The sparse clade problem. Humans belong to a very sparsely populated clade. Our closest living relatives, the chimpanzees, diverged from us a long time ago, and there are few other great apes. This phylogenetic history makes any systematic study of the pattern of relationships between phenotype and environment in this group difficult. (Note: Branch lengths are not to scale.)

mon ancestor that lived something like 10–14 mya, and maybe another few species of "lesser apes" (the gibbons) that share a common ancestor with the great apes that lived on the order of 18 mya (Goodman et al. 1998; Gagneux et al. 1999; fig. 7.1).

The problem with such a sparsely populated clade is that the basic phylogenetic techniques that permit one to figure out how putative adaptations relate to selective regimes (Harvey and Purvis 1991; Martins 2000) are difficult to apply without fairly large numbers of reasonably close relatives. To get good evidence that a trait was subjected to strong selection, and to figure out the nature of such selection (and hence what the trait might be an adaptation for), one must compare the phylogenetic histories of the populations and traits in question with the fitness consequences of the evolved traits, as determined by ecological field studies or experimental manipulations (see, e.g., Griffiths 1996; Larson and Losos 1996; Sinervo and Basolo 1996; Leroi, Rose, and Lauder 1994). Such phylogenetic analyses are possible only if there is enough variation in the traits under investigation to test against the known histories of the relevant lineages.

Recall that this is exactly the kind of evidence used in the untangling of the history of the sword in the sword-tailed fish of the genus *Xiphopho-*

Humans, along with their extinct ancestors and all other species derived from that line after their divergence from the chimpanzees, will be referred to as the tribe "Hominini" (the hominins). The group containing the gorillas, chimpanzees, humans, and all their extinct ancestors, after their divergence from the orangutans, will be referred to as the subfamily "Homininae." The larger group containing all of those species plus the orangutans, but not the gibbons, will be considered the family "Hominidae." We are not, however, committed to these levels, nor, a fortiori, to these particular names.

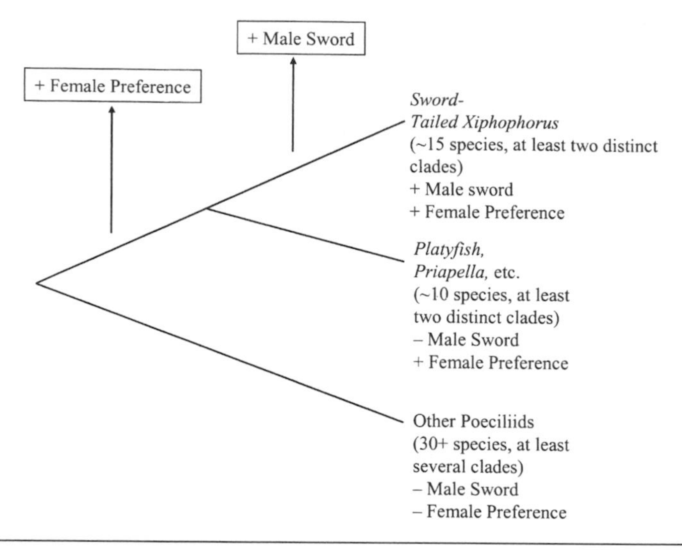

Figure 7.2. How the sword-tailed fish got its sword. Extensive phenotypic manipulation, when combined with information about the phylogenetic relationships of *Xiphophorus* (the sword-tailed fish), *Priapella,* and other poeciliids, generated the following hypothesis about how *Xiphophorus* became sword-tailed: A female preference for swords arose sometime after the common ancestor of *Xiphophorus* and *Priapella* diverged from the other poeciliids; then, sometime after the ancestors of *Xiphophorus* and of *Priapella* diverged, the *Xiphophorus* line acquired its distinctive sword. (Adapted from Sinervo and Basolo 1996.)

rus. It was the finding that the females of some closely related genera (such as *Priapella*) share the preference for swords, even though none of the males of their species are sworded, that suggested that female preference for swords and swordless males are the primitive traits in that clade; as a result, the adaptive hypothesis is well supported (Sinervo and Basolo 1996, 172; fig. 7.2). The problem for evolutionary psychology, and for other fields focused on the adaptive history of human psychological traits, is that this sort of evidence is simply not available in the case of putative universal human adaptive traits; we have no extant relatives that are suspected of sharing similar selective regimes with us and which can therefore be used to test the fitness consequences of our supposed adaptations. If all or most of the estimated dozen or so extinct species in the tribe Hominini still existed (that tribe, consisting of all the species that share a common ancestor with us, but not with the chimpanzees, comprised perhaps two or three genera), phylogenetic studies would certainly be easier and might well be useful for distinguishing between competing hypotheses about the spread and maintenance of behavioral and psychological phenotypic traits of interest. Unfortunately for evolutionary psychologists, all the other members of the tribe Hominini are extinct, and so comparisons of cognitive and other interesting traits between groups, with special attention to the fitness consequences of differences in key traits, are not feasible.

Another difficulty is that, despite many evolutionary psychologists' claims to the contrary, little is known about the details of the environmental (and especially social) conditions under which human evolution took place; without such details, even formulating reasonable adaptive hypotheses becomes difficult. Optimization analyses are therefore hampered both by a lack of information about the environment in question and by the fact that the traits of interest are generally quite distant from any reasonable measure of relative fitness. Optimization analyses of the comparative type are additionally hampered by the lack of reliable phylogenies in the tribe Hominini after its separation from the rest of the subfamily Homininae; it has been argued that without such phylogenetic information, linking changes in phenotypes to changes in the environment will prove difficult (see Collard and Wood 2000 and citations therein).

The best traits for generating testable hypotheses about human adaptations, then, are either those that are widely shared in the family Hominidae (or an even larger clade, such as the primates) (hence, not uniquely human), or those that are of local adaptive significance (such as the *HbS* or the *CCR5-Δ32* mutations or phenotypic traits such as skin color). While many evolutionary psychologists, like the sociobiologists before them, are fond of pointing toward other species to provide *examples* of the kinds of behaviors they are referring to (Thornhill and Palmer's [2000] material on "rape" in other species, including insects, comes to mind), these are *not* meant to be phylogenetic arguments; these authors do not mean to imply that the traits in question are homologous. On the other hand, some researchers have been working with behaviors that humans share with other great apes, but not with other primates less closely related to us; these behaviors, those researchers suggest, might well constitute homologous traits. Examples include comparisons of the emotional repertoire of humans with those of the other great apes and other primates (see Griffiths 1997 for discussion of what the traits in question here actually are) and comparisons of the child care practices of humans with those of the other great apes and other primates (see Maestripieri and Roney, in press; see Jones 1999 and citations therein for a review of some similar work).

The other plausible avenue of research, searching for psychological or behavioral traits of local adaptive significance in humans, is problematic because there is very little known systematic diversity within the species *Homo sapiens,* either at the level of meaningful phenotypes or at the genetic level. Indeed, current DNA evidence points to humans having far less genetic diversity than the other great apes, despite the relatively old age of our clade (Gagneux et al. 1999, 5081).[4] This lack of genetic diversity may mean that there are relatively few significant local variations to be found; however, significant local variations can be the result of a very few genetic differences, and so an overall lack of genetic diversity may not be a particularly important measure of phenotypic diversity. In any event, little systematic work has been done to look for such geographically delimited human variants, and until far more thorough searches are completed, it would be premature to speculate further on this matter (see Kaplan and Pigliucci 2003).

WHAT'S LEFT FOR EVOLUTIONARY PSYCHOLOGY?

Insofar as it is the goal of many of the most vocal proponents of evolutionary psychology to explain "human nature" as a collection of adaptive traits unique to humans (see Cosmides, Tooby, and Barkow 1992, 5), the unfortunate structure of our clade, combined with the ethical and practical constraints on human research, imposes very strict limits on such research programs.[5] Evidence relating to other—perhaps less glamorous—potential human adaptations, however, is more plausibly obtainable.

Wilson (1994), for example, has noted that evolutionary psychology has not dealt seriously with the possibility of studying adaptive phenotypic variation between human populations; he argues that there are both theoretical and empirical reasons to suspect that such variation might exist. Some of the critical difficulties in gathering evidence relevant to claims regarding universal, uniquely human adaptations do not apply to gathering evidence relevant to more local, population-level adaptations. For example, when population-level psychological adaptations are suspected, adoption studies (if properly executed) might provide "natural" transplant

4. It has been suggested that the most likely explanation for the low level of genetic diversity in *H. sapiens* is that at some point, the lineage leading to humans experienced a lower genetic effective population size, though how this played out remains unclear (Gagneux et al. 1999, 5080). If this is so, the implications for nonadaptive explanations of widely shared human traits are intriguing: small effective population sizes could easily result in the fixation of particular phenotypic traits through nonadaptive genetic drift (see Futuyma 1998). Again, this possibility suggests a cautious approach toward attempts to argue for universal, uniquely human *adaptations*.

5. Cosmides and Tooby have been most active in arguing for the strong program in evolutionary psychology, focused on explaining "human nature" as a set of universal adaptations (see Tooby and Cosmides 1989, 1990, 1992, 2000); others include Buss (1995), Symons (1995), and Pinker (1997). Miller (2000) accepts the strong evolutionary psychology approach with respect to "ordinary" adaptations, but rejects it for those traits involved in sexual selection, which are his primary interest.

experiments. It is possible that studies of immigrants to new locales could also shed some light on hypothesized population-level adaptations. And regression analyses done on populations in which the particular kinds of physical processes associated with informal fitness differences are known (or suspected) could provide quantitative assessments of the strength of the (formal) fitness differences associated with particular traits and perhaps even "catch" adaptation in action (see Schlichting and Pigliucci 1998, 166ff, but see our chapters 2 and 4 for a critical analysis of the use of these techniques). Of course, gathering this sort of evidence would probably be difficult.[6] It would require not only cross-cultural fieldwork, but also careful work on adoption and immigration practices, in order to attempt to deal with the large number of confounding (for example, cultural) factors.

Another approach is to deal not with hypothesized adaptive traits that are uniquely human, but rather with adaptive traits shared by at least some of our close relatives (always keeping in mind the limitations imposed by the sparsity of our clade). For example, Sterelny and Griffiths (1999, chapter 14) pointed out that it should be possible to do good work on the origin of those emotions widely shared within the Hominidae, and they noted that this was already a focus of Darwin's attention (1872). While little of that work has been done, hypotheses about the origins and adaptive significance of some aspects of our emotions might be tested by comparisons with related species with which we share problems (and solutions?) of varying similarity. In addition, the work done to test hypotheses about sperm competition shows at least a good (albeit still problematic) start in the same direction (see Futuyma 1998, 359, 588–89 for an introduction to some of this research).[7]

Evolutionary psychology has not yet developed the tools necessary to

6. Indeed, while it seems possible in principle to test population-level adaptive hypotheses in the human case, it may in practice prove to be difficult in many cases due to our inability to control for confounding factors or to gather sufficient evidence to adequately support either the acceptance or the rejection of adaptive hypotheses. However, insofar as this is a serious problem for population-level adaptations, it is even more of a problem for universal ones. The best bet for such studies, as our examples have suggested, may be particular local physical adaptations. See, e.g., Kurbatova, Botvinyev, and Altukhov 1990 (on stabilizing selection on human birth weight) and Kirk et al. 2001 (on so-called life history traits). It is also possible to gather convincing evidence on selection favoring certain allelic variants of human genes (see, e.g., Balter 2005).

7. See Miller 2000 (231–32) for some very speculative suggestions regarding the implications of this work for the human case. Note that so far, many standard techniques, such as phylogenetic correction analysis (see Bohning-Gaese and Oberrath 1999 and citations therein), have not been employed, nor has sufficient attention been paid to the likely ancestral condition, or to the different implications that homologous and homoplasious traits would have for the adaptive hypothesis considered. But these observations do not detract from the fact that this topic presents a plausible research program.

uncover our "shared human nature" (if such there is—see Dupré 1998), any more than physical anthropology has been able to uncover the evolutionary path of even such clear human adaptations as bipedalism. It is obvious that our brains were subject to selective pressures during our evolutionary history; it is not at all obvious what those pressures were. Rather than overreaching by attempting to uncover the historical causes of psychological features shared by all and only humans, evolutionary psychology might be better off attempting to develop ways of adequately testing possible adaptive traits of a more tractable nature, either because they are widely shared outside our species or because they are shared only by specific "ecotypes" within our species (Pigliucci and Kaplan 2003). However, this new evolutionary psychology would look and feel much less sexy than the proto-scientific brand currently available on the market.

8 Slippery Landscapes

The Promises and Limits of the Adaptive Landscape Metaphor in Evolutionary Biology

THE PROBLEM OF HOW A POPULATION CROSSES AN ADAPTIVE VALLEY ON ITS WAY FROM ONE ADAPTIVE PEAK TO ANOTHER . . . MAY BE NON-EXISTENT.
—SERGEY GAVRILETS, "EVOLUTION AND SPECIATION ON HOLEY ADAPTIVE LANDSCAPES"

The metaphors of fitness and adaptive landscapes have played a central role in evolutionary theory and practice ever since they were introduced by Sewall Wright in the 1930s. They provide biologists with powerful imagery that can aid them in thinking about situations that would otherwise require extremely complex mathematics to be handled. However, it turns out that the concepts of fitness and adaptive landscapes have been used in very different ways by various biologists, and that these different uses do not always cohere well with one another or with other aspects of evolutionary theory. Further, recent theoretical work shows that some major problems that have kept the research community occupied for decades—such as finding ways in which populations can "shift" from one adaptive peak to another—may disappear once one adopts a metaphor that better models the kind of phenomenon one is interested in, such as, in this case, macroevolution.

A MUCH-USED METAPHOR

The concepts of fitness landscapes and adaptive landscapes, as well as the related one of adaptive peaks, are part and parcel of all modern training in evolutionary biology, from introductory texts (such as Price 1996) to graduate-level treatments (e.g., Hartl and Clark 1989 or Futuyma 1998). Furthermore, even a cursory glance at papers published by the major evolutionary journals in recent years shows that research on these topics has been consistently at center stage (e.g., Houle 1994; Gilchrist 1995; Whitlock 1995, 1997; Coyne, Barton, and Turelli 1997; Niklas 1997; Pal 1998; Svensson and Sinervo 2000; Wilke et al. 2001; Hadany 2003, and see additional references below). Indeed, as we saw in chapters 2 and 4, the approach most commonly used in attempts to measure natural selection within populations is based on the very idea of fitness landscapes (Lande 1979; Lande and Arnold 1983; Manly 1985; Endler 1986), and discussions have been going on about the most effective statistical approaches to carry out the task (e.g., Mitchell-Olds and Shaw 1987; Crespi and Bookstein 1989; Kingsolver and Schemske 1991; Rausher 1992; Schluter and Nychka 1994; Brodie, Moore, and Janzen 1995; Scheiner, Mitchell, and Callahan 2000; Stinchcombe et al. 2002; but see chapters 2 and 4 on the limits of these approaches). But what exactly *are* fitness or adaptive landscapes? Where do these metaphors come from, how has their presentation changed over time, and how do they influence the way we conceive of fitness and natural selection—among the key concepts in evolutionary biology? How, in other words, is the landscape metaphor currently being used, and how is it being misused?

Before we proceed, we need to clear up a potential source of confusion and make a distinction between *fitness* landscapes and *adaptive* landscapes. As we shall see, failure to separate the two and understand how they are related injects a lot of confusion into this area. The two ideas were distinct in Wright's (1932a) original work (despite his ambiguity of usage), and careful modern authors do make sure to use the appropriate language, depending on the context (e.g., Lande 1976; Kirkpatrick 1982), but Gavrilets (1997a, 1999, 2004)—whose work in this field we will discuss at length—does not, and most practicing biologists are not familiar with the distinction. So: *fitness* landscapes represent the fitness of individual genotypes, not populations (these are also sometimes called fitness *functions*); *adaptive* landscapes, on the other hand, represent the mean fitness of populations.

The landscape metaphors have been in use since the 1930s and have perhaps, in that time, provided some valuable insights into the process of evolution. However, our conceptions of what fitness and adaptive landscapes are (what they are meant to represent) and of what they might actually look like have undergone, and are still undergoing, major changes. These changes force us to consider the possibility that substantial research programs have been based on rather fuzzy thinking—that the intuitive appeal of a rather vague image has encouraged research into problems that may not really exist. In this chapter, we first examine the basic ideas behind the metaphor, as well as the metaphor's history, and then discuss some of the difficulties that contemporary approaches to the concept have revealed. Finally, we suggest that while recent work has pointed toward various ways to improve the metaphor, it may in the end be extremely difficult to articulate it in a way that is both coherent and conceptually fruitful.

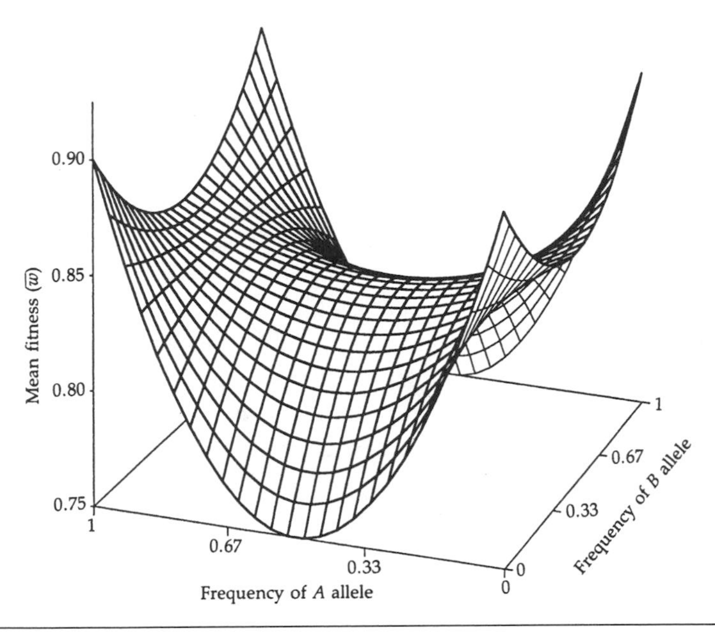

Figure 8.1. A typical textbook depiction of an adaptive landscape: a three-dimensional surface with the base plane identified by the frequencies of two alleles and the vertical axis representing a measure of (mean) fitness. Notice that in this representation the points in the diagram (not shown) would be populations, not genotypes (as in a *fitness* landscape). (From Hartl and Clark 1989, 215.)

THE BASIC IDEA, AND TWO INITIAL PROBLEMS

The concept of adaptive (i.e., population-level) landscapes as usually presented in textbooks, papers, and professional meetings is actually rather simple: Typically, the mean fitness of a population is diagrammed against one or two axes representing a measure of genetic differentiation among populations, usually the frequencies of certain alleles (fig. 8.1). The resulting two- or three-dimensional landscape is meant to show the presence of peaks and valleys, corresponding respectively to the fact that certain genetic combinations have higher or lower fitness. A population finds itself in a region of the landscape determined by its particular genetic makeup; the population is represented by a point on the landscape corresponding to the allelic frequencies within that population. What happens to the population then depends on the structure of the landscape in that region. To put it simply, the intuition is that natural selection should push a population (i.e., alter the allelic frequencies of that population) so that it moves uphill toward whatever peak happens to be nearby (see chapter 1 for a critique of the force metaphor as it appears here). Such movement would continue while there is genetic variation in the population. Lack of genetic variation, or strong stabilizing selection to maintain the population as close as possible to the peak, would slow down or halt the movement of the population and tend to maintain the status quo.

The appeal of visualizing the process of evolution in such simple and intuitive terms is strong. The more one thinks about it, the more one develops the feeling that it is possible to gain deep insights into evolution by thinking of adaptive valleys and peaks: the metaphor rapidly grows on you and begins to redirect your thinking. Indeed, we suspect that many practicing evolutionary biologists may find themselves reversing the relationship between imagery and reality: instead of visualizing complex realities through the simplified lenses of low-dimensional adaptive landscapes, they may succumb to the temptation to project the simplicity of the latter into their predictions of how the real world actually works. The problem is that the tempting imagery conceals several pitfalls of the metaphor, which we examine throughout this chapter.

The first problem worth noting with this articulation of the metaphor is that, at least as it is usually presented, it is overtly simplistic. The idea that changes in the relative fitness of a population are driven by changes in the frequencies of alleles in that population is misguided because, as we noted in chapter 1, there is not a one-to-one mapping of allelic frequencies in a population onto anything like mean population fitness; rather, the (relative) mean fitness of an ensemble of individuals can vary depending on how the particular alleles are distributed in the population (see box 3.2 and fig. 3.1). Indeed, this aspect of the metaphor depends on an uncritical acceptance of the standard textbook definition of evolution as within-population change in allelic frequencies (e.g., Hartl and Clark 1989). If one rejects this definition, either the axes have to be redefined in terms of the various available kinds of heritable variation (not all of them genetic), or one has to restrict the use of the metaphor in some way. And, as we argued in chapter 3, the definition of evolution that makes it simply changes in allelic frequencies ought to be rejected, because it is now well established that, for example, different *heritable* epigenetic mechanisms can result in particular genotypes having very different fitnesses in particular environments (see especially box 3.2 and box 5.5).

Wright's original conception referred to *fitness* landscapes, and therefore did not involve mapping populations with particular allelic frequencies and mean fitnesses onto the axes. Rather, Wright imagined that individual organisms with certain combinations of genes would have particular fitnesses; populations, then, consisted of clouds of points (the individual organisms) on the adaptive landscape (see box 8.1 for a summary).

Box 8.1 What Is Being Graphed on a Fitness or Adaptive Landscape?

As noted in the main body of the text, there have been, and continue to be, serious ambiguities about just what is being represented on graphs of fitness or adaptive landscapes. Here, some of the major possibilities, and some of the difficulties with these interpretations, are presented.

Standard/textbook. In the traditional textbook interpretation of *adaptive* landscapes (based roughly on Dobzhansky's interpretation of the landscape metaphor), various axes representing allelic frequencies in a population are graphed against the mean fitness of a population possessing those allelic frequencies (see fig. 8.1). Points on the graph, then, correspond to populations of organisms with a particular mean fitness.

This interpretation is problematic because (a) the mean relative fitness of a population does not map neatly onto the allelic frequencies of that population, (b) it ignores the complexities of moving between genotypes and phenotypes, and thence to fitness, and (c) the "movement" of populations on the landscape is simply a restatement of the formal (predictive) fitness of the population, and hence the metaphor is not explanatory. Further, environmental variation is difficult to incorporate into these models.

Wright's original interpretation. Wright originally conceived of the axes of *fitness* landscapes as representing allelic substitutions, and of points on the landscape as representing individual organisms with particular genotypes and associated relative fitness values (see fig. 8.2). This conception could be expanded to take account of current molecular genetics understanding of mutations, and the axes could represent molecular changes (or even epigenetic changes) more generally.

Even revised, however, this interpretation runs into several problems: (a) it ignores the complexities of moving between genotypes and phenotypes, and thence to fitness, and (b) it ignores the difficulty of generating a coherent articulation of the overall relative fitness of an individual organism (see chapter 1). And again, environmental variation is difficult to incorporate into these models.

Gavrilets's holey landscapes graph genotypic axes against a fitness axis, and hence are closer, at least in terms of what the axes represent, to Wright's fitness landscapes than to any others.

Simpson's and related interpretations. Simpson's paleobiological use of the landscape concept demands that the non-fitness axes represent variation in phenotypes, not genotypes. Points are therefore organisms with a particular phenotype and an associated relative fitness value. This is, incidentally, what Lande and Arnold's (1983) analyses of natural selection attempt to measure (see chapter 2 and the discussion concerning multiple interpretations of selection analyses in terms of landscapes in Phillips and Arnold 1989).

This interpretation shares several problems with Wright's: (a) it ignores the difficulty of generating a coherent articulation of the overall relative fitness of an individual organism (see chapter 1), and (b) environmental variation is difficult to incorporate into such models. Obviously, this version avoids the genotype-fitness mapping problem, but at the cost of introducing another difficulty; namely, that the relationship between the location and the movement of populations (clouds of points) on the landscape will be mediated by the heritability of the phenotypic variation in question, which will vary with environmental variation as well as with changes in population structure.

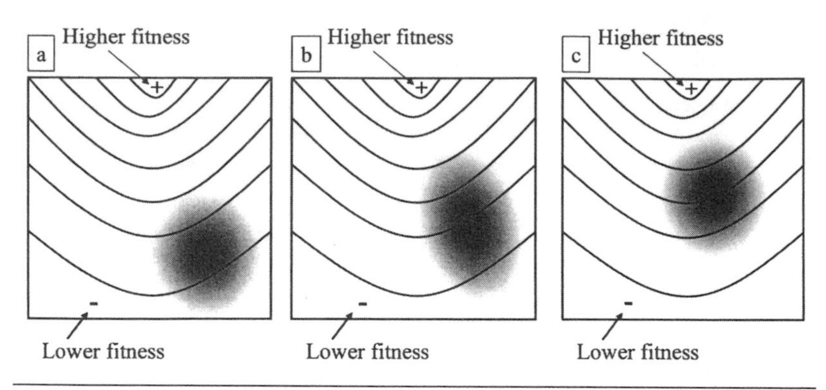

Figures 8.2. If individual organisms are interpreted as points on a fitness landscape, populations (represented by clouds of points) "move" (the a > b > c sequence) uphill because those members of the population that are on the uphill (higher-fitness) side of the landscape tend to outreproduce those that are on the downhill (lower-fitness) side. The population is then "shifted" uphill with each generation. Note that there are no labels on the *x* and *y* axes of these graphs; see the discussion of Wright's initial landscapes and figure 8.3. (Loosely adapted from Wright 1932b, fig. 4, 362.)

While the movements of these clouds (evolutionary changes in the population structure) would usually involve changes in the allelic frequencies of the populations in question, as discussed above, it would not be necessary that they do so because, for example, changes in the distribution of alleles could change the structure of a population without altering the frequency of alleles in that population. But more important, this way of understanding the metaphor makes it explanatory in a different way than the population interpretation. If the points are taken to be individuals with particular genotypes, then the movement of populations on the landscape is explained by the different relative fitnesses of those genotypes. Genotypes that find themselves uphill from the rest of the population will tend to outreproduce those on the downhill side, and the population (the cloud of points) will therefore move, on average, higher on the hill (fig. 8.2). Such an explanation is impossible if the points are taken to be populations with particular allelic frequencies; in that case, all one can say is that natural selection will tend to drive the populations (which are points on the *adaptive* landscape) uphill, which is clearly not an explanation, but merely a restatement of what the metaphor is supposed to represent.

The second difficulty with the landscape metaphor that we wish to draw attention to is the habit of thinking of the landscape as fixed and the population as changing. As Lewontin (1978) pointed out, one of the fundamental problems with thinking of evolution as a climb on adaptive hills is that the landscape does not, in fact, remain fixed. If the non-fitness axes of the landscape represent allelic frequencies in a population, then the shape of the landscape itself has to be altered by ongoing evolution. Organisms, Lewontin noted, are not independent of the environment they occupy; rather, they at least in part define and continuously modify that environment by the very acts of surviving and reproducing. Part of their environment, in other words, is the ensemble of other organisms that are present in the population and with which they interact. As Lewontin put it, evolution is not best viewed as a process of organisms "solving" problems posed by the environment; rather, organisms and environments (which, of course, include other organisms) coevolve—both the "problems" and the "solutions" emerge from the interaction of the two components. It is vanishingly unlikely, in other words, that a population could climb to a fitness peak that remained unaltered during the climb. It is for this reason that Lewontin proposed a change in the landscape metaphor: instead of thinking of a fixed landscape and of populations moving on it, we need to visualize a rubbery landscape whose very features are altered by the entities (populations) that exist on it.

While we have expressed the above problem using the traditional formulation in which points represent populations, it is no less serious if we treat the points as individuals with particular genotypes (and populations as clouds of such points)—that is, if we switch from adaptive to fitness landscapes. One advantage of thinking in terms of individual organisms, however, is that it makes obvious another fundamental difficulty with the metaphor; namely, the relationship between genotype and fitness. If one questions the simplicity of the genotype-phenotype mapping function generally assumed in population genetic models, the problem of how to think about the relationship between the organism, its fitness, and its environment quickly becomes quite complex, as it should be clear that the relative fitness of a genotype depends on the interaction of the overall phenotype and the environment. Further, the phenotype of the organism depends not just on the genetic resources available during development, but on all the available resources, including, of course, those in the developmental environment more broadly. Because the latter is one of the things that evolution will tend to change, the relationship between particular genotypes and the relative fitnesses that characterize the organisms carrying those genotypes is obviously going to be much more complex than the model usually envisioned (see, e.g., Alberch 1991; van Tienderen and de Jong 1994; Schlichting and Pigliucci 1998; Mezey, Cheverud, and Wagner 2000).

In any case, the mathematics necessary to describe Lewontin's rubbery landscapes is vastly more complicated than that used to study the properties of fixed landscapes. Ironically, as we shall soon see, Sewall Wright originally introduced the landscape metaphor precisely because the equivalent math was too complex to be grasped by the majority of his evolutionary biology colleagues. Toward the end of this chapter, we discuss the possibility that Wright's archrival, Ronald Fisher, got closer to a realistic appreciation of the evolutionary process by dismissing the adaptive landscape metaphor altogether while proposing a formulation that is similar in some ways to Lewontin's ideas.

A BRIEF HISTORY OF EARLY LANDSCAPES

It is instructive to consider where the idea of fitness and adaptive landscapes came from, why it caught on and was developed, and how it has changed through time. Ruse (1996) and Skipper (2004) provide an excellent introduction to these topics, as well as to the ongoing debate about what, exactly, Wright was up to; here we consider only the highlights that are essential to our story.

The first representation of fitness and adaptive landscapes (and the beginning of the confusion between the two) can be found in Wright's contribution to the Sixth International Congress of Genetics held at Cornell University (Wright 1932b). He was asked to talk about his shifting balance theory of evolution, which had been the topic of a lengthy paper recently published in the journal *Genetics* (Wright 1932a). The shifting balance theory, as we shall see, still plays a major role in the discussion about landscapes, and it requires a brief explanation. The goal of the theory is to account for rapid evolution in small populations; Wright believed that evolution in such small populations was vitally important to speciation and to the development of new adaptations, whereas Fisher was stressing a process of mass selection in large populations (this was the beginning of the apparently endless Fisher-Wright controversy). The three phases of the shifting balance are usually summarized in the following fashion (e.g., Hartl and Clark 1989):

Phase I: Exploratory phase. Random genetic drift alters allelic frequencies in small populations, which allows an exploration of the topography surrounding the current position of a given population (notice the use of the landscape metaphor and the importance of being clear whether one is referring to individual organisms or population means as points on the landscape).

Phase II: Mass selection. Natural selection increases the frequency (within a population) of favorable gene combinations created by phase I.

Phase III: Interdeme selection. The last phase consists of selection among different populations (demes), in which those characterized by the best

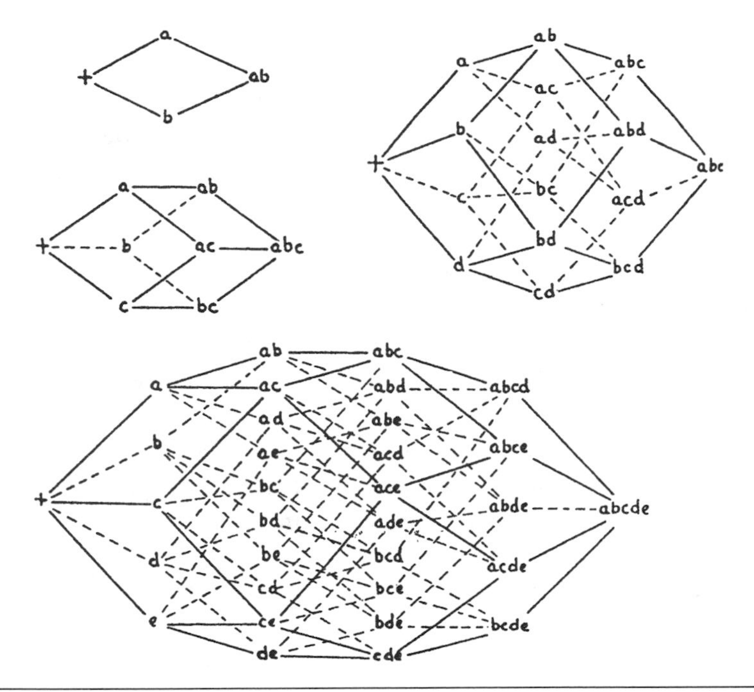

Figure 8.3. Wright's fitness (not adaptive!) landscapes become increasingly complex as one considers more alleles at one locus (or more loci, for that matter). It is obvious that visualization soon becomes unmanageable for any landscape with a realistic dimensionality of gene combinations. (From Wright 1932b, fig. 1, 357.)

combinations of genes grow fast and start exporting migrants to nearby populations. Those populations' genetic constitutions—and positions on the adaptive landscape—are therefore altered and the new combinations of genes spread further.

According to Ruse (1996, 330) the term "shifting balance" (obscure to the large majority of evolutionary biologists who use it) comes from the fact that Wright was a follower of Herbert Spencer and of his idea that evolution consisted of a dynamic equilibrium imposed by natural selection. The shifting refers to the change from one equilibrium (one adaptive peak) to another.

Be that as it may, Wright was asked (by his advisor, E. M. East) to present a simplified version of his *Genetics* paper. East was concerned, apparently, that unless Wright was able to present a nontechnical version of the paper, most of the audience would not understand what he was talking about, and he would lose a formidable opportunity to present his theory. Wright therefore replaced the math with drawings such as the one in figure 8.3, representing his nonmathematical conception of a fitness landscape (notice that there is no explicit representation of fitness in these diagrams, and that they do not define a continuous space, but rather a set of multidimensional polyhedrons).

There are two things that are important to understand about Wright's initial drawings of fitness landscapes. First, as we noted above, he started out by representing possible combinations of alleles such that the points on the landscape were individual genotypes, *not populations* (i.e., these were not adaptive landscapes). The space being visualized, then, consisted of a discrete set of points representing possible combinations of genes; each point had associated with it the mean relative fitness of the genotype represented by that point (presumably, in a particular environment). In the interpretation more common today, in which populations are the smallest units represented on the landscape, the adaptive landscape itself is viewed as a continuous function consisting of allelic frequencies.[1]

Our second point may explain how Wright's landscapes came to be interpreted as being about allelic frequencies in populations: Wright's diagrams did not actually have labels on the axes (see figs. 8.3 and 8.4). Note that in figure 8.3, each "point" on the polyhedron is surrounded immediately (e.g., at a distance of one "step") by all possible single-allele substitutions, but there is no visualization of fitness values, and the result hardly looks like a "landscape" at all. If we treat the points in figure 8.4 as individuals with particular genotypes, each "axis" might represent the possible alleles at each (possible) locus, but the noncontinuous nature of these axes makes this interpretation awkward, and the "scale" of the axis remains obscure. In fact, it seems clear that Wright intended the distance between any two points (representing genotypes) on the fitness landscape in figure 8.4 to represent the distance between those genotypes—the number of one-step substitutions necessary to get from one to the other (see especially Wright 1932b, fig. 4). Wright assumed that in collapsing many thousands of dimensions into two, nothing of importance would be lost; while the distances between every genotype could not be accurately represented, Wright's use of this kind of diagram suggests that he believed that the overall structure of the relationships between genotypes could be adequately represented. Notice especially that it is hard to see how Wright could have labeled the axes on figure 8.4 in a way that was both an accurate reflection of his stated intentions and visually compelling. Forced to choose, the landscape he drew was visually compelling, but is, at best, difficult and awkward to interpret properly.

1. Of course, since all real populations are finite, the possible allelic frequencies in real populations are in fact a set of discrete points, rather than a continuous function. But the basic point remains.

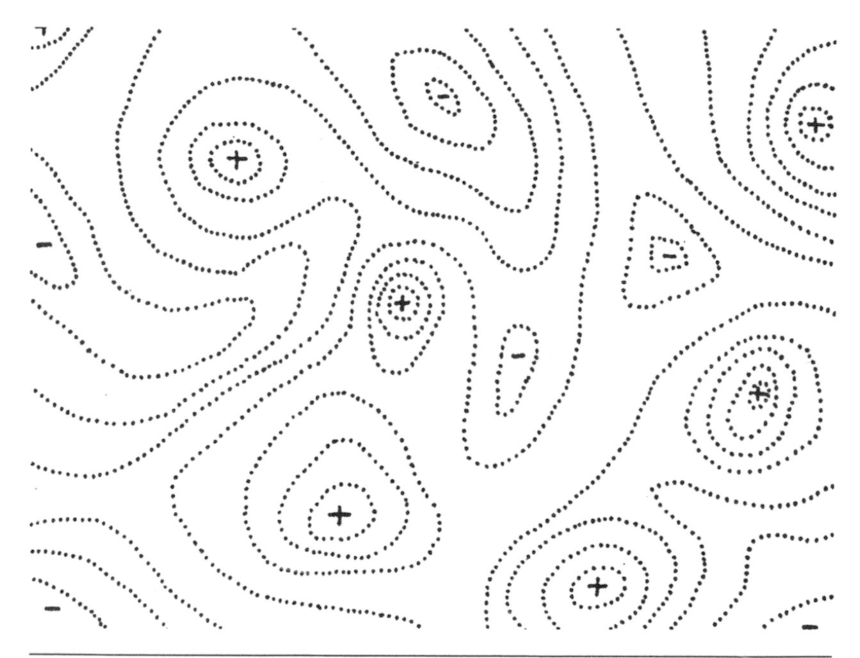

Figure 8.4. One of the first representations of Sewall Wright's fitness (not adaptive!) landscape idea, a field of n-dimensional gene combinations (represented here in two dimensions) and the relevant adaptive topography; note that the topography is not based on empirical evidence or computational models, but rather represents what Wright thought such landscapes probably resembled. Note too that there are no labels on the axes. (From Wright 1932b, fig. 2, 358.)

Despite these ambiguities, Wright's presentation was very successful. Ruse (1996) reports that not only did Wright receive a large number of reprint requests, but Dobzhansky, one of the fathers of the modern synthesis that dominated twentieth-century evolutionary thought, soon referred to the idea of adaptive (not fitness) landscapes as a crucial key to the understanding of evolution, and made use of it in his influential book *Genetics and the Origin of Species* (Dobzhansky 1937). However, whereas Wright thought of populations chiefly as clouds of points on a fitness landscape, adapting to their environment and differentiating on their way to becoming new species, Dobzhansky thought of points as representing populations on adaptive landscapes, and of populations on different peaks as being separate species. In this way, both Dobzhansky and Wright linked the study of landscapes to the problem of speciation, though of course they used different landscapes (adaptive landscapes in Dobzhansky's case, and fitness landscapes in Wright's) and used them in rather different ways. It is worth emphasizing that, despite the fact that many contemporary papers continue to couple adaptive evolution (the origin of adaptations) with speciation (the origin of species), there is no necessary link between the two. Of course, selection may, under certain circumstances, facilitate speciation (Macnair, Macnair, and Martin 1989; Hatfield and Schluter 1999; Doebeli and Dieckmann 2000), but it need not. It is obvious that speciation can occur without selection, but more important, selection does not necessarily lead to speciation. Gavrilets, Li, and Vose (1998) have made this point, and West-Eberhard (2003) has put much emphasis on the demonstrable fact that novel adaptive phenotypes can (and often do) arise within species in the course of evolution without such processes having to result in speciation (*pace* the theory of punctuated equilibria: Eldredge and Gould 1972).

Another grand figure of the modern synthesis (sometimes referred to incorrectly as the "neo-Darwinian synthesis"), the paleontologist G. G. Simpson, also made use of the landscape metaphor, while at the same time helping to popularize it and again altering its meaning. In his *Tempo and Mode of Evolution* (1944), he showed diverging adaptive peaks over macroscopic time scales, reinforcing the connection with the process of speciation, even though, of course, the (still unlabeled) axes of such diagrams can no longer be thought of as any measure of genotypic combinations at all, nor even the frequencies of alleles in a population, since these cannot be ascertained from the fossil record (fig. 8.5). One might interpret the axes as measures of *phenotypic* differentiation. If one represented individual organisms as points based on their fitness (of the "overall" sort we have suggested is so hard to determine—see chapter 1) and their location in phenotypic space, one would have a phenotypic fitness landscape. If, on the other hand, one graphed populations as points based on the frequency of particular phenotypes within each population, one would have a phenotypic adaptive landscape. Insofar as the phenotypic variation mapped on the axes is heritable, the scenario explored in figure 8.2 would, presumably, still obtain.

Ruse (1996, 317) asks whether the pictures of adaptive landscapes (fitness landscapes are rarely pictured—or even mentioned—in papers or textbooks) are part of evolutionary *theory* or just evolutionary *thought*—that is, the way in which evolutionary biologists tend to think of their subject matter. He argues that the answer is "both," and we tend to agree. Since the math used by Sewall Wright was too hard for most professional biologists to follow, they had to use the metaphor and the imagery in order to think about the problems of adaptation and speciation. Indeed, the metaphor is still in wide use, at least in part because the mathematical tools necessary to go beyond the metaphor continue to be out of reach for most practitioners. But while the use of the metaphor led to a consider-

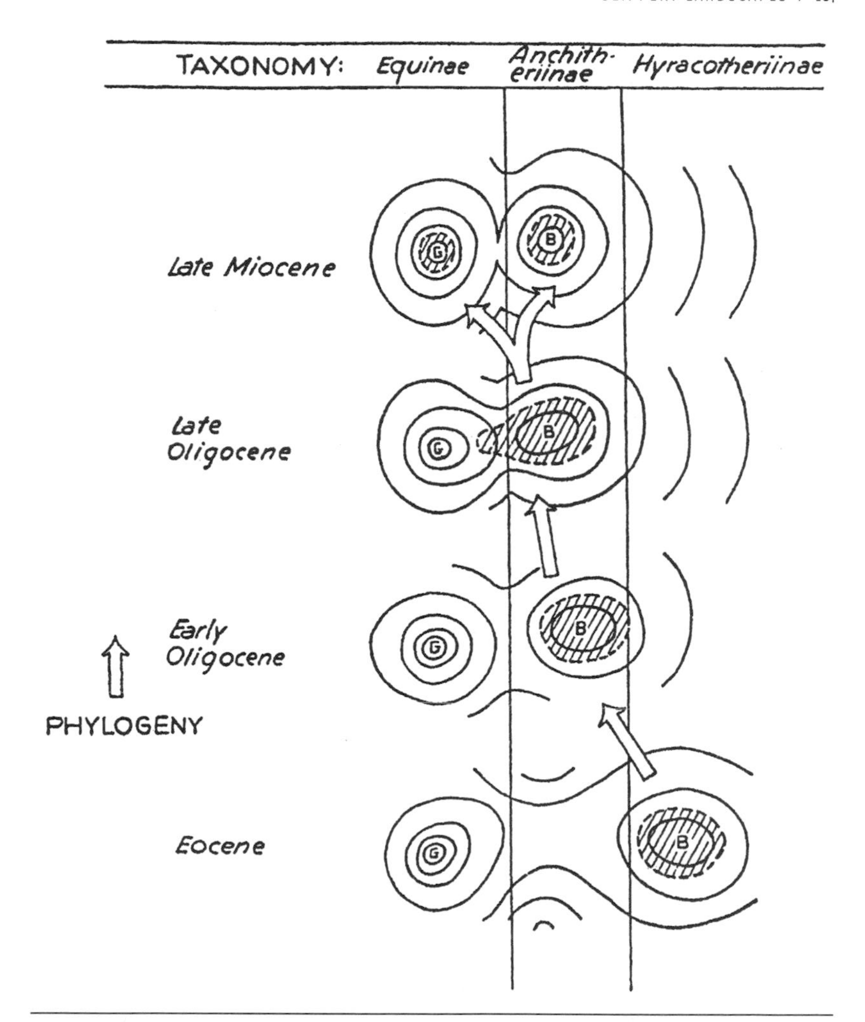

Figure 8.5. Phenotypic landscape showing speciation and macroevolution according to Simpson (1944). Notice that the (unlabeled) axes of the adaptive topography cannot be either gene combinations (Wright's original meaning) or allelic frequencies (the later, population-level evolution of the concept), since that sort of information is not available in the fossil record. The *x* and *y* axes of the landscape must be interpreted as indicating *phenotypic* combinations (either of individual organisms if this is a phenotypic fitness landscape, or of populations of organisms if this is a phenotypic adaptive landscape—on this issue, Simpson is unclear) and their fitness correlates. (From Ruse 1996, 315.)

able amount of both theoretical and empirical research in the decades following Wright's original paper, it seems likely, in light of recent developments, that the imprecision of the metaphor has resulted in much of that research leading to dead ends. Skipper (2004) suggests that Ruse considers the role of landscapes to be limited to illustrating particular evolutionary ideas; he argues instead that adaptive landscapes have been used extensively as a "theory evaluation heuristic" (1185). But if this is true, it is even worse news for the role played by adaptive landscapes in evolutionary theory. As we argue in the next section, visualizations of adaptive landscapes provide a very unsatisfactory heuristic for thinking through the complex dynamic behavior of populations on realistically complex high-dimensional landscapes, and such a use should therefore be resisted.

KAUFFMAN'S OUTLINE OF A GENERAL THEORY OF FITNESS LANDSCAPES

According to Gavrilets (1997a), the next big step in the development of the landscape metaphor was marked by Kauffman's work on applications of complexity theory in evolutionary biology, and in particular by a paper he co-authored with Levin (Kauffman and Levin 1987), in which they attempted to sketch a general theory of adaptive walks on what they called "rugged" (fitness) landscapes.

The problem tackled by Kauffman and Levin is as old as the Darwinian theory of evolution itself. As Darwin (1859) had pointed out, if evolution by natural selection is a natural process of adaptive modification of organic forms, it has to be able to proceed via a series of intermediate steps, each of which has at least the same, or better, fitness than the preceding one. Modern evolutionary biologists think of this requirement very much in terms of adaptive landscapes (at the population level): natural selection cannot favor "downhill" movement, even if that is the only way to reach a higher peak farther in the distance. Because selection is not teleonomic in nature (i.e., it cannot foresee a future advantage), the only way to move from one peak to another is by moving upward (or at least sideways) on the landscape.[2] The problem can be restated in terms of individual genotypes, and therefore of fitness landscapes. What interested Kauffman and Levin, then, was how one would go about outlining a theory that explains evolutionary paths of increasing fitness that use one-step variants (i.e., genotypes differing by a single mutation).

Both mathematically and empirically, answering this question is far from trivial. Mathematically, the description of realistically complex fitness landscapes may require an unwieldy number of simultaneous equations—and hence there may be no general analytical solution to the problem. Empirically, it is difficult even to imagine how one would go about quantifying the characteristics of "real" landscapes, not to mention the

2. As briefly discussed by Frank and Slatkin (1992 and references therein), however, theoretical work shows that selection can in fact cause a temporary decrease in a population's mean fitness when acting on linked loci characterized by epistatic interactions.

phenotypic differences of the universe of one-step mutants when starting from a particular (arbitrarily chosen) genotype (and environment).

Kauffman and Levin therefore first addressed the relatively simple problem of deriving the properties of one-step fitter variant paths in uncorrelated fitness landscapes. In this special category of landscapes—which we will encounter again below while discussing Gavrilets's "holey" landscapes—the characteristics of the landscape at any given point are statistically unrelated to those at any other point. Biologically, this means that the fitnesses of even closely related genotypes are themselves uncorrelated—a rather unrealistic assumption. Indeed, this lack of "quasi-continuity," to use Lewontin's language, is often thought to make adaptive evolution difficult or impossible (see Lewontin 1985, 79–80). Nevertheless, Kauffman and Levin developed their general theory starting from this case, which they consider as a baseline,[3] and then moved on to sketch a general theory for the biologically more relevant case of correlated landscapes.

What Kauffman and Levin found in the case of uncorrelated landscapes is interesting and, in part, still sets the agenda for the field. They identified two major limits to natural selection in these settings: first, when the organisms under selection increase in complexity, the optima that can be reached on the landscape are characterized by fitness peaks that tend to be similar to the mean for the entire landscape. In other words, natural selection is not likely to be a very efficient mechanism for increasing mean population fitness across the landscape. The second limitation is that—again, as the complexity of the entities under selection increases, and assuming a constant mutation rate—it becomes difficult for selection to maintain organisms on adaptive peaks that are connected by one-step fitter mutations.

The limitations to natural selection highlighted by Kauffman and Levin are actually in sync with the thinking of at least some evolutionary biologists (e.g., Van Valen 1973), who see selection not so much as a mechanism to achieve ever-increasing fitness, but as one that minimizes the

3. It is worth noticing that this approach of starting from an unrealistic, but tractable, situation and considering it as a baseline is rather typical of theoretical science, and of theoretical biology in particular. While this is a logical, and often successful, approach, it is also true that what constitutes a reasonable baseline from a mathematical standpoint may not represent a useful starting point in biology, a distinction that is sometimes overlooked (but see Wimsatt 1987 on the ways in which false models can be useful). In this particular case, the lack of fitness correlation in the initial model may not be a severe limitation. It turns out that correlated and uncorrelated landscapes display relevantly similar characteristics as far as the question at hand is concerned—provided that the dimensionality of the landscapes is sufficiently high (see Gavrilets 2004, esp. 90–94).

loss of fitness caused by the continuous deterioration of the environment. This decrease in environmental quality has been attributed by Lewontin (1978), and earlier on by Fisher (1930), to the very existence of living organisms themselves, and Van Valen (1973) referred to its general dynamics as the "Red Queen effect" (named after Lewis Carroll's character in *Through the Looking Glass,* who had to keep running just to stay in place). And again, keep in mind that if we consider that the landscape itself can, and in general will, change in response to the evolution of the populations involved, the situation becomes even more fascinating.

THE CURRENT REVOLUTION: "HOLEY" ADAPTIVE LANDSCAPES

Research on the properties of fitness landscapes (and their relationships to adaptive landscapes) is currently undergoing a revolution, the impact of which is still unclear. What does seem likely to us is that theoretical evolutionary biology is at a potentially crucial crossroads, and that the paths being explored now (and those that we anticipate will be explored in the next few years) will reveal that Wright's original metaphor was far more misleading than most biologists had suspected before. These new interpretations of the landscape idea may relegate most research on "rugged" landscapes (both rugged fitness landscapes and rugged adaptive landscapes) to a minor role and thereby take theoretical evolutionary biology in a different direction entirely.

It turns out that fitness (i.e., genotype-level) landscapes taking account of a biologically realistic dimensionality may have a *qualitatively* different structure than that envisioned by Wright and biologists following his example. Fitness landscapes may best be thought of as "holey" (as in having holes); that is, as characterized by loosely connected networks (or areas) of high fitness separated by areas of very low fitness. This structure is often explicitly contrasted with the rugged landscapes of Wright and Kauffman, in which peaks of high fitness are surrounded by valleys of low fitness. The idea of holey landscapes has apparently been around since Dobzhansky (1937), and has been elaborated on by a number of people at different times, including Muller (1942), Bengtsson and Christiansen (1983), and Nei, Maruyama, and Wu (1983), to mention just a few (for a brief history, see Gavrilets 2004). In the following discussion, however, we comment largely on the work of Gavrilets and his collaborators (1997a, 1997b, 2004; Gavrilets, Li, and Vose 1998), as it is their work that has been chiefly responsible for a revival of the concept and has produced a large body of theoretical results with philosophical implications.

Gavrilets (1997a) set the idea of holey landscapes in the context of the standard problem of how populations (notice the higher level of analysis here) can shift from one adaptive peak to another. Classically, two solutions have been proposed. On the one hand, Wright himself suggested his shifting balance theory as a solution to this problem; recall that for him, the landscape metaphor was a way to make clear how this theory could explain speciation and rapid adaptive evolution (see the discussion above). On the other hand, several authors (e.g., Templeton 1981; Carson and Templeton 1984) contend that a founder event—that is, the establishment of a new population by a small number of initial genotypes—may overcome the gulf between two peaks and eventually lead to speciation. Notice that both solutions rely on drift (better thought of as the relatively wide distribution of probable evolutionary outcomes in smallish populations; see chapter 1) to alter the genetic constitution of small populations, thereby rapidly moving them to a different area of the adaptive landscape. (But see below for another "solution" to the peak shift problem that appeals instead to the failure of landscape models to take environmental change and organismal development into consideration.)

Gavrilets (1997a) correctly points out that there is quite a bit of dissatisfaction with both the shifting balance and the founder effect solutions to the problem of adaptive peak shifts. While several authors have historically considered the shifting balance to be a promising avenue of research, Gavrilets (1996) himself has convincingly shown that the conditions necessary for phase III of the shifting balance (the process of interdeme selection) to occur are actually quite restricted. In particular, he argued that previous authors have analyzed models that do not actually reflect the biological situations they were meant to represent, because such models focus on the specific case of only two populations and assume that the migration rates between the two are constant and unequal. This is not at all what Wright himself was envisioning, since he thought of rates of migration out of a given population to be proportional to the "excess" output of individuals in that population, and hence as variable rates linked to population growth. When Gavrilets studied the properties of a model that more accurately reflects the biological reality of interest, he found that phase III of the shifting balance can take place only under the following conditions:

1. Migration is neither too strong nor too weak when compared with selection (this condition can be made mathematically precise, of course)
2. There is a large difference in "height" between the two peaks
3. There is little recombination (which can otherwise halt the process altogether)
4. Peripheral populations are involved (the central ones not being likely candidates for the process)

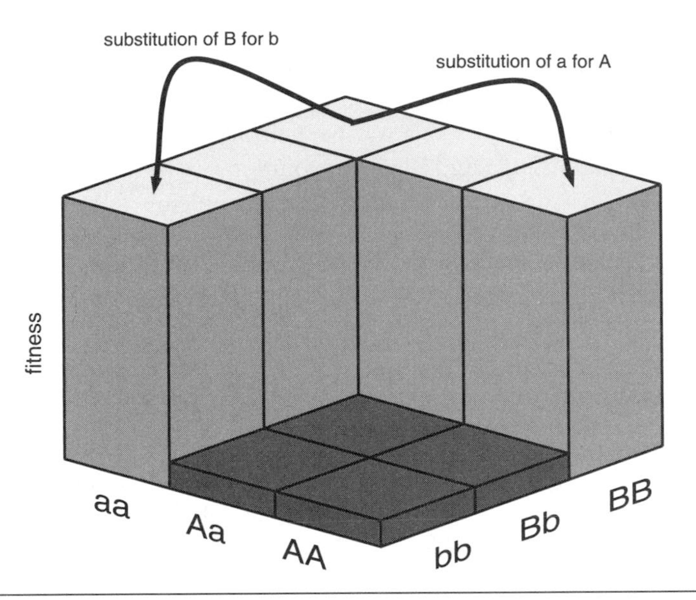

Figure 8.6. Dobzhansky's simple two-locus–two-allele model that originated the idea of "holey" fitness (i.e., genotype-level) landscapes. Notice how it is possible to move across the landscape, by allelic substitution, along a ridge of high-fitness genotypes. (From Gavrilets 1997a, 308.)

These findings, of course, do not mean that Wright's explanation of peak shifting never obtains, but they do cast a significant shadow on its general applicability.

As far as founder effect speciation events are concerned, the discussion in the literature is vehement, with no general agreement on the relevance of these phenomena (see, e.g., Rice and Hostert 1993 vs. Templeton 1996). What is most important for our purposes, however, is that Gavrilets and Hastings (1996) have provided a theoretical reassessment of founder effect speciation that makes it a likely candidate for a (dis-)solution to the peak shift problem once it is recast in terms of holey landscapes. What, then, are holey landscapes actually like, and how might evolution work through them?

The idea of a holey landscape is an elaboration of Dobzhansky's (1937) relatively simple two-locus–two-allele model, in which certain combinations of genotypes (notice the shift to the lower level of analysis: populations here are clouds of points, not individual points graphed as means!) have high fitness and others have very low fitness (the latter correspond to the "holes" in the "holey" landscape). Figure 8.6 illustrates the model, in which five of the nine genotypes that originate from all possible combinations of the two alleles at both loci have high fitness. Notice that there is a continuous path—a "ridge"—on the landscape that connects high-

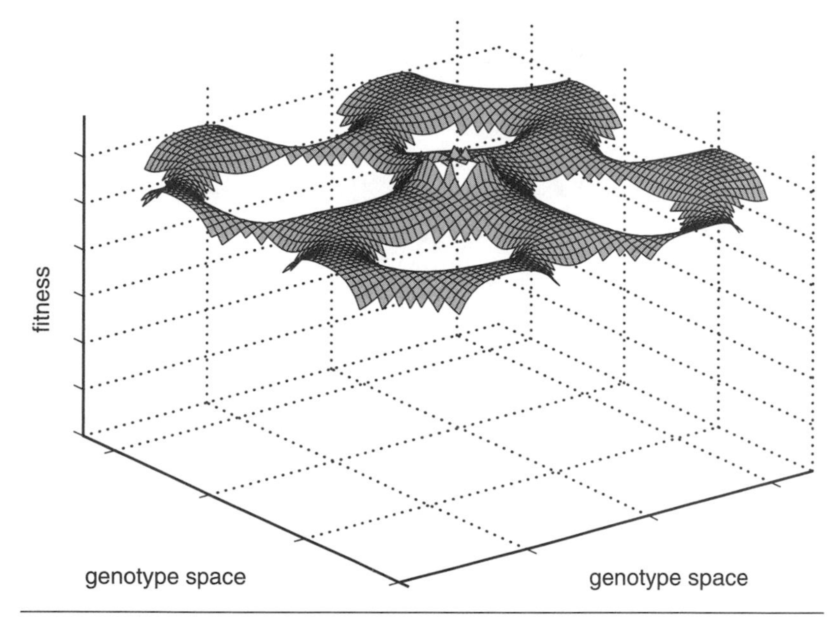

Figure 8.7. A low-dimensional representation of a "holey" landscape—one in which high-fitness genotypes form a large, connected network. The existence of a large network connecting most high-fitness genotypes depends on the probability that a (random) genotype is viable, the degree to which genotypic fitnesses are correlated, and the dimensionality of the landscape; hence, the "shape" of the landscape is determined by the parameters built into the model. Notice the low-fitness areas (holes) and, in particular, the fact that high-fitness genotypes are connected by continuous paths; when translated to the population level, this makes it possible to cross the landscape without having to "jump" from one peak to another or to cross an adaptive valley. (From Gavrilets 1997a, 309.)

fitness genotypes by one-step changes (allelic substitutions), a crucial characteristic of holey landscapes and an important distinction from the Kauffman-like models discussed above. Dobzhansky's idea was to represent hybridization and the fact that different combinations of the recombining genes could have dramatically different levels of fitness; in other words, Dobzhansky wanted to show how it was possible to find combinations of the recombining genes, after hybridization, that have dramatically different levels of fitness.

Gavrilets's holey landscapes are a multidimensional generalization of Dobzhansky's model, and analyzing their properties requires sophisticated mathematical tools borrowed from percolation theory in physics. However, Gavrilets suggested that a three-dimensional projection (fig. 8.7) could help us understand the basic concept.[4] To begin with, the "real"

4. We hope the reader will not miss the irony that one needs a low-dimensional picture of a multidimensional space in order to grasp some of its properties, even though one of Gavrilets's major points—and indeed a point of this very chapter—is precisely that high-dimensional landscapes have very different properties from low-dimensional ones.

landscape, of course, does not have only zero/one fitness levels (for more on what one might mean by the "real" landscape, see below). However, Gavrilets (1997a) argued that it is important to have a mathematically tractable version of Dobzhansky's intuition that there are combinations of genotypes that are clearly fit to a particular environment (and would therefore be favored by selection) and others that are clearly not (and would therefore be selected against).[5] The remarkable feature of the landscape is the presence of large, continuous areas (of very high dimensionality) connecting high-fitness genotypes to one another: these "ridges" (or, better put, paths in N dimensions) of high fitness allow smooth evolution in genotype space. At both the level of populations (adaptive landscapes) and the level of individual genotypes (fitness landscapes), this model eliminates the problem of adaptive valleys, therefore solving—or rather dis-solving—in one fell swoop the problem of peak shifts that has vexed evolutionary biologists for half a century. Simply put, in a high-dimensional landscape, there are no peaks that need to be shifted between.

Things are not quite so easy, of course, because there are several assumptions that go into Gavrilets's models, as well as limitations that emerge from them. For example, these landscapes are generated at random, which is mostly a reflection of our empirical ignorance of what fitness landscapes *really* look like. It is not clear how and when this problem will be overcome. In Gavrilets's early models, fitness values were generated independently, which means that these are "uncorrelated" landscapes, a problem we have already encountered with some of Kauffman and Levin's work discussed above. More recently, Gavrilets has shown that the main results regarding networks of high fitness hold for several types of correlated landscapes (given high enough dimensionality); these results, however, are based on computational simulations rather than analytical solutions, so it is not known how general they are (Gavrilets 2004). Most important, it is not known whether or not these results hold for many or most biologically realistic kinds of correlated landscapes (in part because it is not known what kinds of correlations would be biologically realistic).

Another important aspect of holey landscapes is that they are characterized by two qualitatively different kinds of behavior, depending on a crucial parameter in the models. The parameter, referred to as p, is the probability of a genotype having "high fitness" on the landscape of inter-

5. In addition, Gavrilets (1997a) cites work (e.g., Gavrilets and Hastings 1995) demonstrating that for multi-locus systems like those relevant in the case of holey landscapes, genetic drift "smooths over" fitness differences when they are small, accentuating the distinction between the more extreme classes of "high" and "low" fitness genotypes.

est (in models that have only two values for fitness, high fitness corresponds to "viable" and low fitness to "not viable"). If *p* is low (below what is referred to as the "percolation threshold"), then there are many disconnected high-fitness paths of small size across the landscape; in other words, high-fitness (viable) genotypes may turn out to be isolated on "islands" that cannot be reached from many departure points in the landscape. However, if *p* is greater than the percolation threshold, then the landscape is characterized by paths connecting most of the extant high-fitness genotypes, so that it is increasingly easy to move via high-fitness one-step variants across the holey landscape. Gavrilets notes that the percolation threshold decreases rapidly with increases in the dimensionality of the system of networks considered; given the very high dimensionality of the genotype space considered, he argues that we should expect the percolation threshold to be very small in biologically relevant cases (2004). However, the value of *p* remains, at best, a difficult empirical matter to determine, and further, probably varies with the particulars of the kind of genotype considered (as will the dimensionality, of course); how real values of *p* compare to the percolation threshold, then, is still unknown.

The theory of evolution on holey landscapes is tightly connected with the problem of speciation, which was one of Dobzhansky's (and Gavrilets's) original concerns, although it has played a relatively secondary role in discussions of Wright-type landscapes, where the focus has been on adaptive evolution (recall that adaptive evolution and speciation are not necessarily coupled). For one thing, by following a ridge of high fitness within the high-dimensional space of a holey landscape, populations (as clouds of genotypes) can eventually find themselves on the opposite sides of a hole of low fitness, which means that their hybrids will be inviable; the low fitness of hybrids may in turn promote the spread of mechanisms of reproductive isolation at the population level. By almost any definition of species (see chapter 9 for discussion), such populations would belong to different species. Given a landscape with a particular structure, it is possible to estimate the maximum number of species that can coexist. Indeed, the claim is that what makes speciation possible to begin with is the fact that not all gene combinations are viable; in fact, it is usually thought that *p*, the probability of a random genotype being viable, is fairly low. It is the "holes" in holey landscapes, the large areas that are not part of the connected network of high-fitness "ridges," that drive the dynamics of macroevolution (Gavrilets 1997a, 309).

Perhaps the most crucial and innovative aspects of work on holey landscapes are the ways in which they have been connected to the more traditional representation of fitness landscapes hypothesized by Wright (see

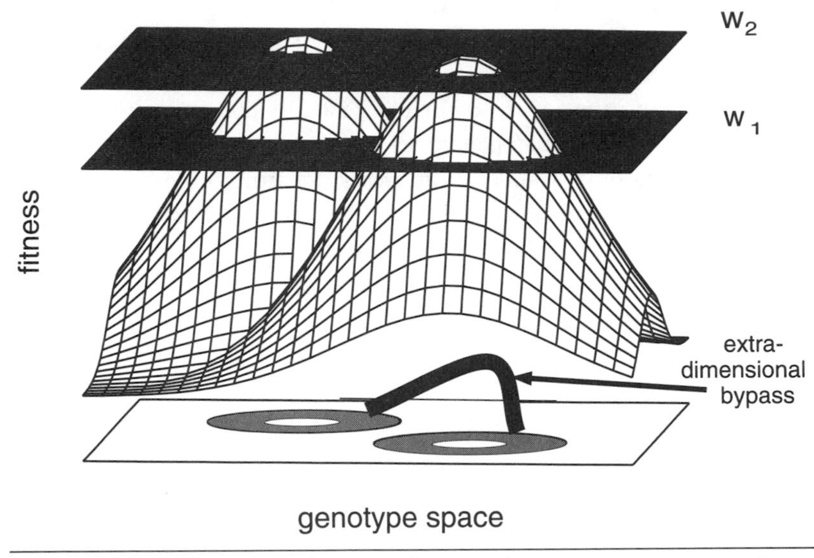

Figure 8.8. How holey (Dobzhansky-type) and rugged (Wright-type) *fitness* landscapes are conceptually connected. Microevolution can be thought of as a process of populations (as *clouds* of points, not as means, as in adaptive landscapes) climbing on Wright-type landscapes (which can bring the population to a connected network of high fitness—that is, to a holey landscape), while macroevolution is a matter of sliding across holey landscapes (with populations "moving" by allelic substitution on the network of high fitness). Notice the possibility of "extra-dimensional" bypasses connecting holey landscapes. (From Gavrilets 1997a, 311.)

Wagner, Wagner, and Similion 1994). Figure 8.8 depicts Gavrilets's suggestion: two rugged fitness landscapes provide the background for microevolutionary change of the classic hill-climbing type. However, macroevolutionary change (and therefore speciation) occurs because of movement (by allelic substitution) across holey landscapes.[6] The two peaks—two points of high fitness—may appear disjointed in low-dimensional representations, but, assuming that p is above the percolation threshold, they are very likely to be connected in high dimensions. It is this high-dimensional connection—part of the nearly neutral network of high fitness that connects all (or most) high-fitness genotypes on such landscapes—that the "extra-dimensional" bypass in the figure is supposed to represent. This concept is important, because it dramatically enlarges our picture of evolution on adaptive landscapes, while at the same time altering the very meaning of the metaphor. It is crucial to clarify what this means for how biologists may come to view the whole microevolution-macroevolution debate: what is emerging here is a view of microevolution

6. For the connection between microevolution and macroevolution in terms of landscapes, see also Arnold, Pfrender, and Jones 2001.

as primarily a matter of local maxima being explored via allelic substitutions on a network in which non-neutral, fitness-increasing, one-step moves are generally available (with respect to fitness). Macroevolution, on the other hand, is the result of higher-dimensional "wandering" on areas of generally equal fitness within a large connected network of high fitness, surrounded by vast areas of very low fitness (the "holes" on the landscape).

Going from low to high dimensionality, then, is not just a matter of "more of the same," but rather introduces radically new population dynamics. An appreciation of those dynamics makes old problems (such as peak shifts) disappear, because—again—in high-dimensional landscapes there are no peaks or valleys, but rather extensive networks of quasi-equal fitness ridges. It is also important to realize that if the connection between Wright-type and holey landscapes is a reflection of biological reality, then microevolution and macroevolution truly are (usually) different kinds of processes, characterized by very different dynamics (as argued at exhausting length, but from a very different perspective, by Gould [2002]). This conclusion is contrary to the standard view of the modern synthesis, and it vindicates several attempts at providing an expanded or alternative synthesis that have been published on and off through the second half of the twentieth and the very beginning of the twenty-first centuries (e.g., Goldschmidt 1940; Lewontin 1983; Rollo 1995; Schlichting and Pigliucci 1998; West-Eberhard 2003; Müller and Newman 2003; Jablonka and Lamb 2005).

FROM THEORY TO EVIDENCE: CAN IDEAS ABOUT LANDSCAPES BE TESTED?

This is all very appealing from a theoretical perspective, but what about the empirical evidence? Is it even possible to find empirical confirmation of the existence of holey (as opposed to other kinds of) landscapes? Intriguingly, Gavrilets (1999, 2) makes the philosophically interesting point that—contrary to the general belief among his colleagues—in evolutionary biology, the testability of the predictions of a model is not necessarily its main attraction. Gavrilets suggests that many models are not testable, either because it is impossible to get precise enough models to derive significant quantitative predictions, or because it is not feasible to implement experimental conditions that would meaningfully test the model (Gavrilets 1999, 2). In this sense, according to Gavrilets, models (such as his holey landscapes) are more useful as *metaphors;* that is, as tools to help us frame our thinking on certain problems and perhaps to move empirical research in a general direction.

Nevertheless, Gavrilets (1997a) does attempt to list some possible examples of empirical tests of the holey landscape model (see also 2004, 95–100). One important thing to realize here is that if holey landscapes adequately represent a general feature of the biological world, then it should be possible to observe cases in which high-fitness, reproductively isolated genotypes are connected by a continuous series of intermediate forms or genotypes, also of high fitness. In other words, since speciation generally occurs by evolution going around the hole (rather than jumping across it), there will probably be cases in which the "intermediate" genotypes (and the populations that contain them) have not gone extinct, but continue to exist along the high-dimensional ridge of high fitness. Possible cases of this sort mentioned by Gavrilets include hybrid inviability governed by epistatic interactions among genes (Dobzhansky's original case); certain kinds of hybrid zones between closely related species; ring species (i.e., series of populations in which neighboring ones are reproductively compatible, but those at the extremes are not); at least one artificial selection experiment in which the genetic makeup of a population was dramatically altered without affecting its mean fitness; studies of RNA sequences and their corresponding folding structures; and the observation of intermediates between two distinct forms in the fossil record.

Of these possible examples, RNA sequences are perhaps the easiest to study from an experimental standpoint, since they constitute a relatively simple system in which one can obtain enough information to characterize in detail the structure of the landscape on which these sequences evolve (after all, the genotype-phenotype mapping here is rather direct, and the range of relevant environmental conditions is limited). Indeed, even the theoretical literature in this area has developed in parallel with the main evolutionary literature on fitness landscapes, and many of the same ideas have emerged in a context in which empirical verification is much more accessible (see, e.g., Schuster et al. 1994; Fontana and Schuster 1998a, 1998b; Reidys, Stadler, and Schuster 1997; Stadler et al. 2001).

As for the other cases listed by Gavrilets, ring species are a classic example of incipient speciation in the neo-Darwinian literature, and several cases have been uncovered in recent years (see Gavrilets 1997a for references). The prediction about hybrid zones is interesting because it discriminates fairly well between rugged and holey landscapes: In the former case, one would expect two fitness peaks (genotypes of the two parental species) and a large valley of inviable intermediates. In the second case, there should be several hybrids with fitness comparable to that of the parentals, all occupying the same ridge of a holey landscape. Of all the examples mentioned by Gavrilets, the least promising kind seems to be evidence from the fossil record, given how fragmentary that evidence typically is as well as the obvious impossibility of getting genetic information from the samples.

Gavrilets acknowledges that there are other major limitations in the work done so far on holey landscapes, some of which would be particularly bothersome to ecologically minded evolutionary biologists. For one thing, there is no ecology in the models (other than what is subsumed in the fact that, presumably, it is the ecology of a species that determines whether a given genotype is associated with a phenotype that has, on average, a high or low fitness). One consequence of this failure to incorporate ecological considerations is that these models assume a fixed landscape; they are therefore inadequate for modeling coevolution. As we have seen, these are the very same things that worried Lewontin (1978) about all the fuss over Wright-style landscapes, and despite all the conceptual progress outlined so far, it seems that in some respects we have not moved very far from Wright's 1932 approach. Indeed, an argument can be made that this is why Ronald Fisher (1930) stayed out of the whole business of producing a dynamic theory of evolution, and incidentally hated the very idea of adaptive landscapes. It is to this aspect of the Fisher-Wright controversy, and some alternative ways of thinking about landscapes and peak shifts, that we now turn.

FISHER VERSUS WRIGHT, AGAIN

Most evolutionary biologists have at one point or another sat through a version of the famous Fisher-Wright debate about the nature of evolutionary change. The standard version of the debate focuses on a small number of recurrent points (Wade and Goodnight 1998; see, in particular, their table 1): Fisher envisioned evolution as happening mostly in large populations in response to natural selection, while Wright emphasized small populations and the role of random genetic drift; Fisher considered phenotypes to be underlain by the action of a large number of genes, each with "additive" (i.e., largely independent) small effects, while Wright saw pleiotropy (the simultaneous effect of a gene on many traits) and nonadditive epistasis (the effects of genes on one another) as rampant and of fundamental importance. The consensus has swung back and forth several times over the past few decades, leaving an increasing number of people simply confused as to what the controversy is about and, especially, how it translates into modern terms: remember that both Wright and Fisher did their work well before the modern molecular revolution, not to speak of the advancements of "evo-devo" (i.e., the evolutionarily informed study of development: Wilkins 2002) and the explosion of studies on genotype-environment interactions (Pigliucci 2001).

Much of this discussion hinges on the interpretation of Fisher's famous "fundamental theorem of natural selection" (FTNS) and how it describes the process of evolution (see Crow 2002). Frank and Slatkin (1992) have argued that the misunderstanding about the controversy stems at least in part from some common misperceptions about Fisher's theorem. Let us briefly review Frank and Slatkin's arguments, as they are directly related to the limits of the adaptive landscape metaphor in evolutionary biology.

The FTNS was stated by Fisher (1930, 37) as follows: "The rate of increase in fitness of any organism at any time is equal to its genetic variance in fitness at that time." This statement was taken by Wright (see references in Frank and Slatkin 1992) to support his own model of evolution on adaptive landscapes: according to Wright, and following the landscape metaphor, the rate of change in the allelic frequencies of a population depends on how steep the gradient of the landscape is. This gradient is measured as the change in mean fitness over the change in allelic frequency. But Fisher rejected the idea of adaptive landscapes and claimed repeatedly that Wright had misinterpreted the FTNS. Indeed, as pointed out by Frank and Slatkin (1992), Fisher (1930, 41–45) argued that the average fitness of a population must fluctuate around zero: if it were positive, the species would overwhelm the planet, and if it were negative, it would go extinct.[7]

Work by Price and Ewens, summarized by Frank and Slatkin (1992), makes clear where the misunderstanding lies: Fisher thought that the total change in fitness in a population depends on two broadly defined components. The first component is the action of natural selection against a given environmental background; the second component accounts for changes in the environment of the population (which, for Fisher, included not just the physical environment, but the changing genetic makeup of the population, and hence pleiotropy, epistasis, and so forth). The common interpretation of the FTNS confines itself to the first component, apparently ignoring Fisher's contention that the second component is usually *negative* because the environment deteriorates. Adding a negative second component to the equation describing change in the average fit-

7. It seems to us, however, that Fisher's and Wright's interpretations here are not necessarily incompatible: the first one can be taken to describe a general state of quasi-equilibrium of populations, while the latter may be useful to describe temporary positive or negative growth of populations. After all, new species are likely to experience periods of very rapid growth, and of course species do go extinct from time to time!

ness of a population means that such change tends to be close to zero, on average, just as Fisher maintained, and contrary to Wright's landscape-driven interpretation. To put it in other terms, and in a fashion that is close to Lewontin's (1978) own objection to the landscape metaphor, Wright's theory assumes a constant landscape, and therefore focuses on only half of the game. It may correctly describe what would happen if the environment (biotic, physical, and genetic) were not deteriorating or otherwise changing. But the environment changes constantly, leaving populations condemned to eternally catch up with an ever-shifting landscape (see the discussion of the "Red Queen" effect above; Van Valen 1973). Those that fail to catch up go extinct.

According to Frank and Slatkin (1992), Price and Ewens were "disappointed" when they realized what Fisher actually meant. The disappointment comes from the fact that it is not possible—as Fisher acknowledged—to produce a mathematical theory of the total variation in fitness (i.e., including both components) because the dynamics of environmental deterioration are too complex and local. Frank and Slatkin (1992), however, still think that Fisher's partition of the fitness change into two components is very useful as a conceptual tool for evolutionary biologists, and we agree. Their example is the study of clutch size in birds. Without entering into too many details, researchers often find a "paradox" when studying clutch size: they frequently observe additive genetic variance for clutch size, which means that—according to the standard version of the FTNS—the trait should change in response to natural selection. One should then observe larger clutch sizes until the genetic variation is exhausted. But such changes are rarely observed. The solution lies in the second part of Fisher's equation: an increase in clutch size brings a deterioration of the environment of the chicks in the form of increased competition for resources. This second component eventually balances the first one, neatly explaining why one observes additive genetic variance and apparent selection for a trait, and yet the trait itself does not change in response to natural selection (see Seger and Stubblefield 1996 for discussion). Frank and Slatkin (1992) remind us of the opening words of Fisher's 1930 book: "Natural selection is not evolution." But if one thinks in terms of Wrightian landscapes, one may be mistakenly led to believe that it is.

OTHER SOLUTIONS TO THE "PEAK SHIFT" PROBLEM

We have already noted the persistent interest that evolutionary biologists have taken in the problem of "peak shifts." The difficulty of figuring out ways for populations to go from one adaptive peak to another across a valley of low fitness is unavoidable if one adopts the Wrightian adaptive landscape metaphor. However, there are more creative solutions in the literature than the standard drift-based answer linked to Wright's shifting balance or to the founder effect. Here we briefly discuss one of those alternatives, suggested by Kirkpatrick (1982), because it again challenges the classic conception of adaptive landscapes and forces us to rethink an old problem from a different perspective.

Essentially, Kirkpatrick noticed what, in hindsight, may seem to be the obvious alternative: instead of the population having somehow to "jump" from one peak to another, something can alter the phenotypic variance of the population itself; that is, it can enlarge the cloud of points on the (phenotypic fitness) landscape. If the cloud becomes spread out enough that its border touches an area of the landscape close to the new peak, then natural selection will result in the population being shifted up the new peak in the standard way (given certain assumptions about heritability). A peak shift, then, can occur under classic Fisherian conditions of large populations and mass selection, without requiring the Wrightian scenario of genetic drift in small populations. But what is this "something" that enlarges the phenotypic variance of the population? Kirkpatrick identified two possible mechanisms: a shift in the external environment,[8] or a change internal to the population itself, such as an alteration of the mutation rate or an increase in "developmental noise" (i.e., in the phenotypic variation produced by any individual genotype). Kirkpatrick (1982) thought that a change in the mutation rate was a rather far-fetched possibility, though in fact such changes are now well documented (Maley 1997; Sniegowski, Gerrish, and Lenski 1997; see discussion in Jablonka and Lamb 2005), and an analogous outcome could be obtained by a change in the recombination rate, which is also known to evolve (Korol and Iliadi 1994; Trickett and Butlin 1994; Otto and Michalakis 1998).

Alterations in the degree of developmental noise of a trait have also been the target of much research (Freeman, Graham, and Emlen 1993; Gavrilets and Hastings 1994). Perhaps more important, as we noted in chapter 6, recent studies of systems that act to buffer the developmental

8. Recall from chapter 2 that in quantitative genetic theory the phenotypic variance in a population is the sum of a number of components, usually including at least the genetic, the environmental, and the genotype-environment interaction variance: $V_P = V_G + V_E + V_{G\times E} + V_{err}$. To make V_P (the phenotypic variance) larger, one can increase the genetic variance, V_G (which can occur by the Wrightian mechanism of genetic drift, or by mutation, or by recombination), the genotype-environment interaction variance, $V_{G\times E}$ (which can occur for a number of reasons), or the environmental variance, V_E (which can be the result of a change in the external environment). (V_{err} is the so-called error, or residual, variance, which basically includes everything else that was not explicitly accounted for during the experiment or in building the model.)

process (such as the system of which the heat shock response protein Hsp90 is a part) have suggested that the buffering process permits the buildup of "hidden" variation in developmental resources, including both genetic and epigenetic inheritance systems (Rutherford and Lindquist 1998; Queitsch, Sangster, and Lindquist 2002). Disruptions in these systems (caused by things such as exposure to novel environmental conditions) can then release the previously buffered variation, radically increasing the range of phenotypic variation in the population. This newly released phenotypic variation may then become available to selection, which can stabilize new variants of traits such that they remain after the disrupted buffering systems become functional again (this is true of variation attributable to both genetic and epigenetic mechanisms; see Queitsch, Sangster, and Lindquist 2002). In this way, buffering systems can act as "capacitors" of phenotypic evolution (but very likely do not have the *function* of doing so; see chapter 6).

It is interesting to note that changes in genetic or phenotypic variance are probably not independent of the other category of causes underlying adaptive peak shifts identified by Kirkpatrick (1982): changes in the external environment. The studies mentioned above point out that it is often a change in the environment that triggers increased developmental noise, levels of recombination, or mutation rates. Such changes can be rapid, as Kirkpatrick pointed out in the context of his discussion on punctuated equilibria, and as Pigliucci and Murren (2003) have emphasized in reference to the role of phenotypic plasticity in facilitating peak shifts to novel environmental optima (fig. 8.9).

It seems that one of the biggest challenges of future research in this area will be to somehow bring together all these threads to derive a coherent theory of the mechanisms by which populations can evolve in the face of ever-changing environmental conditions. Will such a theory necessarily have to make use of the metaphor of adaptive landscapes? If so, in which form will landscapes be used? The rugged, Wright-type ones, the holey, Dobzhansky-type ones, or some combination of both?

DO ADAPTIVE LANDSCAPES EXIST?

Are adaptive landscapes real? In the more philosophical sense of this question, one might be wondering about the ontological status of scientific models and laws (see Cartwright 1983). Without getting into those issues, we note merely that the goal of the adaptive landscape metaphor is to model particular aspects of evolutionary change. Wright (1932a) originally introduced the idea as a way to help us think about evolution, natural se-

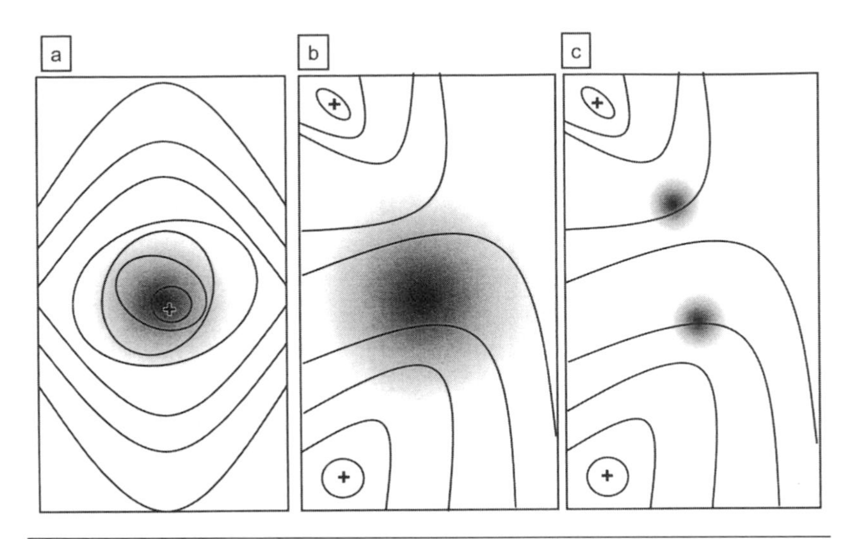

Figure 8.9. An example of "peak shifting" by environmental change (on a phenotypic fitness landscape). In the scenario envisioned here, an environmental disruption results in the expression of previously suppressed genetic variation, resulting in the exposure of novel phenotypes to the new environment. While most fail, some novel phenotypes are stabilized by selection. Note that as in figure 8.5, the axes of these adaptive landscapes must be interpreted as phenotypic rather than genetic variation. (a) A population near equilibrium around a peak. (b) A radical environmental shift and an attendant breakdown in buffering; note the increase in phenotypic variance. (c) Some of the new phenotypes are stabilized by selection, but most are eliminated.

lection, and genetic drift, among other things; making particular kinds of processes more perspicuous is the role of many kinds of models in science. Wright proposed the metaphor because the underlying math was abstruse. While the metaphor has continued to change in a variety of ways over time, it has remained part of modern evolutionary thought and parlance.

But fitness and adaptive landscapes are certainly not a necessary tool in evolutionary theory in general, as evidenced by the fact that plenty of researchers before and after Wright have not made any particular use of them. Furthermore, there may be good reasons to go with Fisher's antipathy for the whole concept and reject the metaphor of the landscape as too deeply misleading to be of value. So, in a less philosophical vein, we might ask whether landscapes model any aspect of evolutionary change with enough fidelity to be of value *as a model.* Indeed, even if one wishes to adopt the metaphor, one is then immediately presented with very different kinds of landscapes (rugged, holey, uncorrelated, correlated, stable, shifting—and, of course, the crucial distinction between fitness and adaptive landscapes), each characterized by peculiar properties, each channeling our thought in certain directions rather than others. So it seems likely that if one particular kind of landscape models certain aspects of evolutionary change well, the other landscape models will not do as good a job with those same aspects.

The claim that fitness or adaptive landscapes may well reflect real features of the world and accurately model some aspects of the world of interest to theoretical biologists is supported by those who suggest that it may be possible, and indeed valuable, to find ways to test particular versions of landscapes against those aspects of the world they are supposed to model (some such methods, proposed by Gavrilets 1997a and 2004, are discussed above). But as yet we are some distance from being able to carry out any such test in a sufficiently robust way, and even Gavrilets (1999) readily admitted that the usefulness of models in evolutionary biology may lie more in the general framework they provide than in their actual empirical predictions.

Could we have developed a theory of adaptive evolution and speciation without using the landscape metaphor? The answer is probably positive. Such a theory might have originated as an elaboration of Fisher's initial intuitions, if Wright's visual rendition of his equations had not so rapidly caught the imagination of major players in the modern synthesis. Ironically, Wright himself may not have introduced the metaphor if the average level of understanding of math had been a bit higher among the leading evolutionary biologists of the time (Ruse 1996). Nevertheless, the use of the landscape metaphor in evolutionary biology has been both fertile and misleading in important ways. It has been fertile because it has generated a significant number of research efforts and papers, one of the few quantitative indices of scientific success. But it has also been misleading, because if Gavrilets, and indeed Fisher, are right, then a lot of people have put a lot of time into the solution of problems (such as peak shifts) that are not really problems at all, or at least are not problems of the sort they are usually thought to be.

We have also seen in this chapter that the whole discussion of peak shifting and evolution on fitness and adaptive landscapes often becomes entangled with the conceptually distinct issue of speciation and the origin of species more generally; the landscape metaphor makes these two problems appear more related than they in fact may be. Which, if any, version of the landscape metaphor one finds attractive will have implications for the way one thinks about speciation and the kinds of things that make a population count (or not) as a separate species from other populations. But what are these things the origin of which we wish to know? In the next chapter, we turn to the question of what species are, and what kinds of characteristics make two organisms members of the same or of different species. We present a philosophical argument that the so-called species problem actually disappears once one unpacks the concept of species, very much in the same fashion in which peak shifting disappears (or is at least dramatically reformulated) as a research problem once one unpacks the landscape metaphor.

9 Species as Family Resemblance Concepts

The (Dis-)Solution of the Species Problem?

WHAT IS YOUR AIM IN PHILOSOPHY?—TO SHOW THE FLY THE WAY OUT OF THE FLY-BOTTLE.
—LUDWIG WITTGENSTEIN, *PHILOSOPHICAL INVESTIGATIONS*

The so-called species problem—the problem of agreeing on a definition of "species" that is both clear and useful—has plagued evolutionary biology ever since Darwin's publication of *On the Origin of Species.* Many biologists think the problem is just a matter of semantics; others complain that it will not be solved until we have more empirical data. In this chapter, we briefly examine the main themes of the biological and philosophical literatures on the species problem, focusing on identifying common threads as well as relevant differences. We then argue for two fundamental claims. First, we argue that the species problem is not a scientific one, but rather a philosophical question that cannot be settled by empirical evidence, though any solution will certainly have to be informed by data from the real world. Second, we argue that the solution lies in adopting Wittgenstein's idea of "family resemblance" or cluster concepts and considering "species" as an example of such a concept. This solution has several attractive features, including its bringing together of apparently divergent themes of discussion among biologists and philosophers. While there is no definition of "species" that permits one to find a set of necessary and sufficient conditions for two organisms to belong to the same species, we can choose to focus on particular aspects of the species question depending on our aims as researchers; this possibility reveals what is valuable about pluralistic approaches to the species problem.

THE PROBLEM THAT NEVER GOES AWAY

Settling on a definition of "species" that is both unambiguous and biologically useful is a challenge that has been around since Darwin's publication of the not so aptly titled[1] *On the Origin of Species* in 1859. The so-called

1. Some people have suggested that while *On the Origin of Species* provides an excellent explanation of *adaptation,* it fails to provide an adequate account of speciation and diversity more gen-

species problem is unlikely ever to go away, at least in part because biologists have a schizophrenic attitude toward the question. On the one hand, they tend to reach for their guns when species concepts are brought up by colleagues, are the subject of papers, or are discussed at conferences. On the other hand, they simply cannot resist offering graduate seminars on the topic or avidly reading anything that is published on the subject. In the span of just two years, two of the major journals of evolutionary biology have devoted several papers in special issues to the ever-burning question of exactly what species are (*Journal of Evolutionary Biology,* vol. 14, 2001, 889ff; *Trends in Ecology and Evolution,* vol. 16, 2002, 326ff; see also the recent book by Coyne and Orr [2004] to be briefly discussed below).

In this chapter, we make two major related claims. First, we argue that the reason why the species problem has not gone away is that it is not an empirical problem (contrary to, e.g., Hey 2001a), but rather a philosophical one. Indeed, the philosophical literature on the definition of species is as extensive as the biological one, with some biologists contributing to both (see, e.g., van Valen 1988; Ridley 1989; de Queiroz 1992; Mayr 1996). This does not mean that empirical information is not relevant here, but rather that the problem represents a paradigmatic example of a philosophical question that requires empirical information (provided by science) to be settled, not of a scientific problem with unwelcome philosophical characteristics. Second, we argue that the problem does in fact have a satisfying philosophical "solution" based on Wittgenstein's idea of "family resemblance"[2] or cluster concepts (1958, 65–69), as was realized early on by Hull (1965). While we can lay no claim to presenting a novel solution to the problem, we hope that by casting it in more modern terms we can offer contemporary biologists reason enough to be satisfied with a rather unusual philosophical solution to what has often been seen as a scientific problem.

Before moving on to the main body of the chapter, it is perhaps worth taking a moment to reflect on why we are bothering to spill any more ink on this issue. Part of the answer should be obvious to any biologist: species are generally considered to be the fundamental level of organization of the biological world (though this too is hotly disputed; see box 9.1),

erally; see below (and chapter 8) for more on decoupling the explanation for adaptation from the explanation of biological diversity/disparity.

2. It is of course interesting that the term "family resemblance," as Wittgenstein developed and used it, refers to a fundamental biological fact: members of a group (for example, a human family of kin) can be clearly more similar to one another than they are to members of any other group, and yet there may not be any characteristic that is either necessary or sufficient (let alone both) for an individual to be recognized as a member of that group.

Box 9.1 Adaptation, Speciation, and Divergence

It is often claimed that species are the fundamental category in the biological world—that they are the "units of evolution" (the title of a 1991 book edited by Ereshefsky) and the "units of biodiversity" (the title of a 1997 book edited by Claridge, Dawah, and Wilson). Further, species are often thought to be the primary units in which adaptation takes place; while it is individual organisms that end up possessing (most) adaptations, it is the species that becomes adapted through evolution (individual organisms, of course, do not evolve in this sense—but, the idea goes, species do). If, as we argue in this chapter, species are not best thought of as the kind of thing that can always be unequivocally picked out, where does this leave the idea that species are the fundamental way in which the biological world is organized?

We suggest that a focus on species as *the* fundamental unit of organization is misplaced. Populations both smaller and larger than species can be units of evolution, biodiversity, and adaptation (see our discussion of de Queiroz's proposal of species as evolving metapopulation lineages in the main body of the chapter). Indeed, it is unwarranted to think that adaptation, diversification, and evolution more generally are closely related phenomena that take place via the same mechanisms in the same populations; it should be no surprise, then, that attention to different aspects of population structure, and to different populations, will be important in dealing with these different phenomena.

For example, adaptation can, and verifiably does, take place without speciation, as does nonadaptive evolution more generally. But more to the point, a population can become adapted to local conditions and yet remain phylogenetically (reproductively) part of a larger population not so adapted. There can be extensive gene flow between ecotypes (i.e., locally adapted populations) as long as the fitness differences between the populations are sufficient to maintain the genetic (or other heritable) differences between those populations. So adaptation can, and verifiably does, take place in groups below the species level. For similar reasons, if we take evolution to be heritable changes in a population, then species cannot be the fundamental units of evolution, either, because populations smaller than what are usually picked out as species can have local heritable differences from the population at large. There is, however, a great deal of terminological confusion in the literature regarding such populations. As noted above, ecotypes are usually defined as populations that differ genetically and are adapted to local conditions, whether or not there is gene flow between the different populations. Similarly, varieties of a type of organism may or may not be reproductively isolated from one another, but also may or may not be different because they are adapted to particular local conditions. Incipient species are populations that are usually thought to be on their way to becoming reproductively isolated (or are already partially isolated), but the populations may or may not be different for adaptive reasons. The term "subspecies" seems to be able to refer to any of the above.

A better strategy than arguing over these terms, it seems to us, is to carefully state what biological features of a population one is interested in and how that population reflects those features. Indeed, in an earlier article (Kaplan and Pigliucci 2003), we suggested that the human species is made up of multiple, overlapping ecotypes, despite the relative lack of overall genetic diversity and the relatively high degree of gene flow between popu-

(*continued*)

Box 9.1 (*continued*)

lations; the perennially volatile question of whether there are biological human "races," then, comes down to what one thinks a "race" is. If biological races are incipient species, then there are not now, nor were there probably ever, biological races of humans; if "races" is just another name for ecotypes, then there are lots of biological races of humans, though they do not, for the most part, correspond to our "folk" racial categories.

When thinking about the units of biodiversity, it matters whether one is interested in species diversity just for the sake of species diversity itself, or whether one is interested in biological diversity more generally. If the former, then of course species will be the fundamental units of biodiversity (by definition), but we can think of no good reason that one should care about species qua species (however they are defined). To take an example from the main body of the chapter, the fact that the plants *Phryma leptostachya* var. *leptostachya* (native to eastern North America) and *P. leptostachya* var. *asiatica* (native to eastern Asia) are phylogenetically very distant from each other and show extensive molecular divergence in (apparently) noncoding regions of their DNA seems, to us, to be of little interest from the standpoint of biodiversity; far more important is the fact that the two plant varieties are all but indistinguishable in practice. And as Sterelny and Griffiths note, the loss of the kakapo (a large, flightless, very endangered parrot with a unique morphology and lifestyle) would seem to many people to represent a loss of biodiversity that the loss of the Chatham Island black robin (a rather endangered but otherwise fairly dull robin, remarkable only for being not quite as good a flyer as other robins) would not, however distant these birds are from their nearest living relatives! Gould's controversial claim that biological disparity peaked in the period immediately following the Cambrian explosion is relevant here (see Gould 2002); whatever one thinks of the status of Gould's claims regarding that period, the intuitive idea that a diversity of forms with very slight variations is different from a diversity of truly disparate forms is surely important.

a status that suggests that one should have a good idea of what (kind of thing) they are. Indeed, the existence and structure of species—the basic organization of the biological world—is itself an interesting biological problem. Why, in other words, are there a large number of relatively discrete species rather than, on the one hand, a continuous gradation of biological variation (see Sterelny and Griffiths 1999) or, on the other hand, a very few very plastic species (see Scheiner and DeWitt 2004)? For this question to make sense, one must have a relatively good grasp of species—what they are and how to identify them. So, while at times Darwin seemed to suggest that the mutability of populations spoke against the reality of species (e.g., 1859, 484), a view articulated in more detail and defended by a very few modern authors (see Mallet 2001), more often biologists (and philosophers) have been convinced that species are a basic feature of the biological world, and as such, are pivotal to several fields of investigation as well as applications of evolutionary biology. Researchers interested in the study of the process of speciation (obviously), evolutionary geneticists, evolutionary ecologists, systematists, and conservation biologists all deal directly with problems for which—it would seem—understanding what constitutes a species is of paramount importance in order to make progress.

Or is it? One can argue that progress in all of the above areas (including, paradoxically, our understanding of the process of speciation: see chapter 8 and, e.g., Gavrilets 2003) has actually been achieved *despite* all the discussion and lack of agreement on what species are. To paraphrase an American judge,[3] many biologists seem to agree that—as in the case of pornography—it may be impossible to define species, but it is easy enough to recognize them when you see them. Indeed, some even go so far as to suggest that it is because of a stubborn tendency to get entangled in such "semantic" (a pejorative term among many scientists) issues that evolutionary biology has not achieved the recognition as a science that, say, physics has (see, e.g., Noor 2002).

But even if this is so, understanding the debate and getting a better fix on how particular species concepts are actually used can steer attention in more, rather than less, fruitful directions. If we understand where particular species concepts are most appropriately used, and where they are less appropriate, we will be in a better position to judge the evidence put forward in making claims about the number of species of a particular genus in a particular area, or whether a purported endangered "species" is really a species at all. Even if there is no single "right" answer to what constitutes a species, getting a better sense of what the various reasonable answers are like is still a worthwhile endeavor.

THE BIOLOGICAL SIDE OF THE CONTROVERSY

It is not our intention here to provide a comprehensive review or history of either the biological or the philosophical literature on the species question. While that would no doubt be a fascinating endeavor, at this point it would require at least a book-length treatise. But if we are to argue for our preferred solution, we must at least sketch the general contours of the debate.

From the perspective of the biological literature, besides the two recent special issues of major journals mentioned above, there have been several

3. Potter Stewart, of the U.S. Supreme Court, who wrote, in the Jacobellis vs. Ohio case of 1964: "I can't define pornography, but I know it when I see it." See Woodward and Armstrong 1979, 193–200, for some of the amusing results of this view.

Table 9.1. An incomplete list of some of the major species concepts that have been proposed in the literature and the biological factors they consider to be most relevant to determining whether two populations of organisms belong to the same species

Species concept	Major focus
Biological	Reproductive characteristics; focus on genetic similarity relevant to reproduction
Cladistic	Phylogenetic relationships
Cohesion	Reproductive and ecological characteristics; genetic similarity
Ecological	Ecological characteristics
Evolutionary	Ecological characteristics and phylogenetic relationships
Genetic	Genetic similarity
Morphological	Genetic similarity as manifested in phenotypic similarity; focus on key characters
Phenetic	Genetic similarity as manifested in phenotypic similarity; all measurable traits considered
Phylogenetic (several variants)	Phylogenetic relationships
Recognition	Reproductive characteristics and genetic similarity; focus on behavioral traits

books (e.g., Otte and Endler 1989; Hey 2001b; Coyne and Orr 2004; Gavrilets 2004), and of course countless articles. A brief history of the species concept has been provided by Grant (1994); an example of how easily different concepts of species fail to mesh, even for the same group of organisms, has been provided by Gleason, Griffith, and Powell (1998); and Mayden (1997) provides a (still incomplete!) list of a whopping twenty-one species concepts. It is instructive, however, to look at the major species concepts that have emerged so far to seek common themes. These concepts are not difficult to find: table 9.1 lists what one might consider the "top ten" species concepts (in alphabetical order—we are not taking sides here), together with the emphasis they place on specific biological factors that are considered essential to the definition of species. We can see that there are, broadly speaking, three factors entering into the equation: phylogenetic relationships, genetic similarities (sometimes specifically concerned with reproductive traits, sometimes more broadly defined), and ecological similarities, broadly construed. These factors are, of course, reflections of the different interests biologists have in using species concepts, and may go far toward explaining why different practicing biologists express sometimes strong preferences for one or another species concept. To complicate matters further, it should be obvious upon reflection that the factors invoked to define species are not all-or-nothing criteria, but rather quantitative assessments that are bound to vary to some extent in real populations (fig. 9.1).

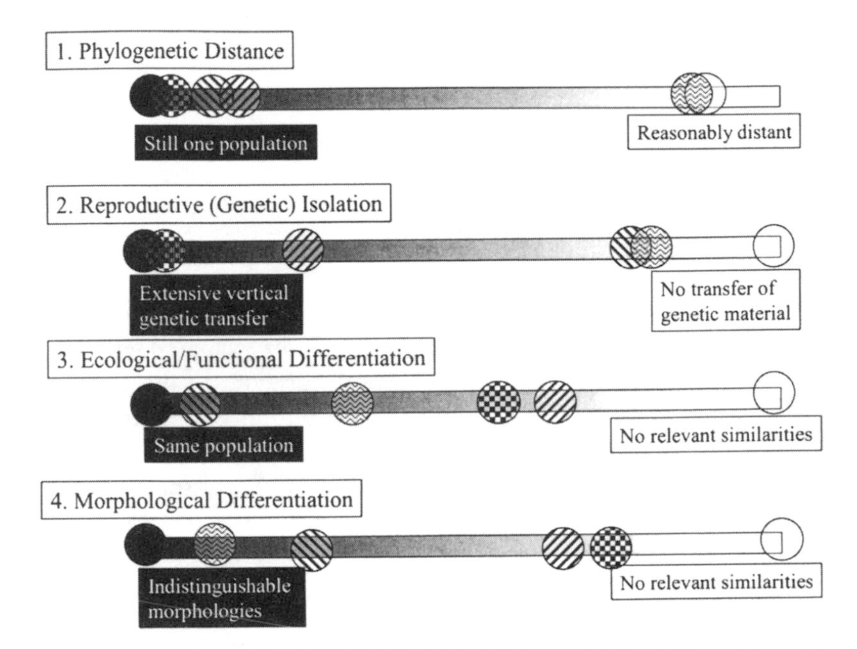

Figure 9.1. There are several criteria by which species can be identified; all admit of degrees. Different criteria are relevant for different purposes, as are different cutoff points on the various axes shown. One can think of the "species-ness" of a population as being given by its location in the multidimensional space determined by the four axes below (and perhaps others). Most "classic" species (e.g., those species of animals students in introductory courses tend to think of) would be clustered toward the origin in all four dimensions; other plausible species candidates would be close to the origin on some, but not all, of the axes. Those populations that are close to the origin on few or no axes are probably not species on any reasonable view.

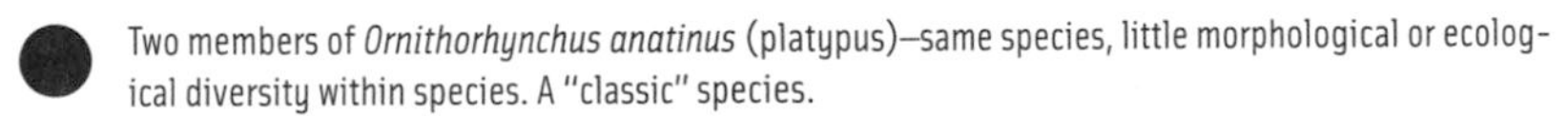

Two members of *Ornithorhynchus anatinus* (platypus)—same species, little morphological or ecological diversity within species. A "classic" species.

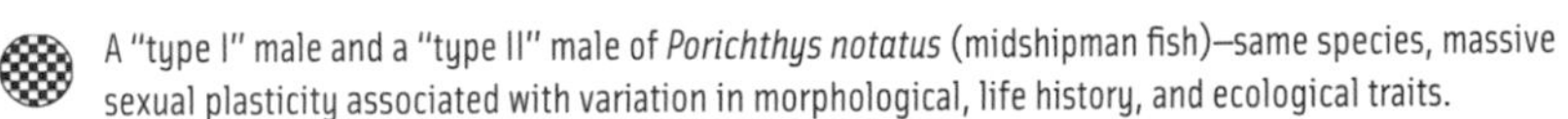

A "type I" male and a "type II" male of *Porichthys notatus* (midshipman fish)—same species, massive sexual plasticity associated with variation in morphological, life history, and ecological traits.

A member of *Homo sapiens* (modern human) and a member of *Gorilla gorilla gorilla* (lowland gorilla)—two species of primates.

Anemone quinquefolia and *Anemone rivularis*—a recent polyploidy event in a flowering plant (see table 9.4).

Drosophila mercatorum (from slopes of mountains) versus *Drosophila mercatorum* (from saddle between mountains)—significant ecological differentiation, but with significant gene flow between populations (see Templeton 1999).

Phryma leptostachya var. *leptostachya* (eastern North America) versus *P. leptostachya* var. *asiatica* (eastern Asia)—phylogenetically very distant, but morphologically and ecologically very similar flowering plants (see Wen 1999). (Some botanists, apparently focused on a phylogenetic species concept, have argued that, far from being two varieties of the same species, these plants should actually be in separate *genera*!)

Moreover, these three factors are not independent: genetics, phylogeny, and ecology are usually tightly intertwined when it comes to determining the fate of any population of organisms. And, as most biologists in practice recognize most of the same species, various interpretations of the species concept are generally trying to pick out the same entities (the same populations of organisms, that is). So it ought not to be surprising that there are more commonalities among the various species concepts than one might at first suspect. Generally (although not always), any population that is genetically distinct from other related populations will be so in part because of phylogenetic separation and/or ecological separation; similarly, populations that are phylogenetically distant or ecologically distinct will tend to have separated genetically as well. It is for these reasons that, in general, species are relatively easy to recognize (even if the concept is hard to pin down). Indeed, Templeton's "cohesion" species concept (Templeton 1989; Templeton, Maskas, and Cruzan 2000) works by capturing a compound of genetic, phylogenetic, and ecological aspects; by doing so, it may be the best proposal that has come our way so far from biologists.

THE PHILOSOPHICAL SIDE OF THE CONTROVERSY

Like the biological literature, the philosophical literature on the species problem is vast and complex, and we cannot possibly do it justice here. However, it is interesting to note the main themes, particularly as they partly overlap with the concerns of biologists (some of whom, as we noted above, have participated actively in this side of the debate as well) and are partly characteristically philosophical in nature (which is, of course, not to say they are irrelevant to science).

Like its biological counterpart, the philosophical discussion can best be understood as focusing on a small number of basic themes, which are summarized in table 9.2 (which, like table 9.1, is certainly not to be taken as an exhaustive list). First, as in the biological literature, there are discussions of species concepts proposed by biologists, mostly focusing on the "biological" species concept[4] and, to a lesser extent, the phylogenetic species concept. These are the discussions that, understandably, have seen the major contributions by biologists to the philosophical literature. Giray (1976) produced one of the early criticisms of morphological species concepts and proposed a synthesis of the different variants of the biological

4. As Theodosius Dobzhansky once aptly remarked about the so-called biological species concept, what species concept would not be biological?

Table 9.2. Some of the major themes dominating the philosophical discussions of the species problem (the list of themes and the references cited are not meant to be comprehensive; rather, they are meant to give some sense of the terrain and to provide an entry into the relevant literature)

Theme	Representative references
Discussions of species concepts proposed by biologists, mostly the biological and phylogenetic ones	Ruse 1969; Giray 1976; Mishler and Brandon 1987; Splitter 1988; Sterelny 1994; Horvatch 1997
What kind of thing are species? Are species individuals or sets? Are species natural kinds?	Giray 1976; Kitcher 1984; Hull 1978; Splitter 1988
Pluralism: Do we need more than one species concept? Does the inherent heterogeneity of purposes to which the species concept is put in the biological sciences imply that pluralism is our best option?	Kitcher 1984; Dupré 1993

species concept that had been put forth by biologists up to that point. Horvath (1997) examined the consequences of the phylogenetic species concept, while Sterelny (1994) is unusual in defending the evolutionary species concept as opposed to the two dominant ones (biological and phylogenetic). Ruse (1969) suggested early on that biologists prefer one well-defined species concept (and are therefore very unlikely to go for any of the pluralistic proposals briefly outlined below) because they would prefer to find *the* answer to the problem, and because they would like this answer to be embedded within general laws and to be derived logically from first principles of biology (an approach likely to be especially problematic for a largely historical science such as evolutionary biology: Cleland 2001, 2002).

An issue closely related to how species are defined is just what kind of "thing" species are. Mishler and Brandon (1987), for example, suggest that species are individuals, something they claim is necessary if, for example, species selection is possible (see Stanley 1975; Gould 2002).[5] Kitcher (1984), however, has argued that species are not individuals at all, but rather sets that have individual organisms as their members; if this is so, he suggests, we should expect there to be limitations on the usefulness of the biological species concept.[6] Splitter (1988) goes further and suggests

5. One might suggest that this line of reasoning has the ontological priorities reversed, and that one should investigate the possibility of species selection *after* one has convincingly made the case for species as individuals, not the other way around.

6. Kitcher argues that if species are sets, rather than individuals, they are sets whose members belong to them because of biologically interesting properties they share; since the members of many species do not reproduce in a way to which the biological species concept can apply, that concept cannot be uniquely useful for identifying those species (Kitcher 1984, 317ff).

that species are not natural kinds at all, an idea that—with an interesting twist that includes arguments concerning the human brain's physiology—has been revisited by biologist Hey (2001b).

We then come to pluralism, the contention—put forth in various fashions by authors such as Mishler and Donoghue (1982), Kitcher (1984), and Dupré (1993)—that the reason there are many species concepts is that biologists are legitimately interested in a heterogeneous group of questions, each of which requires logically independent, and yet equally valid, concepts of species. Kitcher argues that there are at least two such components to biological endeavors, which he terms the historical/evolutionary and the structural/functional inquiries. These two (equally valid) approaches to biological questions translate for Dupré into a genealogical and an ecological conception of species, neither one of which has logical priority. In other words, if one wishes to emphasize historical relationships, one must adopt a phylogenetic concept of species and speciation; however, one could also equally validly be interested in the ecology and function of organisms, in which case something along the lines of the biological or ecological species concepts would serve the purpose better.

As we noted above, these kinds of criteria will usually pick out the same entities, but they need not, and sometimes verifiably fail to do so. For example, traditionally the plants *Phryma leptostachya* var. *leptostachya* (native to eastern North America) and *P. leptostachya* var. *asiatica* (native to eastern Asia) have been classified as varieties of the same species on the basis of their morphological, ecological, and reproductive similarities; however, studies of the molecular divergence (at apparently neutral sites) between the two populations reveal that they are phylogenetically very distant, so much so that some botanists have argued that they should be classified as belonging to separate genera (Wen 1999) (see fig. 9.1). To a pluralist about species concepts, this state of affairs is unproblematic; the plants are varieties of the same species if one is interested in the ecology of the plants, but belong to different species if one is interested in the plant's genealogical relationships.

Dupré's pluralism extends to the question of what kind of thing species are, as well. His "promiscuous realism" permits him to claim that the different entities, and even different kinds of entities, picked out by various species concepts are all equally real. Hence, species (picked out by whatever criteria are most useful for a particular project) can be treated as individuals or sets, depending on the interest of the researcher. To avoid confusion, one must simply be clear about how one is identifying species (and for what reason), and what kind of thing one is treating the entity so identified as (and why).

But of course, for those who believe that there must be *an* answer (a single, unequivocal answer) to the question of whether or not two organisms belong to the same species, pluralistic approaches will be deeply unsatisfying, and so any approach that avoids the pitfalls of most monistic solutions without embracing full-blown pluralism should be welcome. For example, while Templeton's (1989) "cohesion" species concept, incorporating as it does both a genetic/phylogenetic and a structural/ecological component, has some of the advantages of the pluralistic approaches defended by Kitcher and Dupré, it is not, strictly speaking, a pluralistic theory. Templeton is not interested in the possibility that the same group might be divided up into species in a number of different ways depending on the interests of the biologists doing the dividing. While the approach we argue for below has many similarities to Templeton's, it is distinguished in part by our openness to pluralistic answers (at least in some cases).

While Templeton's cohesion species concept does not point unequivocally toward a particular conception of the kind of thing species are, it does cohere well (no pun intended) with the position that species are a kind of "homeostatic property cluster kind" (see Boyd 1999; Griffiths 1999). Homeostatic property cluster (HPC) kinds are, roughly speaking, kinds by virtue of sharing a number of properties (none of which are necessary or sufficient) that can be used in prediction and explanation and which are reliably maintained by causal mechanisms; in the case of species, the causal mechanisms maintaining the shared properties are such factors as the mechanisms of reproduction (via gene flow, etc.), the ecological behavior of the organisms (via stabilizing selection, etc.), and the reliability of the developmental processes more generally (via buffering, etc.). Again, while the HPC approach shares some key features with the approach we favor, it tends to point away from pluralistic answers to species questions in cases in which our view is more open to those possibilities.

WHY THE PROBLEM HAS NOT GONE AWAY

We come now to the first of the two points we wish to make in this chapter: The species problem has not yet been solved because it is not the sort of empirical problem that can be solved by biologists alone. Insisting that a purely scientific solution can be found not only goes against the obvious fact that generations of biologists have tried and failed, but also reflects the not uncommon refusal of scientists to concede that philosophy can make important contributions to scientific inquiry.

The species problem is not, we maintain, the sort of problem that simply gathering more data or improving experimental techniques could permit one to solve. This is not, however, to say that the problem is not solvable, nor that it is not a real problem (i.e., contra Noor 2002, it is not "just" a matter of semantics). On the contrary, the species problem is a prime example of the sort of philosophical question that requires input from empirical science in order to be solved and whose solution can be of value to that science (in this case, to evolutionary biology, as well as, for example, conservation biology and ecology). Our position here runs counter to some views of the relationship between science and philosophy, but, as we argue below, there are good reasons to reject those views. Indeed, some of those reasons will emerge directly from our analysis of the species problem.

The relationship between philosophy and science is a complex one, and entire books have been written on its turbulent nature by authors on both sides of the so-called cultural divide (see, e.g., Sorell 1991; Levitt 1998). As we mentioned in the prelude, Nobel Prize-winning physicist Steven Weinberg's essay, aptly titled "Against Philosophy" (1992), provides an excellent example of the kinds of charges that some scientists have leveled against philosophy. In it, Weinberg accuses philosophers of not having contributed to the advancement of science in a single instance; he claims that the influence of philosophy has in fact retarded the advancement of science, arguing that the logical positivists' rejection of unobservable quantities slowed the acceptance of quantum mechanical theory. In this context, it is instructive to ask a simple question: if (part of) the aim of philosophy of science is to understand what science is and how science works, why *would* we expect philosophy to have contributed to answering scientific questions? After all, it is not the goal of philosophy to solve scientific problems, but rather to understand the nature of such problems. This effort may include discussions of what ought to *count* as a solution (philosophy of science can, after all, have a normative component), but that is still a far cry from recommending any *particular* solution.

Returning to the specific difficulties posed by the problem of species concepts, an article by Hey (2001a) is instructive. Hey suggests that the main reason why the species problem is still unsettled is that "we [biologists] are not acting like scientists. We are acting like some philosophers." It is unclear, however, what Hey might mean by this (other than the obvious slight against the philosophical profession). The straightforward interpretation is that further biological research, rather than further conceptual analysis, is necessary. But what additional empirical research could possibly solve the problem? In the same article, Hey himself asks, "How could our knowledge, upon which the species debates have been built, be missing something?" We have, after all, been studying species and speciation for decades now, and if anything is clear, it is that more information will not unequivocally point toward one species concept rather than another. Hey suggests (2001b) that the heart of the species problem is that we evolved as pattern-recognition animals, but this ability is far too crude to match the sophistication of nature when it comes down to the makeup of species. The result, he claims, is that what we perceive as species have little correspondence with what species "really" are; Hey conjectures that the ways in which we pick out species do not accurately reflect the factors that actually delineate species. He is, of course, assuming here that there is a fact of the matter about what constitutes a species, and further, that the real criteria (which have not yet been discovered) are monistic: that there is one and only one answer to the question of whether two organisms belong to the same species.

Even if Hey's analysis is on the right track—and the presence or lack thereof of correspondence between natural categories in biology and what the human brain picks out as categories in biology is an empirical question that awaits further study[7]—it cannot be the entire explanation for why the species problem has been so difficult, nor, obviously, does it solve the problem. For the most part, biologists have been able to function very well without full agreement on what species are or by what criteria they are identified. This is true even for empirical studies of species and speciation, something that would be difficult if we were really so bad at picking species out (recall that, for the most part, we seem to know them when we see them!). This kind of independence of scientific progress from the solution of a conceptual problem is one of the hallmarks of questions that are philosophical and not (just) scientific in nature. Analogously, research on the neurobiology of consciousness is proceeding at a fast pace (see, e.g., Gazzaniga 2000; de Quervain et al. 2004) despite strong disagreements about what consciousness really is (a conceptual/philosophical problem; see, e.g., Block, Flanegan, and Guzeldere 1997).

Again, this does not imply that biologists should not be interested in philosophical questions, nor that the answers to philosophical questions will be useless to practicing scientists. Rather, it means that the species problem is, at least in part, philosophical in nature, and therefore that more empirical research is not the answer. But again, it is worth stressing that there is, of course, an empirical aspect to the problem. As our knowledge of the biological world has increased, so has our stable of counterexamples to the approaches to defining species generated by biologists and philosophers alike. These counterexamples are the places where phylogenetic distance, ecological similarity, reproductive isolation, and the

7. But see a summary of the available evidence in chapter 1 of Coyne and Orr 2004.

Table 9.3. What is a game? This table shows some possible criteria for something being a game and applies these criteria to some activities; activities 1–13 are typically recognized as games, whereas activities 14–18 are typically not considered games. Note that there is no set of necessary and sufficient conditions for something being a game. Any set of criteria that picks out all the activities generally considered games will also capture some activities generally considered non-games; any set that captures only games will not capture some activities normally considered games. (Note too that the Yes/No treatment of the criteria in the table is oversimplified, as the criteria generally admit of degrees.)

	1. Chess	2. Checkers	3. Monopoly	4. Diplomacy	5. Foosball	6. Solitaire	7. Bridge	8. Magic	9. Poker	10. Tennis	11. Baseball	12. Football	13. Catch	*14. Plumbing*	*15. Criminal law*	*16. Reading fiction*	*17. Finding parking*	*18. Sex*
a. Competition	Y	Y	Y	Y	Y	N	Y	Y	Y	Y	Y	Y	?	N	Y	N	?	N
b. Winners	Y	Y	Y	Y	Y	N	Y	Y	Y	Y	Y	Y	?	N	Y	N	?	N
c. Luck	N	N	Y	N	N	Y	Y	Y	Y	N	N	N	N	?	?	?	Y	?
d. Skill	Y	Y	Y	Y	Y	Y	Y	Y	Y	Y	Y	Y	Y	Y	Y	?	Y	Y
e. Amusing	?	Y	Y	Y	Y	Y	Y	Y	Y	Y	Y	Y	Y	N	N	Y	N	Y
f. Physical	N	N	N	N	?	N	N	N	N	Y	Y	Y	Y	?	N	N	N	Y
g. Teams	N	N	N	N	N/Y	N	Y	N	N	N/Y	Y	Y	?	N	?	N	N	N
h. Official rules	Y	Y	Y	Y	?	Y	Y	Y	Y	Y	Y	Y	N	N	Y	N	?	N

like fail to line up—where something that looks like one species according to one definition looks like two (or more) species according to another (see fig. 9.1 and table 9.4 below). As it is hard to see what could tempt us to give up on the criteria we now use to identify species in practice, these counterexamples are not going to go away, so it is toward a different approach to dealing with the species concept that we now turn.

"SPECIES" AS A FAMILY RESEMBLANCE CONCEPT: A (DIS-)SOLUTION?

Our second major claim in this chapter is that the species concept is best viewed as what Ludwig Wittgenstein referred to as a "family resemblance" concept (see, e.g., 1958; for discussion, see Simon 1969); as we noted above, this claim was defended by David Hull (1965) back in the mid-1960s and then more or less ignored by both philosophers and biologists.

Let us briefly look at Wittgenstein's concept as it applies to games (table 9.3) and then explore how it might be applied to the species problem (table 9.4). In *Philosophical Investigations,* Wittgenstein addresses the question of what complex concepts mean—how our words acquire meaning, how they are used, and so forth. As an example, Wittgenstein considers the question of what we mean by "game." It is worth noting that

Table 9.4. Species and the family resemblance concept. Cells in the matrix show whether the criteria (rows) distinguish pairs of populations (columns) as different species. Note that there is no criterion or set of criteria such that two organisms are members of different species *if and only if* they differ with respect to that criterion or set of criteria. As figure 9.1 makes clear, there is also room for disagreement about whether, for example, a particular degree of reproductive isolation or ecological differentiation is sufficient for two organisms to be assigned to different species, so rather than providing a simple "yes/no" answer, these criteria should perhaps be viewed as being met (or not) by degrees (i.e., one ought to apply fuzzy, rather than binary, logic).

	Anemone quinquefolia and *A. rivularis*[a] (two plants separated by a recent polyploidy event)	*Cyprinodon diabolis* and *C. nevadensis*[a] (recent speciation event in desert pupfish)	*Phryma leptostachya* var. *leptostachya* (eastern North America) and *P. leptostachya* var. *asiatica* (eastern Asia)[b] (two phylogenetically very distant plants that are remarkably similar)	*Homo neanderthalensis* and *H. sapiens* (extinct versus extant)	Bacteria: *Pseudomonas putida* KT2442 and *Planctomyces maris*[c] (phylogenetically distant bacteria that exchange plasmids)
Phylogenetic distance	N	Y	Y	?	Y
Reproductive isolation	Y	Y	Y	?	N
Ecological/ functional differentiation	N	Y	N	?	Y
Morphological differentiation	N	Y	N	Y	Y

[a] From Futuyma 1998.
[b] From Wen 1999.
[c] From Dahlberg, Bergström, and Hermansson 1998.

"game" is not a terribly sophisticated concept, nor is it one whose application we are generally inclined to argue about (unlike, say, truth or justice). Indeed, our use of the game concept is straightforward, despite some wrinkles.[8] This is, of course, similar to the way the species concept is used by practicing biologists; except for a few wrinkles, the application of the concept rarely produces much controversy. But it rapidly becomes clear that, despite our comfort with the application of the game concept, any attempt to find a set of necessary and sufficient conditions for something being a game is doomed to failure. There is no feature (or set of features) shared by all games and only games—any definition broad enough to capture such disparate things as board games, card games, ball games, and sporting games will capture many non-games as well. Similarly, any definition narrow enough to exclude all non-games will also exclude many typical games (see table 9.3).

Rather than a set of necessary and sufficient conditions that can be used to demarcate games and non-games, Wittgenstein suggests that there are many threads that crisscross the multidimensional linguistic landscape occupied by the concept of "game." Some of these threads connect several types or instances of games, others connect additional instantiations of games, and yet other threads run through some (but not all) examples of different classes of games. "Game," in other words, is defined by a relatively loose *cluster* of characteristics. Wittgenstein refers to cases like these as "family resemblance" concepts, by analogy to the way in which members of a human family resemble one another, but not because they share any particular feature—rather, members of the same family usually share some characteristic or other with some other family members. For example, some members of a family might have similar faces, whereas others might have similar hair, or eyes, or a vague look, or gestures, but there is no feature or set of features that every member and only members of the family share. Some members of the family might not resemble each other at all, but even so, these members can usually be linked by a chain of resemblances to other members.

Wittgenstein addresses one of the practical problems posed by cluster concepts, a problem similar to the question of how we can possibly use the concept of species if we fail to agree on what species are: "How should we explain to someone what a game is?" (69). Wittgenstein's answer is in-

8. Wittgenstein provides the following example: "Someone says to me: 'shew the children a game.' I teach them gaming with dice, and the other says 'That sort of game isn't what I meant.'" Wittgenstein notes that when we use words like "game" we do not necessarily have in mind exactly what instantiations of the concept we mean to exclude, even if we do in fact mean to exclude some (1958, 28e).

structive: "I imagine that we should describe games to him, and we might add: 'This and similar things are called games.' And do we know any more about it ourselves? Is it only other people whom we cannot tell exactly what a game is?" (69). Wittgenstein's point is that our inability to come up with a precise definition of "game" is not a fault of our weakness in providing definitions. It is not that we know precisely what games are—that we have an internal set of necessary and sufficient conditions that allows us to pick out all and only games, but for some reason cannot communicate it. Rather, the concept itself is "fuzzy" or inexact (69–71). This does not make the concept useless, nor even less useful than a precise concept would be. Indeed, for some things (games among them), an inexact concept is precisely (no pun intended) what is needed.

Similarly, as biologists, we teach our students what species are by example until they form a general idea of what we mean by the term and how it is used. Our inability to come up with a set of necessary and sufficient conditions for two organisms to belong to the same species that would work in every case we might be interested in is not the fault of our being sloppy thinkers or lacking empirical data. Rather, it is a feature of the concept itself. As Wittgenstein notes, "But this is not ignorance. We do not know the boundaries because none have been drawn" (69). If for some purpose we need to be able to specify precise conditions, we can do so, but we should not mistake our ability to do so with a general necessity for doing so: "We can draw a boundary for a special purpose. Does it take that to make the concept usable? Not at all!" (69). Generally, we learn and use the species concept without needing any precise definition; indeed, any precise definition would fail us in practice in at least some cases (see table 9.4). But for particular purposes, we can draw one line or another in order to make the species concept precise in a particular arena. For example, for all the dismissal of the morphological species concept, morphological evidence is, by necessity, the only kind that can be employed, for instance, in recognizing species in the field and in paleontology, even if few biologists consider such evidence part of the definition at a theoretical level. And whatever one thinks of Mayr's "biological" species concept more generally, it is obviously useless for most microorganisms and many plant species.

Taking a Wittgensteinian approach to the species problem resolves several issues at once, on both the biological and philosophical sides. Such a solution (or, rather, a dis-solution) has at least the following advantages:

1. We can move from a Platonic "essentialist" view of species (which, despite protestations to the contrary, strikes us as perennially tempting to many bi-

ologists) and the related question of what species *really* are to a cluster concept view that is more nuanced and realistic. It is worth noting that this is a philosophical shift, although, as noted above, it is induced by the wealth of empirical information we have on species.

2. Hey's (2001b) suggestion (discussed above) that part of the problem may lie in a mismatch between the categories recognized by the human brain (and hence language) and whatever natural categories are really "out there" can be rearticulated in a less contentious way. The fact that we are unable to precisely define "game" does not imply that games do not exist, but neither does it imply that there is some real natural category of games that we are unable to recognize. Our desire to precisely define species is different from our ability to recognize species and to use the concept of species without there being any precise definition.
3. Wittgenstein's suggestion that we can draw boundaries around subsets of family resemblance concepts for practical purposes matches Kitcher's (1984) and Dupré's (1993) ideas of a legitimate pluralism of interests among biologists. We can abandon our quest for *the* correct answer to the problem and focus instead on which boundaries we ought to embrace for which purposes.

Before presenting our conclusions, we must address two very recent, and wildly divergent, entries in the literature that are symptomatic of the whole problem of species concepts, how it is perceived by biologists, and what philosophy has to do with it. The first is a book on speciation by Coyne and Orr (2004); the second a paper on species concepts by Kevin de Queiroz (2005).

Coyne and Orr are concerned with speciation, not species concepts per se. However, they recognize the need to address the species concept problem, and they do so in both the first chapter of their book and a lengthy appendix. This is obviously not the place for an extensive review of the book, but we wish to point out how Coyne and Orr come out strongly in favor of the so-called biological species concept, dismissing the alternatives on a variety of more or less reasonable grounds while at the same time minimizing the equally troublesome problems (pointed out by countless authors before them) with their favorite solution. This rather conservative approach, perfectly in line with the mainstream position characteristic of the modern synthesis, is a necessary move for Coyne and Orr because they wish to reduce their main concern, speciation, largely to the equally classic allopatric model of speciation (i.e., speciation catalyzed by geographic isolation of evolving populations). To do that, they must eliminate alternative modes of speciation, chiefly the sympatric one. Sympatric speciation can, in theory, occur by a variety of mechanisms, including the action of ecological factors that can lead to divergence in spite of a certain degree of gene flow, as well as genetic mechanisms such as polyploidy in plants and some animal groups. Coyne and Orr manage to reject sympatric models because they put an overwhelming emphasis on reproductive isolation (notoriously difficult to achieve in sympatry). In doing so, they dispense with an increasingly large and robust body of literature that points to a more nuanced and pluralistic view not only of species concepts, but also—consequentially, one might argue—of speciation modes. The circle, in Coyne and Orr's scheme, is therefore complete, with their favorite species concept buttressing their favorite mode of speciation, and vice versa (indeed, at times they seem to wish to exclude non-allopatric speciation essentially by definitional fiat rather than by empirical investigation). Needless to say, we find such a conceptual straitjacket rather too tight to accommodate the variety of biological phenomena of interest.

Kevin de Queiroz takes a completely different, and much more innovative, perspective on the whole problem. He starts out his paper by suggesting that there are *three* problems, not just one, with species concepts. The first concerns the existence of disparate and apparently incompatible species concepts, the second deals with the variety of *processes* underlying speciation and how they are related to the various species concepts, and the third is the practical problem of reliably identifying species faced every day by practicing biologists. De Queiroz argues, correctly we think, that the third problem already has plenty of good solutions, relying on a host of empirical and analytical techniques devised by biologists especially throughout the latter part of the twentieth century. This is why, as we suggested above, the study of species and speciation has made so much progress in spite of the conceptual quagmire into which discussions of species concepts have plunged biologists. The other two problems are solved by de Queiroz in a rather Solomonic fashion. On the one hand, species concepts can be unified by his proposal that they all have one property in common: all species, regardless of how they are defined, have to be separately evolving metapopulation lineages. On the other hand, de Queiroz finds that our proposal of species as cluster concepts nicely addresses the complementary problem of the different causal threads that lead to speciation, and hence to different subtypes of species. Further work needs to be done on both the biological and philosophical implications of de Queiroz's proposal—for example, as far as the status of pluralism is concerned (it seems that de Queiroz would admit pluralism of mechanisms, but not pluralism for the core concept). We look forward to these developments over the next few years.

CONCLUSIONS: WHAT, THEN, ARE SPECIES, AND WHY SHOULD WE CARE?

With the proviso of de Queiroz's interesting distinction among three different species concept problems, we are suggesting, as Hull did back in 1965, that "species" is best viewed as a family resemblance concept whose underpinnings are to be found in a series of characteristics such as phylogenetic relationships, genetic similarity, reproductive compatibility, ecological characteristics, and morphological similarity. These traits take on more or less relevance depending on the specific group one is interested in, as well as the particular problem one wishes to explore (see fig. 9.1).

Biologists can benefit from this recognition in a variety of ways: First, they no longer need to try to empirically solve a problem that emerges from a philosophical stance. The desire to find *the* answer to the species problem emerges from what is, in the end, a philosophical assumption—that we can uncover the necessary and sufficient conditions for the application of the concepts we use. Accepting that this assumption is unwarranted permits us to move on to exploring how the concept is actually used, despite there not being precise criteria for its application. But this recognition does not imply that the species problem is irrelevant to the aims of working biologists—that it is just a semantic issue for philosophers to quibble about. The recognition that there are a number of different ways of making the species concept precise, none of which is the right way, but many of which will be useful in particular circumstances, points toward important empirical projects—projects aimed at determining which articulations will be most valuable in which situations and how to properly use them. The family resemblance view of the species concept emerges from a particular philosophical perspective, but it is not therefore irrelevant to working scientists.

This kind of interplay between science and philosophy may well represent the best model for the relationship between science and the philosophy of science. Some problems in the philosophy of science are at least in part empirical in nature, and attention to the empirical work of the fields involved can influence the kinds of solutions regarded as philosophically adequate. At the same time, the empirical sciences make use of philosophical assumptions, whether they are recognized or not; being aware of such assumptions can sometimes have a significant payoff in terms of the empirical research suggested and the kinds of inquiry considered fruitful.

10 Testing Biological Hypotheses

The Detective versus the Statistician

I CONTEND THAT THE GENERAL ACCEPTANCE OF STATISTICAL HYPOTHESIS TESTING IS ONE OF THE MOST UNFORTUNATE ASPECTS OF 20TH CENTURY APPLIED SCIENCE.
—MARKS R. NESTER, "A MYOPIC VIEW AND HISTORY OF HYPOTHESIS TESTING"

Here, at the end of this book, we take a broader look at what it means to test hypotheses in a complex, partially historical science such as evolutionary biology, a theme that has been present in virtually every chapter so far. We begin our discussion by examining the limitations of classic hypothesis testing in statistical analysis, the common currency in the quantitative biological sciences. This approach is closely related to a rather outmoded simplified Popperian philosophy of falsificationism and has provided the conceptual basis for much theory evaluation in evolutionary biology throughout the twentieth century. After discussing two major examples (the Hardy-Weinberg principle in population genetics and Kimura's neutral theory of molecular evolution), we conclude by sketching what better alternatives are available to evolutionary biologists in the realm of hypothesis testing and theory evaluation. These alternatives also provide better ways of approaching the various problems discussed in previous chapters, from measuring natural selection to studying genetic drift and evolutionary constraints. Evolutionary biologists need to move beyond physics as a model for the sciences as a whole, and past naive falsificationism, to embrace the beautiful—if often frustrating—complexity of their discipline.

WHAT IS THE GOAL OF HYPOTHESIS TESTING IN ORGANISMAL BIOLOGY?

Throughout this book, we have tried to make sense of the various complex concepts and methods that constitute the theory and practice of modern evolutionary biology. As the reader will surely have noticed, there seems to be somewhat of a disconnect between what evolutionary biologists are interested in finding out and what they are actually studying, given the techniques and approaches they employ. For example, we have seen in chapter 2 that the Lande-Arnold multiple regression models aim to study natural selection in action, but—because of what natural selection actually *is*—end up doing something quite different and only partially related to the stated objective. Such methods provide useful summaries of patterns of correlations between characters and fitness, but those patterns may—or may not—reflect something about natural selection at the ensemble level. Similarly, in chapter 4, we have considered the concept of genetic variance-covariance matrices and what they are supposed to tell us about evolutionary constraints and trade-offs. But a closer look, especially after the work of Houle (1991) and Gromko (1995), reveals that the link between processes and patterns is much more complex and indirect than quantitative genetics theory would like to assume. As a consequence, biologists obtain (again, useful) statistical summaries, but all too often write as if they were actually studying causal relationships directly. In this chapter, we delve a bit more deeply into this recurrent disconnect between questions and methods, and we examine more closely what it might mean to test hypotheses in organismal biology.

We suggest that there are two major components of the problem: On the one hand, questions in evolutionary biology are extremely complex and are unlikely by their very nature to be settled in a clear-cut manner.[1] On the other hand, organismal biologists are being hindered in their research by a widespread approach to hypothesis testing, stemming from the use of concepts such as null hypotheses in statistics, that reflects a philosophically unsophisticated Popperian attitude of simple falsificationism. There is little biologists can do about the first problem: nature was not designed with the convenience of the scientist in mind (indeed, as far as we know, it was not designed at all), and sometimes we just have to accept that the progress we make on some interesting questions has been, and will continue to be, slow and tortuous. The second problem, however, is a matter of training and attitude, both of which can be changed for the benefit of future generations of evolutionary biologists. Since we cannot alter the complexities of nature, it is to this second problem that the remainder of this chapter is devoted.

NULL HYPOTHESES: STATISTICAL AND PHILOSOPHICAL CONSIDERATIONS

One of the common insider's jokes in quantitative biology is that all biologists who have ever used statistical analyses think of themselves as experts

1. This is, of course, not true of scientific endeavors in general (think of the rather precise answer to the question of what the charge of the electron is) or of biology in particular (for example, there is an unequivocal answer to the question of what amino acids particular nucleotide triplets code for).

in statistics. Furthermore, some have a tendency to become zealous about it, pounding on colleagues who fail to stick to the established canons of the discipline. Conversely, statisticians with whom we have interacted seem by and large more casual about their own profession, thinking of it as providing a convenient toolbox for exploring data as well as some aid in hypothesis testing. Indeed, a debate has been going on for decades about the usefulness of the two basic tools of statistical analysis in the quantitative sciences: null hypotheses and "*p* values" (the latter being numbers that allegedly tell researchers with what confidence they should reject or accept a given null hypothesis). The point we wish to reinforce here (since, as we shall see, it has been made by others several times before) is that many quantitative scientists, and evolutionary biologists among them, have developed an apparent confusion (in their practice, if not conceptually) between causal and statistical hypothesis testing, and an uncritical use of null hypotheses and *p* values is among the chief culprits. Many statisticians and some scientists who have approached these problems from outside the traditional statistical framework have argued that this confusion is significantly hindering scientific progress; the resulting debate has the potential to change the way peer review and editorial decisions about publications and grant proposals are conducted, which makes it much more than a purely philosophical or methodological issue.

In a nutshell, the problem is that many quantitative biologists take for granted the value of null hypotheses and *p* values as key concepts of their statistical practice; this is hardly surprising, as these are the concepts they have been taught in graduate school and which are used almost universally in technical journals. And yet, the debate on the meaning and usefulness of such concepts has intensified to the point of influencing the actual practice of researchers in fields such as statistics, psychology, sociology, and some applied disciplines that rely heavily on decision-making procedures (Cronbach 1975; Matloff 1991; Loftus 1993; Cohen 1994; Gregson 1997; Dixon and O'Reilly 1999; Anderson, Burnham, and Thompson 2000). For the purposes of this chapter, we will define the "standard," or classic, approach to statistical analysis as a mixture of Fisher's and Pearson and Neyman's concepts of hypothesis testing as presented in most statistical textbooks for biologists (e.g., Sokal and Rohlf 1995). The main problem with this classic approach is rooted in the fact that it involves testing the probability of the observed data given an often unsatisfactory "null" hypothesis (Matloff 1991; Dixon and O'Reilly 1999; Anderson, Burnham, and Thompson 2000). That is, biologists (and social scientists) typically set up statistical tests to tell them whether, for example, the correlation coefficient between brain size and body size in a sample of vertebrate species is zero (null hypothesis) or different from zero (any other value). They then collect an appropriate set of data and ask what the probability of getting that data (or an even greater departure from the null expectation) would be if the null hypothesis were correct (the so-called "*p* value"). If that probability is low, they "reject" the null hypothesis and "accept" the alternative one.

There are a number of problems with this procedure:

1. What scientists are, or ought to be, actually interested in determining is the probability of a series of alternative hypotheses being correct given the data, not vice versa (i.e., scientific inference is about comparing hypotheses on the basis of data, not the other way around: Howson and Urbach 1991; see also Howson and Urbach 1989, but see the critical discussion in Chalmers 1999).
2. The null hypotheses tested by standard procedures based on *p* values are often of limited interest to the researcher.
3. The standard approach does not take into account any prior knowledge one may have about the system (such as the expectation that not only will there be a correlation between body and brain size, but it will be positive and linear). If science is to be a progressive enterprise, then surely scientists do not—and should not—start each project with a blank slate.
4. The so-called alternative hypothesis is, in reality, an infinite family of hypotheses, and concluding that one of these is more likely than the null hypothesis is not very informative.
5. The best science proceeds by considering several alternative hypotheses, not just one (as was clearly pointed out by Chamberlin [1897; see also Chamberlin 1890]). The more alternatives, and the better defined, the better scientists can design informative experiments and make progress.

A more formal way of thinking about the problem posed by the use of null hypotheses can be based on the concept of conditional probability: scientists are interested in $P(H|D)$ (the probability of a hypothesis given the data), not in $P(D|H)$ (the converse), but the standard approach comes closer to testing the latter. One of the most common fallacies in statistical reasoning is the assumption that the two are related: that if the probability of the data given the null hypothesis is low (i.e., we reject the null), then the probability of the null given the data must also be low. It can be shown, both mathematically and by example, that this is not the case; one might, for example, pick a very unrealistic hypothesis, or the sample size might be inadequate (leading to what is known as a type I error, the incorrect rejection of the null).

In typical applications of the standard approach, *p* values themselves are also often misinterpreted in a variety of ways. The most important thing to understand about the p value is that it is neither a measure of $P(H|D)$ (again, what we are actually interested in) nor (exactly) of $P(D|H_0)$ (where H_0 is the null hypothesis). In fact, it is the cumulative probability of the exact null hypothesis and of all values more extreme than the one observed (Sokal and Rohlf 1995). While this probability is related to $P(D|H_0)$, it is clearly not the same thing. Furthermore, it is absolutely incorrect to interpret a p value as an estimate of the strength of the effect under investigation (another practice that emerges from perusing the literature); while it may be the case that "given a small p value the results are highly significant" in a statistical sense, this is quite different from the results being significant in a broader scientific sense (see Giere 1979; Dixon and O'Reilly 1999; see also McCloskey and Ziliak 1996 and Ziliak and McCloskey 2004 for evidence that there remains rampant confusion among economists between *statistical* significance and the *economic* significance of the effect). The p value is a function of the sample size for any effect of fixed size (large *or* small). One can repeat the experiment with a much larger sample size and obtain a much smaller p value associated with the same null hypothesis, and yet the effect size (e.g., the strength of a correlation coefficient: Cohen 1992) can be unchanged (Loftus 1993; Gregson 1997). Since it is the effect size that is of interest insofar as our goal is to understand underlying causal structures through the statistical "shadows" they create (see the prelude), it ought to be clear that statistical significance is not the kind of significance we are most interested in.

It follows from all of this that null hypotheses, as usually presented, are really straw men: their rejection is not very informative, and failure to reject them is even less so. Indeed, the null hypothesis is almost certain to be rejected given a large enough sample size. This is because the null hypothesis is stated very precisely (e.g., $r_{bs,bs} = 0$; that is, the correlation between body size and brain size is exactly zero). In reality, what scientists mean to test is whether a correlation (or whatever other statistic they are interested in) is far enough from zero to be scientifically interesting (fig. 10.1). The effect of any interesting variable on the system is highly unlikely to be exactly zero (because of small, often unaccounted for, systematic deviations), but that does not imply an effect size worth thinking about, nor a causal pathway of any real interest. Part of the problem is that "chance is lumpy," as some statisticians put it (Abelson 1995, 21); that is, it tends to cause clumped distributions of data that make it too easy to "spot" patterns that are not really there. To put it in yet another way, because it is unlikely that if we censused the entire population there would be *zero* effect, as our sample size approaches the population size, we can be sure to find statistically significant effects, even if there is no *causal*

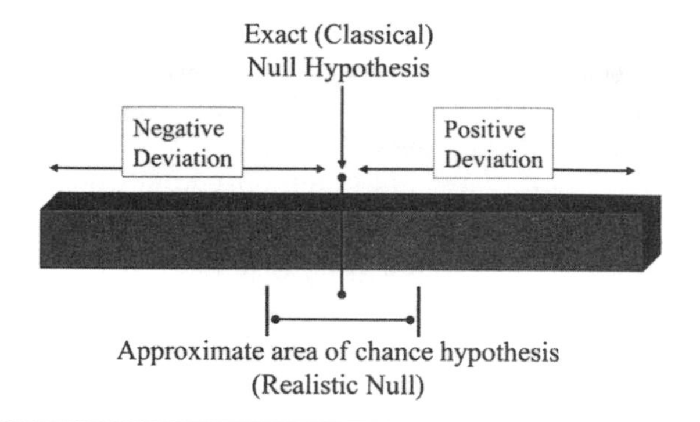

Figure 10.1. The structure of hypothesis testing in a hypothetical experiment. What one is really interested in is not whether the exact null hypothesis can be rejected (which is what standard statistical tests tell us), but whether the outcome of the experiment is far enough from that point to seriously raise the possibility of a systematic effect of biological interest. The problem, of course, is that "far enough" is a subjective concept that requires a judgment call on the part of the scientist. As much as we would like it to be otherwise, this problem is unavoidable.

mechanism that generates the difference (the reader may notice a similarity here to the conceptualization of "genetic drift" as a "force" as opposed to a statistical expectation, discussed in chapter 1).

On the other hand, small effect sizes can be biologically important. Indeed, even effects too small to be revealed as statistically significant given reasonable sample sizes may be biologically important—as in the commonly encountered case of small selection coefficients in natural populations (Kingsolver et al. 2001; see chapter 2). While they often cannot be statistically detected given the size of real wild populations, if sustained over time, such small selection coefficients can nevertheless yield major evolutionary change (Fisher 1930). In these cases, researchers have to rely on their knowledge of the system, not on ready-made statistical cutoff points; in particular, if small deviations can be confirmed by repeated experiments, and if they are in the direction predicted by suitably realistic biological hypotheses, then they may become biologically interesting, regardless of the p values that accompany them.

What we have been calling here the "standard" approach in statistics was developed by Fisher and by Neyman and Pearson (though Bayesian analyses, which ameliorate several of the above-mentioned problems, were actually popular in statistics for decades prior to Fisher's work). Fisher constructed a system of statistical inference that is echoed by the insights of Karl Popper (1963), particularly his ideas about falsifiability. As is well known, in the Popperian view, hypotheses can never be proved; they can only be disproved (falsified). Popper's reasoning was that no matter how much evidence there is in favor of a hypothesis, there is always the possibility that an as yet unknown alternative hypothesis will explain the same evidence and then some, or that the hypothesis in question will fail when faced with a novel situation. On the other hand, a theory that can be shown to have made a false prediction is falsified, and therefore ought to be discarded. This is often taken to be the idea behind the use of a null hypothesis: one is attempting to falsify a given possibility in order to move on and consider the others.

One problem with the standard approach should already be evident from our discussion so far: even according to Popper, the rejection of one hypothesis does not imply the acceptance of another (and Fisher did not advocate such a procedure, though this interpretation has entered common usage). Furthermore, Popper suggested that the most interesting hypotheses to falsify are the most daring ones (because they are the most informative), whereas most null hypotheses actually tested tend to be trivial and uninformative. Finally, philosophers of science have long since moved past what is sometimes referred to as "naive falsificationism" (see the discussion in Chalmers 1999; even Popper himself may not have actually advocated the sort of "naive falsificationism" that is sometimes attributed to him). Indeed, some authors have come to the conclusion that the scientific method is really a continuous competition among alternative models, with some gaining and some losing, depending on the available evidence (a fairly Bayesian view of science: see, e.g., Howson and Urbach 1991). All of this notwithstanding, scientists begin their training, as undergraduate and graduate students, with a rather simplistic Popperian view of science. It may turn out that the connection between the testing of null hypotheses and falsificationism is one of the most profound examples of bad philosophy guiding scientific research ever documented; not only has it resulted in much misguided effort, but, because textbook authors seem to have a hard time catching up with developments in philosophy of science since Popper's work, it is a hard problem to correct.

As has long been realized, an infinite number of hypotheses can be constructed to explain (fit) a certain data set (the so-called problem of the underdetermination of the theory by the data: Quine 1960; Okasha 2000). In general, no matter what statistical approach one uses (standard, Bayesian, or likelihood; the latter two are discussed below), the resulting output should be taken only as a (re)description of the data, not as a decision-making procedure. No statistical procedure is capable of substituting for informed judgment by an expert in the field; if nothing else, given that multiple underlying causal structures are compatible with the same statistical patterns, there are always alternative explanations to be consid-

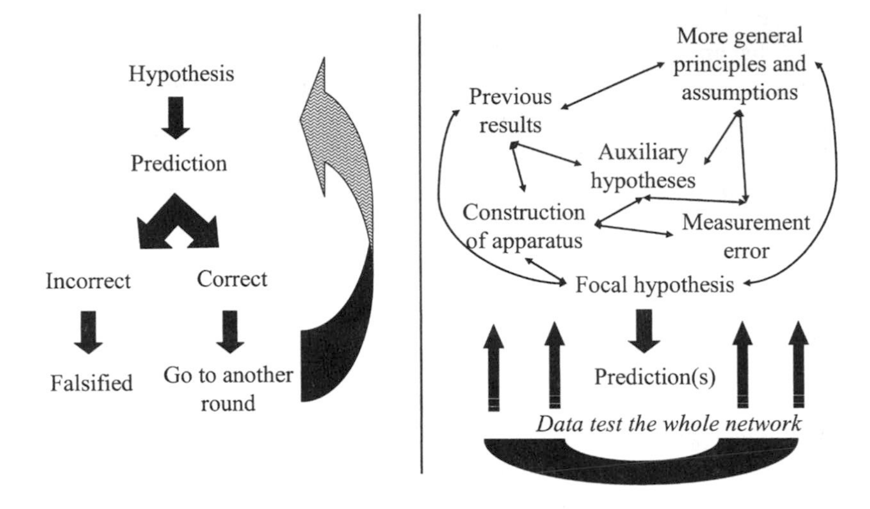

Figure 10.2. Two models of hypothesis testing: (*Left*) Popperian "naive falsificationism" (which is reflected by the standard statistical approach), in which a hypothesis makes falsifiable predictions that are directly tested by data. (*Right*) A complex web of knowledge in which assumptions, hypotheses, and reliability of measurements are simultaneously tested during an ongoing process of research. Note that this should not be taken to imply that it is impossible to test the reliability of such experimental setups; rather, such tests can be performed only by refocusing the research on parts of the experimental setups themselves (see, e.g., Wimsatt 1987; Cleland 2002).

ered and discussed. More generally, the data generated may not be sufficient to be informative, or may not be reliable enough to distinguish between alternative hypothesis. Nor, of course, can statistical analyses account for the actions of factors not explicitly considered by the researcher up to that moment. Finally, no statistical analysis can be used to discount the possibility of problems with the experimental setup or with the instruments of measurement (fig. 10.2).

All this complaining may lend credence to the accusation often directed at philosophers by scientists of failing to help science solve its problems (Weinberg 1992; but see Wilkins 2002) while arrogantly sitting in judgment of what scientists (allegedly) do wrong. Similarly, Stigler notes that criticisms such as the above represent a "string of general caveats" for the use of statistical analyses in biology that, while justified, are nonetheless unlikely to change the practice of working scientists unless "positive principles" are put forward along with the warnings (1987, 148). We therefore provide a primer on several of the alternatives to the use of null hypotheses and *p* values that have been identified in the course of this debate and which are increasingly changing the content of publications, and especially the way of thinking, of practicing scientists (though, so far, mostly in the social sciences: Matloff 1991; Loftus 1993; Gregson 1997; Anderson, Burnham, and Thompson 2000).

One simple change scientists can enact immediately has already been discussed: abandon the practice of pitting an often uninformative null hypothesis against every other possible alternative, which lets the latter, highly heterogeneous category easily win the day (especially if the sample size is large enough). Interesting and informative scientific papers consider two (or more) plausible hypotheses and provide empirical support for an increase or decrease in the likelihood of one (or more) of them. This was realized as early as the end of the nineteenth century by Chamberlin (1897) and was forcefully reiterated in a more recent book on ecological investigations by Hilborn and Mangel (1997; the appendix to that volume includes Chamberlin's original paper). To insist on publishing papers in which the results are presented as if the authors were kicking a very weak opponent (the null hypothesis) makes for the appearance of simplistic thinking and makes advancement in evolutionary and ecological disciplines more difficult (Pigliucci 2002). In reality, of course, most scientists do understand the distinction between statistical and scientific inference. As a result, they write their papers in a rather schizophrenic fashion: The introduction and discussion sections are usually framed in terms of a series of alternative hypotheses whose relative merits are judged on the basis of previous knowledge (e.g., other papers) and the current results. However, these sections of the paper sandwich a quite different middle section that presents the results largely as endless tables of p values, purportedly rejecting a series of often uninteresting null hypotheses.

Thinking in terms of several alternative hypotheses is not only more natural and more effective, but can also be an important way of avoiding a typical occupational hazard of being a scientist—what Jacques Monod (1971) referred to as falling in love with one's own hypothesis. If one considers several viable and interesting possibilities, one can avert more or less unconscious biases in favor of *the* "alternative" hypothesis. Furthermore, this practice might ameliorate the known problem of the reduced publication of "nonsignificant" results, which, even though they may sometimes be highly informative, tend to be underrepresented in the literature (the "file drawer effect": Moller and Jennions 2001) because editors do not want to waste precious journal space to publish a negative conclusion. Again, this is a consequence of the widespread more-Popperian-than-Popper attitude adopted by many practicing scientists in which it is assumed that the only way to make progress in science is by rejecting hypotheses, never by confirming them; accordingly, failing to reject a null allegedly does not advance science. But, provided, at least, that one has chosen an interesting hypothesis[2] to attempt to falsify, failing to reject it can indeed advance the field.

An obvious alternative to the standard statistical approach to hypothesis testing has been mentioned already in passing several times: Bayesian analysis. Essentially, Bayesian analysis asks the right question (what is the probability of the hypothesis given the data?) and attempts to answer it by taking into consideration not only the actual data obtained during a given experiment, but also whatever knowledge was available beforehand, summarized in the form of prior probabilities. The basic equation of Bayesian analysis is

$$P(H|E) = \frac{P(E|H) \times P(H)}{P(E|H) \times P(H) + P(E|{\sim}H) \times P({\sim}H)}, \tag{10.1}$$

where $P(H|E)$ is the probability of a given hypothesis (H) given the evidence (E)—also known as the *posterior* probability of the hypothesis; $P(E|H)$ is the probability of observing the evidence assuming the hypothesis is correct; and $P(H)$ is the so-called prior probability of H being correct (${\sim}H$ is an alternative hypothesis, or a family of alternative hypotheses).

We will not get into the perennial discussion concerning the choice of objective or subjective prior probabilities here (Howson and Urbach 1991; Berger and Jefferys 1992; Malakoff 1999; see also Chalmers 1999 for a critique of the use of Bayesianism as a general philosophical theory of science). Although we agree that it may be impossible in many (but not all) situations to come up with objective priors that are actually useful, we see this as part and parcel of the practice of science: science is a human activity that melds subjective, intuitive judgments with empirical and quantitative means of objective analysis. The fact that the same is true of the Bayesian approach is therefore a reflection of reality, not an admission of weakness (let us remember that numerical quantification does not guarantee objectivity or meaningfulness). The Bayesian approach is becoming more popular within organismal biology, having been applied to ecological modeling (Hilborn and Mangel 1997), selection analyses (Rudge 1998), quantitative genetics (Shoemaker, Painter, and Weir 1999), phylogenetic studies (Sinsheimer, Lake, and Little 1996; Huelsenbeck, Rannala, and Larget 2000), and conservation biology (Wade 2000).

2. In this sense, an interesting hypothesis can be a "null" (or, better put, a *neutral*) hypothesis, as long as it specifies in some detail what is not going on and with what kinds of results it would be compatible; see below for discussion.

Along similar lines, sometimes researchers prefer to compare hypotheses by means of what is known as Bayes's ratio:

$$B = \frac{P(D|H_1)}{P(D|H_2)}. \tag{10.2}$$

That is to say, B is the ratio of the likelihood of the data under the first hypothesis to the likelihood of the data under the second hypothesis, and it does not involve the priors of either hypothesis. This ratio is also called the weighted likelihood ratio, and it is similar (indeed, when priors are uniform, it is identical) to the more familiar (and simpler to calculate) likelihood ratio that we discuss next (see Dixon and O'Reilly 1999).

A more sophisticated alternative to the standard statistical approach, likelihood ratios, combines the insights (although not the formalism) of the Bayesian method (sans the difficulties related to the estimation of priors) and the simplicity of readily available statistical approaches (and computer packages). A likelihood ratio is a quantity that compares the likelihood of the data based on one model with the likelihood of the data based on a second, competing model (comparisons can also be made among multiple models, pairwise or cumulatively). The ratio provides a measure of the relative match of models and data. The likelihood ratio is conceptually (and mathematically; see Dixon and O'Reilly 1999) related to the Bayesian approach, but it is also conveniently close to the familiar standard statistics, which makes it a powerful practical alternative. There are at least three ways to calculate likelihood ratios from standard statistics. We provide simple examples of those calculations for illustrative purposes in box 10.1; a more detailed treatment (and additional examples) is provided by Dixon and O'Reilly (1999).

Three important caveats apply to the use of likelihood ratios, however. First, a likelihood ratio is not a measure of the posterior probabilities of the models (the Bayesian interpretation) because, as the attentive reader will have noticed, likelihood ratios calculate the probability of the data given a certain model, not vice versa. However, a quasi-Bayesian interpretation of the likelihood ratio is defensible because, again, the two are equivalent if the priors for the two models are taken to be the same (i.e., one has no a priori reason to prefer one model over the other) and if the models are of comparable complexity (measured by the number of parameters they employ; see Dixon and O'Reilly 1999).

Second, in contrast to a Bayesian analysis, the models being considered need not be exhaustive of all possible models; they are just the particular subset of models that the experimenter wishes to test (or has thought of testing). This means that just because one model is much better than

Box 10.1 How to Calculate Likelihood Ratios to Compare Hypotheses

One way to calculate the likelihood ratio for a comparison of two models is to obtain the ratio of the standard deviations derived from the two models and raise it to the power of the sample size, n:

$$\lambda = \left(\frac{s_1}{s_2}\right)^n, \tag{10.3}$$

where s_1 and s_2 are the likelihood standard deviations based on the two models.* For example, if model 1 has an associated standard deviation (SD) of 0.65, model 2 has a SD of 0.58, and there are $n = 40$ observations, then the likelihood ratio between model 1 and model 2 is

$$\lambda = \left(\frac{0.65}{0.58}\right)^{40} = 95.4. \tag{10.4}$$

This means that the data fit the second model (the one in the denominator) roughly 95 times better then they fit the first one. One should definitely prefer the second model.

It is even easier to calculate the likelihood ratio from a standard table of sums of squares like those produced by any commercial software that carries out general linear model calculations. In this case, the formula is

$$\lambda = \left(\frac{SSE_1}{SSE_2}\right)^{(n/2)}, \tag{10.5}$$

where SSE is the "error" sum of squares calculated after fitting either model, and n is again the sample size. The SSEs can be obtained either by summing up the SS of various sources of variance within a single model, according to the corresponding sub-hypotheses, or—which may be simpler—by rerunning the analysis using each model and looking up each SSE value.

An even simpler method is to run each model separately and calculate the R^2 (i.e., the amount of variance explained by the model), something that again is part of the standard output of any statistical package. The ratio between models then becomes

$$\lambda = \left[\frac{(1 - R_1)}{(1 - R_2)}\right]^{(n/2)}. \tag{10.6}$$

All three of these methods, of course, give exactly the same numerical results. As Dixon and O'Reilly (1999) pointed out, this approach can also be extended to repeated measures models, to mixed and nested ANOVAs, and even to cases in which the data are not completely independent. In other words, likelihood ratios between competing hypotheses can be calculated for most problems that usually fall under the rubric of general linear modeling, which includes the majority of practical applications for quantitative biologists and social scientists as well as every statistical technique discussed in this book. There really is no excuse, in terms of conceptual sensibility, ease of calculation, and flexibility, not to present results in terms of likelihood ratios among competing models.

*Note that the likelihood standard deviation is calculated by dividing the sum of squares associated with a given model by $n - 1$ instead of n.

another, one cannot exclude the possibility that the winning model is still not adequate for the task at hand; a much better explanation may exist, but may not have been thought of (yet?). This, however, is just part of life for a scientist, and it is hardly a limitation of the likelihood ratio approach. It may seem at first as if model fitting is a hopeless enterprise, given the potentially large number of models one could compare with one another. However, most standard statistical packages come equipped with model-fitting algorithms using maximum likelihood or restricted maximum likelihood approaches that can aid researchers in iteratively exploring the parameter space and finding the best model. More radically, Shipley (2000) has suggested an intriguing new approach to model selection based on fundamental properties of statistical procedures such as path analysis and structural equation modeling (see Magwene 2001 for an application to evolutionary questions, as well as our discussion of path analysis in the context of studying selection in chapter 2).

Third, it is, of course, possible that the data will not strongly favor one model over another. This result is important because it means either that the data do not sufficiently discriminate between the models or that both (or all) models are inadequate representations of the causal structure of the phenomena being studied. The advantage of the likelihood ratio is that in this situation one is not forced to accept a "null" hypothesis based on weak evidence.

Yet another powerful approach to the problem of model selection and hypothesis testing—and one that is favored by an increasing number of statisticians as well as some philosophers and biologists[3]—is provided by the Akaike Information Criterion (AIC). This measure is based on information theory and is derived from the concept of entropy in physics (Anderson, Burnham, and Thompson 2000). Briefly, AIC measures the fit of the data to a given model when the model is penalized in proportion to the number of parameters it employs (a sort of statistical version of Occam's razor). This approach is sensible because, other things being equal, a model with more parameters will always fit the data better than one with fewer parameters, but the improved fit will not always be scientifically relevant; recall again that all-important difference between statistical and scientific inference. Our goal is to understand the underlying causal structure of an observed pattern in nature; while a model with many parameters may more closely reflect the underlying structure, it need not,

3. Elliot Sober, for example, has actively explored and defended the use of the Akaike Information Criterion in biological contexts; see Sober 2002 and Sober and Steel 2002.

and a misguided model with more parameters will tend to fit more closely than one with fewer.

When one engages in model fitting using standard sums of squares (as is often the case in the biological literature), the AIC for a given model can be computed simply as

$$\text{AIC} = n \times \ln\left(\frac{\text{RSS}}{n}\right) + 2K, \tag{10.7}$$

where n is the sample size, ln is the natural logarithm, RSS is the residual sum of squares for that model, and K is the number of parameters (the degrees of freedom taken up by the model).

Once AIC has been calculated for each alternative model, one can rank models by computing

$$\Delta_i = \text{AIC}_i - min\,(\text{AIC}), \tag{10.8}$$

where i indicates a particular model, and *min* is the value of the model with the smallest AIC.

Finally, one can compute the likelihood of a certain model given the data by using the transformation $e^{(-1/2\Delta i)}$. If this quantity is normalized to one (the so-called Akaike weight), it allows a convenient comparison among models:

$$w_i = \frac{(e^{(0.5\Delta i)})}{\Sigma_i(e^{(0.5\Delta t)})}, \tag{10.9}$$

where the summation is over all models.

The Akaike Information Criterion lends itself very easily to the comparison of several alternative models, but is not a "test" at all in the sense of requiring the calculation of associated p values. From a philosophical perspective, it is a particularly intriguing approach because it takes into account not just the fit of the model to the data, but also the complexity of the model itself, and therefore of its underlying hypothesis.

One thing may need to be reiterated at this point, although it should be clear after a moment's reflection: *any* statistical analysis is an exercise in model fitting coupled with some approach to gain a measure of how well the model(s) actually did fit the available data. This is true whether one uses null hypotheses, p values, likelihood ratios, AIC, or Bayesian approaches implemented on regression analyses, analyses of variance, or a host of other exploratory and confirmatory statistical techniques common in organismal biology. The alternative approaches advocated here simply make this more obvious and explicit, thereby helping the researcher to think through the problem at hand more clearly.

When all of the concerns outlined above are seriously considered, one might suspect that we are suggesting that most published literature in organismal biology should be thrown away, since it is based on rather simplified and sometimes even misleading exercises in hypothesis testing. Fortunately, no such drastic step is required. As we mentioned earlier, most published papers do not really test hypotheses and make use of *p* values in the way the sections of those papers devoted to statistical analyses would suggest; rather, scientists compare alternative hypotheses on the basis of the data gathered and what was previously known about the systems in question. Moreover, in many cases, the inferences based on the classic *p* value approach are similar to those that would be arrived at using the alternative methods described here.

Why bother with a new way of doing things, then? For two reasons: First, there are cases in which the two approaches will yield very different results. In particular, *p* values are problematic when sample sizes are either too small or very large (both rather common occurrences in the biological literature, as it turns out). In the first case, when sample sizes are too small, a failure to reject the null hypothesis may be premature because the problem may be the low power inherent in small sample sizes. In the second case, when sample sizes are very large, *p* values may lead researchers to reject the null hypothesis when the difference between effects is scientifically irrelevant, though technically nonzero (as we have seen, this is because *p* values refer to a point null hypothesis, which is never of interest to scientists; see fig. 10.1).

The second reason to opt for alternatives to standard hypothesis testing is philosophically more relevant. Often we are simply not interested in the hypotheses tested with the standard methods, and science would proceed more quickly if we explicitly compared the relative merits (using the data) of a series of reasonable alternative models (hypotheses). That is something that standard hypothesis testing simply cannot do, but that scientists naturally tend to implicitly incorporate into the structure of their research proposals and papers. An explicit acknowledgment of this problem would help them and their students to be clearer about the questions they are asking and how to go about answering them. As noted earlier, a difference of opinion about how to interpret a graph may point toward more meaningful disagreements and possible experiments that would distinguish among several possibilities; a claim that a result is not significant because the *p* value falls above a magic cutoff rarely leads to such discussion.

Why is it that these rather elementary considerations, known to any statistician and to many researchers in other fields, have failed to bring about a change in the way organismal biologists present their results in published papers? As we have seen, statisticians and some scientists in other disciplines have been aware of the problems with the standard approach to hypothesis testing and the concept of the null hypothesis essentially since their inception. Yet discontent has been slow to build, and, perhaps oddly, started to surface first in the social sciences.[4]

There are several reasons why the *p* value and null hypothesis testing approach is still so widespread, some of which have been hinted at above. First, quantitative biologists actually do think about their work, and write the discussion sections of their papers, largely by disconnecting their conclusions from the *formal* results of the statistical tests they present. Most practicing biologists realize that what is important is the comparison of different hypotheses and the extent to which each is favored by the data, not the rejection of simple nulls. That is why the discussion section in any given paper usually focuses on comparing different hypotheses, not on interpreting the formal statistical results. Second, several of the alternative methods are not widely implemented in the available commercial software and so are not familiar to most researchers, even though they are conceptually simpler and more intuitive than the standard approach. Third, and perhaps most important, once a system of textbooks and teachers built around the standard approach is in place, it is difficult to do without it because of simple intellectual inertia. Indeed, many editors and peer reviewers are reluctant to take seriously an article that fails to use traditional statistical analyses and *p* values (one of us has found this out through repeated empirical attempts!). That is why we went into some rather practical details in this section—once people learn to do things one way, and editors and reviewers are happy with the status quo, it is often counterproductive for individual researchers (and especially graduate students) to try to change things. The fourth reason is the philosophically interesting one: as we mentioned above, many scientists learn very little philosophy of science, and what they do read is outdated; nonetheless, most practicing scientists fiercely cling to a naively Popperian view of hypothesis testing, a major contributor to the problems discussed in this and previous chapters.[5]

4. See, e.g., Cronbach 1975 and Meehl 1978; notice especially the latter's direct reference in the title to the symmetry between Fisherian statistics and Popperian philosophy of science: "Theoretical Risks and Tabular Asterisks: Sir Karl, Sir Ronald, and the Slow Progress of Soft Psychology."

5. We would like to note in passing that while this case of philosophy hindering science may seem to buttress Weinberg's "case against philosophy" (see the prelude), it actually does not. The

All of the above reasons notwithstanding, the situation needs to change, for a number of very good reasons, in order to foster better and clearer thinking at all stages of the scientific enterprise, from teaching of undergraduate students to the publication of intellectually more satisfying papers. We continue this chapter with an examination of two examples of research and debate in evolutionary biology that we believe are rooted in the problems caused by inadequate null hypotheses, more generally known as "neutral models": the Hardy-Weinberg equation in population genetics and Kimura's theory of molecular evolution.

FROM NULL HYPOTHESES TO NULL MODELS AND THEORY BUILDING

The use of null expectations is deeply embedded in all of the most common methods of statistical analysis used in the quantitative sciences, and ecology and evolutionary biology are no exception. But how does the idea of null hypotheses (and its tight philosophical connection with the concept of falsification discussed above) relate to the broader issue of theory building in biology? What sort of work are null models supposed to do in the context of a partially historical science such as evolutionary biology? We argue that thinking in terms of testing the empirical world against neutral models is fraught with the same kinds of problems that underlie the use of null statistical hypotheses. These problems have unnecessarily limited the work of evolutionary biologists and ecologists and have led to a disconnect not only between the data presented in research papers and the way those papers are interpreted, but also between empirical research and theoretical modeling more broadly.

At their most basic, we think of null models in quantitative biological theory as the natural extension of null hypotheses in statistical analysis: they answer the question, "What would happen if nothing were going on?" This role is immediately complicated, as Sober (1984, 32) remarked, by the need to consider exactly what is not expected to change under a neutral model. For example, Sober contrasts Aristotelian and Newtonian physics, where in the first case "nothing happening" to a given body means that the position of the body does not change, while in the second case "nothing happening" means that no forces are acting on the body, and hence the velocity of the body remains unaltered. One function of null models, therefore, is to indicate what one can expect to be invariant in a system, as well as to force the researcher to state explicitly what kinds

problem here is caused by scientists not being up to date with the philosophy, not by current philosophical ideas slowing science down.

of things (e.g., the action of mechanical forces in Newton's theory) are not happening.

An additional role of neutral models may be, as we shall see, to provide a realistic description of the system under study—not just a simplified reference situation. Kimura's (1983) neutral theory of molecular evolution, for example, was proposed to account for how evolution usually proceeds at the molecular level, and hence was supposed to explain what really accounted for molecular polymorphisms; this is clearly not quite the same as asking what would happen if nothing were going on. Although these two roles of null models are not logically mutually exclusive, they do fulfill quite different purposes and reflect some of the distinct interests and projects of empirical biologists.

Rather than examining these issues in abstract terms, we build on our preceding discussion of null hypotheses in statistics by considering two important cases of null models that have marked the history of evolutionary biology during the twentieth century: the Hardy-Weinberg equilibrium in population genetics (Hardy 1908; Weinberg 1908) and Kimura's (1968, 1983) neutral theory of molecular evolution. Our intent is not to provide a review of the vast literature on either of these ideas; such reviews are already available and would in any event be tangential to the focus of this book. Rather, our goals are to highlight what, if anything, these particular null models have in common by virtue of being null models, and to highlight the conceptual uses to which they have been put by biologists in the course of the development of the theoretical framework of their fields, including their use in guiding empirical research. Throughout this discussion, the reader should keep in mind that evolutionary biologists study all the concepts discussed so far in this book (natural selection, genetic variance-covariance matrices, constraints, etc.) by using null models as the proper contrast, so that what follows applies mutatis mutandis to essentially all the major components of the currently accepted framework of evolutionary theory. We conclude this chapter with a broader discussion of the usefulness of null models in organismal biology and of some possible alternatives.

The Fundamental Basis of Population Genetics: The Hardy-Weinberg Principle

The Hardy-Weinberg principle (H-W) is so fundamental to population genetics theory (and, as we will see shortly, to its practice) that it is taught in every introductory course on the subject, and often even in general courses on evolutionary biology or genetics. In its simplest form, the principle can be summarized in a few sentences. In essence, it says that—

given certain assumptions—the following equation can be used to predict the genotypic frequencies in a population at equilibrium:

$$p^2 + 2pq + q^2 = 1, \qquad (10.10)$$

where p and q represent the frequencies of two alleles (let's say, A and a) at the same locus in a population, the two quadratic terms are the frequencies of the respective homozygote genotypes, and the middle term is the frequency of heterozygotes. The sum of these three quantities is equal to one by definition (since these are the only three genotypes possible with two alleles at one locus).

The list of conditions under which a population is expected to be at Hardy-Weinberg equilibrium includes the following (Hartl and Clark 1989):

1. The organism is diploid
2. Reproduction is sexual
3. Generations are nonoverlapping
4. Mating is random
5. Population size is very large (technically, infinite)
6. There is no migration
7. There is no effect of mutation
8. There is no natural selection on the alleles in question

The principle can also be extended to a few more complicated cases, such as polyploidy (i.e., more than two sets of chromosomes), multiple alleles at one locus, sex chromosome–linked loci, and two loci. Anything beyond that becomes too unwieldy for simple mathematical modeling and analytical solutions, and we leave population genetics to enter the realm of quantitative genetics (see chapters 2 and 4). Here we will concern ourselves only with the simple formulation of H-W for the case of one locus with two alleles in a diploid population, ignoring the possibility of sex linkage. Conceptually, the extensions of the basic principle mentioned above follow exactly the same logic as the basic version.

The importance that both biologists and philosophers attribute to Hardy-Weinberg is hard to overstate. As Hartl and Clark (1989, 33) say, "It is perhaps hard to believe that so simple a result from so simple a model can hold so widely and be so important." Indeed, this simple result does seem to hold "widely." A brief survey of the population genetics literature using the *Science Citation Index* revealed fifty-seven papers published during the first six months of 2003 that reported tests of the H-W predictions. Of these, 58 percent concerned human populations and the remainder a scattering of other animals (both vertebrates and invertebrates) and non-animals (plants, fungi). The existence of H-W equilibrium for all or almost all of the loci examined was confirmed in 54 percent of the studies. When deviations from the predicted genotypic frequencies were found and a possible reason given, 54 percent were attributed to selection, 38 percent to population substructuring (i.e., deviations from the assumption of population-wide random mating), and 8 percent to gene flow (migration) from other populations.

However, there is an obvious, and well-known, problem with such widespread confirmation of the simple equilibrium. As Hartl and Clark themselves (1989, 36) point out, "it is important to note that the Hardy-Weinberg principle is not very sensitive to certain kinds of departures from the assumptions"—namely, nonoverlapping generations, random mating, and especially large population size, migration, mutation, and selection. That, unfortunately, covers just about every important phenomenon with which evolutionary biology is concerned! Therefore, using departures from Hardy-Weinberg to "test" for these phenomena seems a poor strategy—or at least one likely to generate many "type II" errors (in which one fails to detect something that is actually going on). Indeed, Hartl and Clark put it rather dramatically when they cited the following example (1989, 37), concerned with the effect of selection (w_{ij} are the relative fitnesses of the three genotypes): "The Hardy-Weinberg proportions will hold even if $w_{11} = 1.0$, $w_{12} = 0.50$, and $w_{22} = 0.25$—that is, even if three-fourths of the *aa* zygotes and one-half of the *Aa* zygotes die before reaching adulthood! [exclamation point in original] . . . For plausible values of w_{11}, w_{12}, and w_{22} the deviations are expected to be small and detection therefore requires very large sample sizes." This is nothing new: an early warning of the severe limits imposed by logistics and statistical power on the use of H-W for detecting selection had already published back in 1959 by Lewontin and Cockerham. But what are we to make of an allegedly fundamental principle that cannot possibly hold in reality and that, furthermore, is known not to be sensitive to departures induced by most things that can happen to a real biological population? While models that are known to be, strictly speaking, false can be quite useful, their use must derive from either their heuristic value in guiding research or the ways in which their failures can point toward interesting empirical research (or the creation of new models) (see Wimsatt 1987). So are the actual uses of H-W justified on these grounds?

There are two reasons why evolutionary biologists think that H-W is important despite its extreme simplicity and its well-known lack of sensitivity to deviations. One of these reasons is of empirical concern to evolutionary biologists, though it also has interesting conceptual implications; the other is fundamental to theoretical population genetics and to evolutionary theory itself, and thus has important ramifications for the philosophy of biology.

Let us start with the empirical aspect of the question. Even a cursory look at how empirically minded biologists use Hardy-Weinberg (for example, our limited survey of the *Science Citation Index*) reveals that they do consider it a practical tool to check whether real populations show detectable deviations from the neutral expectation. If that is the case, researchers then examine the distribution of genotypic frequencies more closely, beginning by determining the relative proportions of homozygotes and heterozygotes to establish which class is in excess compared with the H-W prediction. Some well-defined categories of explanation are then available, based on the kind of excess observed and on general knowledge of the biology of the organism in question. For example, a dearth of heterozygotes can be attributed to selection against such genotypes. If the genetic system under study is known to be subject to selection, then the selective explanation is assumed to be correct.

The problem is that factors other than selection can yield the same pattern of deviation from Hardy-Weinberg, and therefore complicate (and often make impossible) a straightforward biological interpretation. Multiple underlying causal structures can, and often do, generate similar observable patterns; failure to take seriously the difficulty of mapping patterns to processes has been a recurring theme in this book. In our example, a well-recognized alternative cause of heterozygote underrepresentation is population substructuring, which yields the so-called Wahlund effect (Hartl and Clark 1989, chapter 6). The idea behind the Wahlund effect is that if an apparently continuous population actually comprises two or more partially isolated groups, there will be less random mating than expected. If one were to calculate H-W proportions *within* each subpopulation, they might be found to be in equilibrium (assuming that no additional factors were causing detectable deviations), but when one pools the data from the population at large, one finds a deficiency of heterozygotes because of the unaccounted missed mating opportunities.[6] In other words, as a consequence of the many-to-many relationship between underlying causes and observable patterns, lack of fit to the Hardy-Weinberg predic-

6. On the other hand, if one were to *physically* "fuse" two populations into one population that could then breed freely, one would instead find an excess of heterozygotes because there would be more random mating than before (Hartl and Clark 1989, 282).

tions cannot be unambiguously attributed to any particular cause. Additional lines of evidence will always be needed in order to figure out what is actually causing the deviations from the null hypothesis. Unfortunately, such supplementary information is seldom provided in published papers, and more often than not, when such deviations from the null model are detected, one of the possible causes is identified as the likely one without sufficient evidence.

This is not merely a case of the classic underdetermination of hypotheses by the data that we flagged above. Rather, the problem, outlined by Lewontin and Cockerham (1959) and reiterated by Hartl and Clark in the quotations given above, is that while finding deviation from Hardy-Weinberg very likely means that *something* is going on, the lack of such deviation is not at all good evidence that nothing is happening. Hence, a lack of deviation leaves the biologist unable to draw any useful conclusion at all. This is a general problem typical of the use of uninformatively exact null hypotheses in statistical analysis, as we saw above. In the case of H-W, it is a notorious feature and, in many circumstances, is essentially impossible to circumvent. To see why, consider what scientists can typically conclude when they are unable to reject a null hypothesis: either they can clearly attribute the result to low statistical power (because of small sample sizes) or, more informatively, they can attribute it to the fact that there really is no interesting systematic effect causing a deviation from the null. While one cannot prove a negative conclusion, the hope is that one will be forced to accept it as real if one can show that the problem is not simply one of low statistical power. But Lewontin and Cockerham (1959; see their table 1, 563) have demonstrated in the case of Hardy-Weinberg that there are many reasonable combinations of relative fitnesses of the various genotypes at one locus for which detection of a deviation from H-W takes sample sizes on the order of tens or hundreds of *thousands* of individuals. This is not only much more than any laboratory can hope to achieve, but in many instances is orders of magnitude more than the actual size of the entire wild population itself!

The theoretical use of Hardy-Weinberg in population genetics and, indeed, in evolutionary theory is even more important than its (rather limited, as we have pointed out) practical application. Textbooks such as Hartl and Clark's (1989) present H-W at the beginning and use it to provide the conceptual framework for all that follows. This makes sense, given that population genetics is concerned precisely with what happens to natural populations—at the level of one or few loci—when one of the many assumptions of Hardy-Weinberg is violated. But there is much more to H-W than this, especially from a philosophical standpoint.

Sober (1984, especially 32), for example, elevates the neutral expectations of population genetics to the status of "zero-force" law for the whole of evolutionary theory. His articulation of this position is the clearest and most comprehensive available, but biologists themselves clearly think of Hardy-Weinberg in the same fashion, as is transparent from the way the word "force" is used in textbooks such as Hartl and Clark's. That is why we began this book with an in-depth discussion and critique of the view of selection and other evolutionary mechanisms as forces (chapter 1). In the context of Hardy-Weinberg, we wish to note that while biologists continue to debate how to test for the different effects of selection and drift in natural populations (e.g., Orr 1998; Roff 2000), philosophers disagree on whether the two are even distinguishable conceptually (e.g., Beatty 1984; Millstein 2002; Walsh, Lewens, and Ariew 2002) and suggest that in any case, their "effects" are not directly comparable because the two are not best thought of as alternative mechanisms pushing populations one way or another (Matthen and Ariew 2002; Walsh, Lewens, and Ariew 2002).

We propose that—if one has to have a null model in evolutionary theory—a more sensible (but mathematically less tractable) alternative to the "nothing happens" of Hardy-Weinberg equilibrium is a random walk (box 10.2). Natural populations are always finite (and therefore always violate at least one of the assumptions of the H-W law) and are thus constantly subjected to stochastic fluctuations in the frequencies of both their phenotypic and their genotypic attributes. Notice that a stochastic process such as a random walk can result in the *appearance* of directional trends because of the cumulative effects of individual fluctuations. This is exactly what happens with (so-called) random drift, and it is an outcome that one can expect in a variety of situations, from the path taken by a drunk in the streets to the appearance of a macroevolutionary "trend" over time (as discussed at length in Gould 1996).

Our recommendation to use a random walk instead of no change as a null model in population biology comes directly from our critique of the usefulness (or rather, the lack thereof) of the Hardy-Weinberg equilibrium, which is in turn a result of our philosophical skepticism about null hypotheses in general. It is also a result of the observation outlined by Sober (1984), mentioned at the beginning of this chapter, that the crucial question with null models is what exactly is *not* supposed to change. This question arises because not all null models imply no change at all in the object of study (again, think of Newtonian versus Aristotelian physics). Our answer is that gene frequencies should be expected to change, following a random walk, often in what appear to be specific directions. The

Box 10.2 Random Walks as Null Models

In statistics, a random walk is defined as "a walk in which the walker's movements are a consequence of a sequence of observations on one or more random variables. For example, suppose that, at each time point an individual walks one step to the left (with probability p) or one step to the right (with probability $1 - p$). This simple Markov process is a one-dimensional random walk. If the individual is allowed at each stage to move one step in any direction in two dimensions, then this is the drunkard's walk" (Upton and Cook 2002).

Random walks can occur in any number of dimensions and have some interesting properties that can be counterintuitive. For example, suppose that in a one-dimensional random walk, $p = 0.5$ (i.e., there is an equal probability of movement in either direction). One might intuitively expect that the times spent on either side of the starting point will be approximately equal. It turns out, instead, that the percentage of time spent by the system to the left or the right of the starting point is likely to be close to zero or 100 percent, because of the cumulative effect of each individual step. Understanding this property is crucial if one wishes to use random walks as "null" models in biology (or other sciences), since the expectation is *not* one of no average movement of the system, unless the system is replicated many times (in which case there will be no directional movement computed over all replicates, but each replicate will show directional trends and apparently intriguing patterns—see chapter 4 for an example concerning the evolution of genetic variance-covariance matrices in *Drosophila*).

Statistical tests to detect random walks (or deviations from them) are available and are often used to discriminate between deterministic and stochastic dynamics (e.g., Kaplan and Glass 1992). Organismal biologists are aware of the use of random walks as more sophisticated null models than the usual "no change" option, although their use is sparse and is probably limited by both the lack of familiarity with the concept and the scarcity of software able to perform the necessary computations. For example, Den Boer (1991) showed that fluctuations in population size in sixty-two species of carabid beetles could be explained by a random walk without having to invoke ecological mechanisms of regulation.

More frequent (though still fairly rare) examples of applications of random walks in evolutionary biology are found in the paleontological literature. For instance, Roopnarine, Byars, and Fitzgerald (1999) discussed the limits of using random walks as null hypotheses in the study of anagenetic evolution (i.e., gradual morphological changes within species) in the fossil record, and pointed out that short data sets that are not generated by a random walk can be statistically indistinguishable from a true random walk because of the fractal nature of random walks. Sheets and Mitchell (2001) examined the power of random walks to provide a null hypothesis for paleontological time series intermediate between the cases of directional and stabilizing selection (see chapter 2). The ability to use random walks as a null hypothesis is of general importance to evolutionary biologists interested in distinguishing between different selection regimes and random drift (see chapter 1). Sheets and Mitchell, however, pointed out possible biases in the failure to reject the null hypothesis of random walks when directional rather than stabilizing selection has been acting.

Our point here is not that random walks are a panacea for null hypotheses. Indeed, in the main body of this chapter, we argue that the whole idea of null hypotheses is not particularly fruitful in scientific research. Rather, it seems clear that more nuanced and interesting alternatives to the classic approach are available, and that they should be explored to potentiate organismal biologists' ability to distinguish between different causal hypotheses competing to explain the patterns they observe in nature.

problem is that while random walks are certainly tractable mathematical entities, they are clearly not as simple as something like Hardy-Weinberg. Indeed, there are intriguing problems for statisticians interested in constructing tests based on random walks, as well as for biologists wishing to implement such tests. Further, even given a test with well-characterized properties, it is often unclear how one ought to interpret the results: as we have seen here, the expectations of Hardy-Weinberg can be met despite the fact that something of interest is actually going on, and certainly the expectations of a random walk can be met under a variety of different circumstances as well. Nevertheless, we think that random walks can provide the basis of a more useful model that can help us clarify (or, better, dissolve) the general issue of how to detect genetic drift versus selection—another long-standing debate in evolutionary biology, as we have seen earlier. As it happens, perhaps the best-known example of the drift-selection debate is also rooted in a deceptively simple null hypothesis: Kimura's theory of neutral molecular evolution, to which we now turn.

The Evolution of the Neutral Theory of Molecular Evolution

The neutral theory of molecular evolution was proposed by Motoo Kimura and some of his close collaborators (Kimura and Crow 1964; Kimura 1968, 1983; Kimura and Ohta 1971). While the mathematical details of the theory are complex, its basic claim is fairly simple: most evolution at the molecular level (specifically, as originally formulated, in terms of protein sequences) is either neutral (i.e., not subject to selection) or strongly deleterious (and hence subject to "purifying" selection).

There are two things we must clarify at the outset of our discussion, as they may otherwise bring much confusion. First, Kimura's theory is *not* a theory of evolution at the phenotypic levels classically contemplated by evolutionary biologists interested in selection. The theory most certainly does not say that phenotypic evolution is neutral, and hence does not impinge directly on the other famous debate about selection—the one between supporters and critics of the so-called adaptationist program in organismal biology (see chapter 5). Of course, the molecular and higher phenotypic aspects of evolution cannot be entirely decoupled; if there were not a reasonably reliable relationship between genes, proteins, and phenotypic traits, the possibilities for heritable variation would be much reduced, and adaptive evolution (at least as we know it) would be essentially impossible. But as long as the genotype-phenotype "mapping function" (Oster and Alberch 1982) is loose and complex enough, the two need not be directly related to each other. This means that we can examine Kimura's theory quite apart from parallel discussions on the relative importance of (so-called) drift and selection in other areas of biology.

Second, calling this theory "neutral" is a bit misleading. As stated above, the theory actually predicts quite a bit of negative selection at the molecular level, which should eliminate severely defective protein variants that may show up by mutation. What makes Kimura's theory "neutral" is its prediction about the causes not of molecular evolution per se, but of two other phenomena often debated in the evolutionary literature: molecular (protein) polymorphism within species and its macroevolutionary equivalent, sequence divergence among species. That is, the so-called neutral theory of molecular evolution actually states that the majority of genetic polymorphisms within species and sequence divergence among species is due to "genetic drift" rather than selection. This is important conceptually, but also historically. Before the era of molecular population genetics, the great debate had been between a view of natural populations as essentially genetically homogeneous (as proposed by some of the very early geneticists, such as Johanssen [1911]), offering little variation for selection to act on, and a view assuming plenty of genetic variation, and hence no impediment to Darwinian evolution (e.g., Dobzhansky and Spassky 1944). While the latter "selectionist" view prevailed during the modern synthesis (Mayr and Provine 1980), we had to wait until the mid-1960s for the relevant evidence at the molecular level to begin to emerge (Lewontin and Hubby 1966). Interestingly, this evidence concerned protein polymorphisms, and so it should not come as a surprise that Kimura's theory was born around the same time.

All of the above notwithstanding, the philosophically most interesting thing about Kimura's theory is the way in which the theory itself, and its use by biologists, developed and changed over time (see, e.g., Wayne and Simonsen 1998; Hey 1999; Ford 2002). In its original formulation, the neutral theory of molecular evolution was quite bold—it made very specific predictions about a host of biological phenomena, from the level of protein polymorphism to that of sequence divergence; from the properties of the "molecular clock" (the rate of interspecific sequence divergence) to the relative rates of evolution of different portions of the same protein (e.g., positions at which a mutation would or would not result in an amino acid substitution). These uses of Kimura's theory are still referred to as "strong inference" (Platt 1964) by some biologists today (e.g., Ford 2002, 1255 and 1256). These early articulations and uses of the theory therefore fit in very well with one traditional view of science; namely, that presented by Popper (1968), for whom the ability to make bold predictions is closely linked to their desirable property of being falsifiable (and hence to the whole idea of null hypotheses, as discussed earlier).

As a result, the neutral theory was seen as a classic case in which empirical evidence could decide the issue—and the debate raged on for almost two decades between "neutralists" and "selectionists" in the widespread conviction that it would soon be settled one way or the other. Despite the apparently clear-cut predictions of the model, the conviction that the debate would be clearly decided was rather odd; most of the combatants would have readily conceded that it does not make sense to ask whether proteins evolve by drift *or* (positive) selection (recall that everybody agreed on the role of negative selection). The real question was rather *how much* protein evolution could be explained by one or the other cause (but see chapter 1 on whether it makes sense to think of drift as a cause at all). That is, the question was inherently quantitative, not qualitative. Nevertheless, a definitive solution was expected and actively sought by various means (Hey 1999): continuous refinements of the theory, invention of more sophisticated statistical tests (see Wayne and Simonsen 1998 for a concise review), and—of course—additional data.

One of the first effects of all this activity was the modification of the theory in a subtle but crucial way by its defenders: it soon became the theory of "quasi-neutral" molecular evolution, meaning that Kimura (1983) acknowledged that *some* positive selection had to be going on at the level of protein sequences. If all selection were really reduced to the elimination of disadvantageous mutations, then protein function could hardly change over time—a conclusion that would create serious repercussions for the whole modern synthesis.[7] Indeed, data were beginning to accumulate that suggested a role for at least weak positive selection at the protein sequence level (and, more recently, much more evidence has emerged from studies of DNA sequences: Ford 2002). Weak selection had already been demonstrated by Fisher (1930) to be sufficient, when acting over long periods, to explain Darwinian (adaptive) evolution, so a quasi-neutral theory of molecular evolution went a long way toward reconciling the neutralist and selectionist camps. Part of the problem had been that while the selective scenario had always implied complex predictions dependent on exactly what sort of natural selection one envisioned as acting on protein sequences, the original neutralist hypothesis had been clear-cut. Not so for the quasi-neutral version of Kimura's theory, which gained realism at the cost of losing the lure of strong inference—and consequently the high degree of apparent "falsifiability" that had made it so tempting.

7. Curiously, some modern proponents of that neo-creationist movement known as "intelligent design" (e.g., Behe 1996) maintain a similar, if rather more absurd, thesis; namely, that selection cannot improve or change protein function. For in-depth critiques of this position, see Fitelson, Stephens, and Sober 1999; Shanks and Joplin 1999; Pigliucci 2003a.

The surprising upshot of the debate (Hey 1999) was that just when a large amount of new data on molecular evolution (at the DNA, rather than the protein, level) became available to biologists, the controversy was not so much resolved, as it simply went away. Hey (1999) pointed out that part of the reason for this surprising (and, to many scientists, disappointing) outcome was that the new data sets—while obviously germane to the understanding of polymorphisms and divergence at the molecular level—were not of the kind envisioned in Kimura's theory. The theory was specifically about selection and drift affecting the evolution of protein sequences, and it is no trivial task to translate the theory's mathematical structure to adapt it to a rather heterogeneous set of DNA-level markers, from sequences of known genes to those of promoters, from short repetitive sequences scattered throughout the genome to random bits of DNA picked up by amplification techniques such as the polymerase chain reaction (PCR).

Hey (1999) also comments that the neutralists had "won" the debate, but in a sense that was rather uninteresting. The new data did show that a surprising amount of DNA was neither coding for proteins nor, apparently, doing much else other than replicating itself. But of course, this was not the sort of genetic material in which selectionists would have expected to detect positive selection to begin with. And, to be fair to the selectionist group, "junk" DNA was not predicted by the neutralists either, again because of their focus on the evolution of protein sequences (which were all that was actually available from the empirical studies at the time of the formulation of the neutral theory).

Other interesting complications have emerged from recent studies of molecular evolution and selection (Hey 1999; Ford 2002). For example, we now know that some polymorphisms in natural populations involve (sometimes large) groups of markers that are not independent of one another because of physical linkage on the same chromosome. Kimura's theory, like the alternative selectionist scenarios, was formulated, for the sake of simplicity, in terms of individual mutations, but it turns out that linkage and selection can interact in complex ways; this indeed constitutes a major area of active research in modern molecular ecology. We need not get into the technical details here, but linked polymorphisms can be explained in at least two ways (which are actually opposite ends of a continuum): by "background selection" and by "selective sweeps." In the first case, the polymorphism is assumed to be created by slightly deleterious mutations at several loci present in genomic areas with low rates of recombination. The selective sweep hypothesis, on the other hand, posits that a few mutations are under intense directional selection, but they also "carry with them" all neighboring loci because of physical linkage (the "hitchhiking" effect). Background selection could be accommodated by a modification of the quasi-neutral theory, while selective sweeps are clearly cases in favor of the selectionist viewpoint. Discovering which kind(s) of selection occur is an empirical matter, but we suspect that researchers will demonstrate convincing cases of both mechanisms, and probably a host of intermediate situations as well.

As if all of this were not enough, one of the major contentions of the neutral model has been eliminated by novel empirical results: apparently, natural selection can "see" (see chapter 3 on the targets of selection) the codon usage that a cell employs to make a protein (Hey 1999). Because of the partial redundancy of the genetic code (i.e., the fact that the same amino acid can be coded for by more than one triplet of DNA, or codon), so-called synonymous substitutions—once considered one of the quintessential cases of neutral evolution—are not synonymous after all. As Hey puts it (1999, 37), "evolutionary genetics has become more difficult than it once seemed." And yet, he goes on to defend the usefulness of the neutral theory as a convenient null hypothesis to be used as "the baseline limiting case for virtually all evolutionary genetic theory," a view shared by Ford (2002). We tend to disagree, and suggest that while Kimura's theory has certainly played a pivotal role in the history of ideas in evolutionary genetics (and we concur with Hey that Kimura deserves to be added to the holy trinity of that field: Fisher, Wright, and Haldane), it is no longer helpful (if it ever was) as a benchmark for empirical studies.

This critical position of ours seems to be shared to some extent by Wayne and Simonsen (1998), who wrote a seminal review titled "Statistical Tests of Neutrality in the Age of Weak Selection." In that review, they discussed the several assumptions of the neutral theory as well as the advantages and limitations (many, as it turns out) of all the major statistical tests used by biologists to accept or reject the null hypothesis of neutral molecular evolution. Again, we refer the interested reader to the original article for the details, but some highlights are pertinent to our discussion. First off, the list of assumptions that make a test of the neutral model possible reads a lot like the comparable list of assumptions underlying the Hardy-Weinberg equilibrium given above, and then some:

1. Neutrality of polymorphic sites (see the discussion above about the interaction between selection and linkage)
2. Random mating
3. No migration
4. A very large population of sequences sampled

5. An infinite number of sites (no "multiple hits" within species)
6. No repeated mutations at the same site (no "multiple hits" between species)
7. Nonoverlapping generations
8. Constancy of the neutral mutation rate

We have seen that one of the problems plaguing the application of the Hardy-Weinberg principle is that it is insensitive to fairly large departures from the assumptions, which makes it a particularly weak null hypothesis. The same problem, and the identical conclusion, applies in the case of all tests for the neutrality of differences in molecular sequences (Wayne and Simonsen 1998, 238).

Furthermore, the two tests most commonly used for distinguishing selection from neutrality in molecular data suffer from an additional problem that has an interesting conceptual underpinning. These two tests are the Hudson-Kreitman-Aguadé (HKA) and the McDonald-Kreitman (MK). The HKA is a goodness-of-fit test, aimed at determining whether there is a correlation between polymorphisms (variation within species) and divergence (variation among species), as predicted by the neutral theory. The MK tests the neutral expectation that the ratio of nonsynonymous to synonymous substitutions is the same within and between species. Both tests require a careful (one might argue too careful) choice of an outgroup—that is, of a species that is phylogenetically distant enough (but not too distant) from the one of interest to make for a useful comparison. The problem (Wayne and Simonsen 1998, 238) is that the test can be biased by the choice of an outgroup that is either too closely or too distantly related to the focal species. The choice of an appropriate outgroup requires a rarely achieved accuracy of phylogenetic estimates, but there is no general guidance on how to make that choice. In fact, the "best" outgroup depends on the rate of neutral and selective evolution in the group of organisms in question, which is, somewhat awkwardly, precisely what the tests are attempting to measure.

A particularly vivid graphic illustration of the problem with testing neutrality, and especially with using it as a "null" model, comes from Wayne and Simonsen's figure representing the differences among several hypotheses of sequence evolution (fig. 10.3). The trees depicted there are genealogies of five molecular sequences as they should look according to (from left to right) the neutral model, the hypothesis of a selective sweep, the prediction from background selection, and a model of balancing selection. The thing to notice is that the shapes (the topologies) of all the trees are identical; all that changes is the relative length of the different branches. This means that the four underlying hypotheses do not make

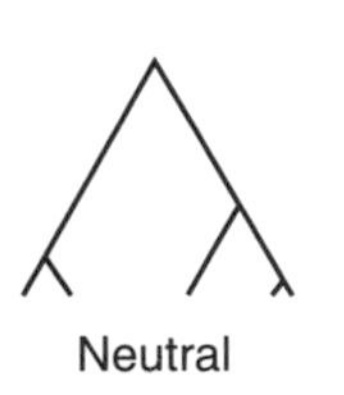

Figure 10.3. A graphic comparison of the genealogies of molecular sequences predicted by different hypotheses concerning the causal factors shaping molecular evolution. Notice that the topology (shape) of the tree is the same in all cases and that the differences among models are quantitative, not qualitative. As such, these are not discrete falsifiable hypotheses that can be subjected to strong inference (i.e., to clear-cut elimination based on crucial experiments), but rather represent a continuum of complex explanations that can be compared only by sophisticated methods, such as likelihood ratios or Bayesian analysis. (From Wayne and Simonsen 1998.)

qualitatively distinct predictions, and are therefore not falsifiable in the simple sense of the term: they do not lead to "strong inferences." Rather, one is faced with a continuum of possibilities that can be tested by empirically estimating a set of parameters for each model and comparing how well the sets account for the available data.

Wayne and Simonsen suggested a better approach to the problem than the more or less arbitrary a priori determination of a particular hypothesis as the "null": the alternative models can be compared using likelihood ratios (or, similarly, a Bayesian approach—both discussed above), which will allow two or more competing models to battle it out quantitatively for their ability to explain the available data. Given a certain data set, more than one model may temporarily emerge as the "victor," which will lead to another round of data collection, focused on the most likely models; this is in turn will be followed by another set of model-data comparisons, and so on (Hilborn and Mangel 1997). Different models will win in different cases, and in some instances the data will simply not be sufficient to pinpoint the one best model (Okasha 2000). This should be the expected outcome for a partially historical science like evolutionary biology, and hence the twin dreams of clear-cut "crucial" tests and a unique determination of the truth ought to be abandoned. Rather, we should relax and take a more realistic view of research as a kind of detective work; it is rare that we will be able to find all the evidence we might want, and we ought to expect to "crack the case" only some of the time.

The recurring theme in our analysis so far has been the usefulness of thinking in terms of null hypotheses to begin with and what—if any—

alternatives might be available. In the final section of this chapter, we broaden the discussion to a more general assessment of the use of neutral models in biology.

SO, WHAT GOOD ARE NEUTRAL MODELS IN BIOLOGICAL THEORY?

The question of the proper use, if any, of null models in biology has occasionally been pursued by practicing biologists, often with an overtly philosophical undercurrent. For example, Hoffman and Nitecki (1987) start out their discussion by invoking Hempel and Oppenheim's deductive-nomological pattern of scientific explanation, in which to explain a phenomenon means to identify a set of conditions that are sufficient for its occurrence. So, for instance, Mendelian genetics and our molecular understanding of mutations "explain" the origin (though not necessarily the maintenance) of genetic variation for quantitative characters in natural populations; while Darwin had good empirical evidence that phenotypic variation sufficient to fuel evolution was usually present in populations, he had no explanation for the phenomenon.

Accordingly, Hoffman and Nitecki present neutral models as "sufficiency models" (1987, 8) that can be used to test for the need for additional causal explanations. For example, Kimura's neutral theory of molecular evolution would be a sufficiency explanation for the evolution of protein sequences (and perhaps of DNA sequences) on the basis of mutation and drift. If the data indicate that the neutral model is not enough, then researchers have to start considering additional causal mechanisms, such as specific kinds of selection (e.g., selective sweeps or stabilizing selection). Wimsatt (1987) expands on this point, noting that models known to be oversimplistic can be used to detect (and estimate the magnitude of) the phenomena they fail to account for (by "factoring out" the phenomena explicitly considered in the model), or as a "limiting case to test the adequacy of new, more complex models" (30–31).

The problem, as we have shown so far, is that too often the null models chosen in biology are difficult to reject, either because of limited power in the data or because the models themselves are too flexible (e.g., compare Kimura's original neutral model with his much more flexible, and hence harder to falsify, "quasi-neutral" version). Indeed, Hoffman and Nitecki recognize this when they point out that there is a big difference between using a null model as a description of the data (i.e., when the data fit the null expectations "well enough") and using it as an explanation of those data (i.e., when one takes a further step and argues that the causal mechanisms used in the neutral model are what accounts for the empirical evidence).

In the same volume edited by Nitecki and Hoffman, Slobodkin (1987) takes a more critical attitude toward null models. He too starts with some philosophical background, observing that null hypotheses were originally used by biologists only as statistical tools (recall the problems even at this level that we pointed out in the first part of this chapter), but what he refers to as "Popperian zealotry" (100) really caused biologists to fall in love with null models. In fact, observes Slobodkin, if one takes Popperian mottoes such as "science consists of refutation of hypotheses" seriously, one ends up demolishing many significant advances in biological understanding because "no data set can survive testing by an infinite battery of null hypotheses." This is, of course, the problem pointed out by Lakatos (1977) and others: strict falsification is not possible because there is always the chance that something else (experimental errors, the assumptions that went into building the measurement apparatus, etc.) went wrong, while the hypothesis allegedly being tested was actually sound. Historically, strict (or naive) falsificationism has probably never been applied by scientists, or fine theories such as the Copernican and the Newtonian ones would never have gotten off the ground (and, again, it is likely that even Popper himself was not really a "naive falsificationist" in the sense described here).

Furthermore, Slobodkin refers us to an earlier study by Colwell and Winkler (1984), who discuss the problem of determining what constitutes a null hypothesis and how to choose the appropriate one, particularly in the field of biogeography. They conclude that building "appropriate, unbiased, null models is extremely difficult" (1984, 358). Slobodkin goes on to pin the problem with null hypotheses on the limitations intrinsic to the logical system that they seem to be based on—rejecting the null means accepting not just *one* alternative view of the world, but an infinite number of unspecified alternatives. The world, Slobodkin notes, simply does not seem amenable to that sort of investigation, and to keep asking binary questions (with one of them winning by default under a wide range of circumstances) does not seem to be the best way to proceed.

Fortunately, there are other, better, ways to proceed. Once one takes to heart the conclusion that null models are not as informative as they are so often assumed to be, other options will seem more tempting. As a real-world example, let us consider Stigler's (1987) exercise on hypothesis testing versus model fitting in the particular case of mass extinctions. Stigler, a statistician, takes a measured approach to the problem of null hypotheses. He does not claim (and neither do we) that they are useless, just that they are often less useful than other ways of proceeding. What is particularly convincing about his approach is that he does not limit himself to "preaching." Rather, he translates his advice into an elegant examination of a major question in organismal biology: are occasional mass extinctions a qualitatively different phenomenon from "background" extinctions? That is, are mass extinctions a different kind of thing than the usual extinctions that occur regularly?

In the few years immediately preceding Stigler's essay, major magazines for the general public published headlines suggesting that paleontologists were uncovering a mysterious periodicity in the frequency of mass extinctions in the fossil record, a periodicity that might hint at the existence of a regular extraterrestrial cause of such catastrophic events. Stigler starts out with a simple hypothesis, which can be framed in the classic mode of null models: perhaps extinction rates reflect a Poisson process [8] (he excludes a priori the hypothesis of a fixed flat rate of extinction as being of interest only to a hypothetical "group of eighteenth-century social philosophers who had rather restricted views as to what could be expected in a divinely ordered universe"). He subjects the Poisson hypothesis to a chi-square test and soundly rejects it. He then moves on, in a perfectly "classic" approach, to consider a nonhomogeneous Poisson process—that is, one in which the main parameter of the Poisson distribution varies with time. This second hypothesis can be tested by a regression model of the square root of the extinction rate versus time, another "null versus alternative" sort of model. This time he does find a monotonically decreasing trend (i.e., a significant regression line), but also a much larger scatter in the data than one might expect if a nonhomogeneous Poisson process were all that were going on.

Here is where things become interesting: Stigler points out that discarding one null hypothesis at a time is not likely to bring us very far because "acceptance may simply indicate an insensitive test, and rejection a test that is sensitive to an extraneous nuisance factor" (153). Rather, he proposes moving on to more sophisticated parametric modeling, in which several distinct hypotheses are made to compete with one another for the best available explanation of the data. He then goes on to set up four such hypotheses (fig. 10.4), two in which only "background" extinctions exist (in one case modeled by a narrow Poisson distribution; in the other case following a gamma distribution with a long tail), and two in which both background and mass extinctions occur (but in one case both with a narrow Poisson distribution, in the other with a mixture of gamma and Pois-

8. The Poisson and gamma distributions discussed here are two of the most flexible and useful distributions used in statistics and modeling. The Poisson is particularly apt to model rare events, while the gamma is flexible enough to approximate many different shapes, including the classic normal (bell-shaped) distribution.

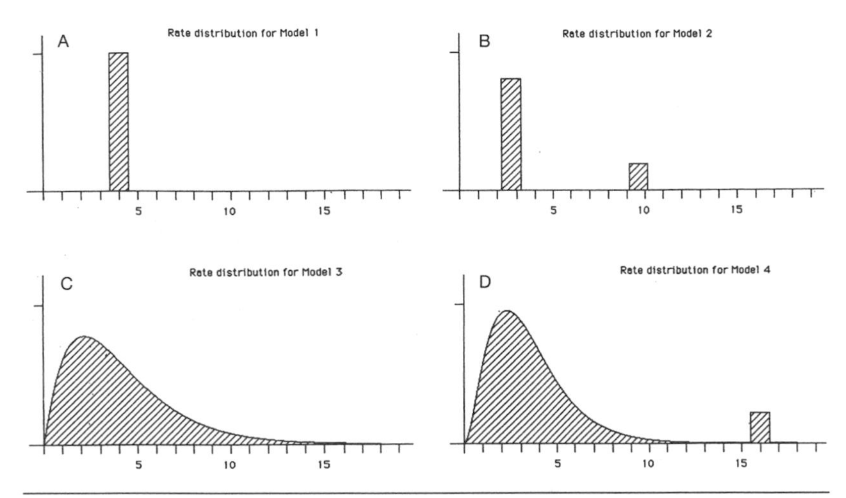

Figure 10.4. Four hypotheses formulated to explore the idea that mass and "background" extinctions are qualitatively distinct phenomena. The four graphs represent distributions of extinction rates predicted by the four hypotheses contemplated by Stigler: (a) low extinction rate governed by a Poisson process; (b) two extinction rates, both governed by Poisson processes with different parameter values; (c) one dynamic underlying extinction rates, with a low mode and a skewed distribution, conforming to a gamma distribution; (d) two dynamics, one following a gamma distribution, the second one the result of a Poisson process. The actual data fit (statistically) both hypotheses 3 and 4 well. (From Stigler 1987.)

son). Stigler then fits the various hypotheses to the data and determines the gain in likelihood of one model over the other (a method similar to the method of likelihood ratios discussed above). Since the different models are characterized by different numbers of parameters (and models with more parameters naturally fit the data better, other things being equal), he also carries out tests comparing models while taking the difference in degrees of freedom into account (a method similar to the Akaike Information Criterion discussed earlier).

The results are interesting because they illustrate quite a bit about the method of multiple hypotheses, the underdetermination of hypotheses by the data, and the tentative nature of scientific conclusions—all themes we have been discussing in this chapter and which underlie many of the criticisms leveled in the previous chapters. It turns out that both models featuring only Poisson distributions (either with only background extinctions or with background and mass extinctions) perform poorly when confronted with the data. The remaining two models, both featuring gamma distributions (but with the second model also incorporating a Poisson distribution for mass extinctions), do much better, but their relative performance is not very different. While technically the two-distribution model does better than the one-distribution alternative (thereby suggesting that mass extinctions are really distinct phenomena), the difference—given the available data—is not overwhelming, and Sterling concludes with a degree of cautious skepticism. Notice that an obvious way to tip the balance between the two remaining competing hypotheses would be to gather more data. To do this, we would not necessarily have to wait for another couple of mass extinctions; rather, we could redo the analysis at a different taxonomic level in order to provide finer resolution. Another way to proceed might be to add additional hypotheses to the mix (e.g., a single gamma distribution, but characterized by different enough parameter values that its tail goes further and possibly covers the range of rates typical of mass extinctions) and thereby continue to learn from the interplay between data and models. We find Stigler's example one of the best, most nuanced, and most useful available examples of how hypothesis testing ought to be carried out in the quantitative sciences. It is so much more intellectually satisfying than the simplistic black and white world of null hypotheses!

The use of null hypotheses has become ingrained in the modus operandi of organismal biologists, from testing hypotheses about selection to studying genetic drift, constraints, and molecular evolution. It should be clear at this point, however, that while there is nothing technically wrong with null models, they leave much to be desired from both conceptual and methodological viewpoints. If our analysis in this chapter and throughout the rest of the book is correct, they are certainly an approach that could profitably be given up and replaced with already available alternatives that are better—if less user-friendly—tools for the exploration of a complex world. We think that evolutionary biologists would do well to abandon both physics as their model of how to do science (Cleland 2001, 2002; Pigliucci 2002) and falsification as their philosophy of hypothesis testing. Organismal biology is a full-fledged science because it is built on empirically driven hypothesis testing, but this testing needs to be carried out in the fashion of a detective attempting to solve a complex puzzle by continuously comparing complex hypotheses with complex data sets. The shortcuts attempted so far are neither good "first approximations" nor render justice to the beauty of the research program started by Charles Darwin almost two centuries ago.

CODA A Philosophical Dialogue

TRUTH SPRINGS FROM ARGUMENT AMONGST FRIENDS.
—DAVID HUME

SKEPTIC: OK, guys, I must admit that I'm a bit baffled by this book. On the one hand, it seems to me that a lot of what you say has been said by others before—though I admit that perhaps there is some value in putting it all together into a coherent whole. On the other hand, you seem to suggest that there are some major flaws in the way evolutionary biologists conduct their business, and even in their conceptualization of crucial ideas in their field. What gives?

M&J: Actually, your summary may be closer than you think to what we intended, with a couple of caveats. Yes, it is true that a lot of what we say can be found elsewhere. This, however, isn't all that uncommon in modern academic books in either philosophy or, especially, science. Philosophy of science is becoming a more and more technical field, in which plenty of smart people have said plenty of smart things. That's why there is an extensive bibliography at the end of this and similar books! However, there is an overarching theme and an underlying argument being laid out here, chapter after chapter. That argument is most certainly *not* that evolutionary biologists don't know what they're doing, nor that the field is in a hopeless state of disarray. Rather, we suggest that there are two recurring problems: First, there is somewhat of a disconnect between the conceptual understanding of phenomena such as, say, natural selection and genetic drift and the way these are studied empirically. Second, and underlying the first point, there seems to be an interesting level of ambiguity in the conceptualization of some crucial ideas—for example, when people think of "constraints" in a variety of not necessarily related ways, from observable patterns of genetic correlations to the intricate molecular details of the genetic architecture of a set of characters.

SKEPTIC: Maybe, but does all this really mean that we need some major rethinking of the modern synthesis? If not, isn't this a tempest in a teapot? It sounds at times as if you guys want to have it both ways. You put forth some pointed criticisms throughout the book, but then occasionally appear as if you wish to hide the hand that threw the ball, perhaps for fear of the consequences.

M&J: There is no question in our minds that the modern synthesis—although extremely useful and historically productive—is in need of some major reworking. We aren't the only ones to suggest this. In the last few years, books by biologists such as Stephen Jay Gould, Mary Jane West-Eberhard, Eva Jablonka and Marion Lamb, and indeed, one of us (Pigliucci, with Carl Schlichting) have repeatedly attempted to outline what might be necessary for the next major developments of modern evolutionary theory. We do see this book in a similar fashion, though coming from a more philosophical perspective, and therefore focused more on conceptual criticism than on biological theory construction.

SKEPTIC: Ah! Precisely! Aren't you afraid that this book is going to be dismissed by biologists as just a lot of armchair speculation, the sort of thing that philosophers do all the time and that has never led to any significant advancement of science? You know, I wasn't too unsympathetic toward Steven Weinberg's attack "against philosophy" . . .

M&J: Yes, we are open (and sensitive!) to that charge. But the charge is, we feel, based on the same misunderstanding of the proper roles of philosophy and science that underlined Weinberg's outburst. Philosophy's role is *not* to solve scientific problems, which is why we refer repeatedly throughout the book to the work of Gould, West-Eberhard, and many others, if one is interested in suggestions about what to do from the point of view of a viable scientific research program. But we maintain that it is an intellectually valuable exercise to engage in philosophical analysis and criticism of what scientists do. Indeed, since scientists are obviously busy enough actually *doing* the science, we hope that some of them will read this book just to see what a different perspective can bring to their own activities. Again, the goal isn't to provide scientific solutions to scientific questions—something for which science itself is obviously eminently qualified and is astoundingly successful at—but rather to provide some food for thought, hopefully to scientists as well as philosophers, about how science works (or doesn't). This may even help to change some scientists' (especially the new generation's) ways of thinking about specific problems and may ultimately, perhaps indirectly, lead to the solution of actual scientific problems. But actually solving those problems isn't the goal of this book.

SKEPTIC: What do you say to the people who will complain that real practicing biologists are already aware of many of the issues you bring up?

For example, let's consider your discussion of studies of **G**-matrices as indicative of constraints on evolution. Surely you realize that people doing the field or laboratory work know their organisms very well and understand that the kinds of organisms, traits, and environments on which they conduct their research will make a difference. In fact, recent work by Steve Arnold and his collaborators is beginning to point out, and test, precisely these notions of context dependence. Nonetheless, biologists reasonably seek some general answers to the broad questions. What's wrong with that?

M&J: We are aware of (and cite) Arnold's and others' works. In fact, remember that our criticism is based on theoretical work (e.g., by Houle and Gromko) as well as on empirical papers detailing the many conceptual and practical problems with the study of **G**-matrices—works that were authored by biologists. Nonetheless, we perceive a clear disconnect between what "everybody knows" and what "everybody does," if you will. If one reads most papers, examines grant proposals, or attends conferences where biologists present their ideas about constraints and **G**-matrices, almost everyone is *talking* as if the problems were minor, the approaches being used really were "first approximations" (they are most emphatically not, at least not to causal inference), and there will be general answers to broad questions. We are very skeptical of the latter claim in particular. **G**-matrices are one of those fascinating beasts in biology in which almost everything is local. We will not be able to uncover general answers (other than to trivial questions) in this area simply because "**G**" refers to a highly heterogeneous set of statistical constructs, which in turn have only vague and indirect relationships to what people would really like to understand: the underlying genetic architecture of multivariate phenotypes.

SKEPTIC: OK, what about natural selection, then? I found your discussion of the two levels rather, how shall I put it, philosophical? I mean, even if one agrees with the analysis you propose as an elaboration of Matthen and Ariew's ideas, why should biologists be interested at all? Does it make any difference in practice for them? One of you is a practicing biologist: is this going to change the way you'll do things in your lab?

M&J: Good questions, and the answers are, as usual, many and nuanced. Yes, the distinction between formal and informal selection has a clear philosophical flavor and may have little appeal to an empirically minded biologist. But then again, remember our comment about the existence of a proper role for philosophy that is independent of the practical consequences for science. Philosophers really shouldn't apologize for not doing science, any more than scientists feel compelled to apologize for not being sufficiently philosophical. If there is a conceptual distinction to make between different ways of thinking about selection, that is intellectually interesting in its own right, period. However, we do think it *should* affect practicing biologists as well. Indeed, one of us (Pigliucci) is in the process of reeducating his graduate students accordingly. This doesn't mean that Lande-Arnold-style analyses of selection should be abandoned, only that they should be seen for what they are: relatively simple (compared, for example, with path analysis) statistical summaries that can be used to suggest (but not test!) causal hypotheses about natural selection. But one cannot *redefine* selection, as some authors have done, as a (partial or not) covariance between fitness and trait variation. It may be appealing in its statistical simplicity, but it is close to nonsensical, or at least uninformative, from a biological perspective.

SKEPTIC: Speaking of path analysis, I thought it was odd that you tried to cut the Lande-Arnold multiple regression approach down to size because it was eminently statistical in nature, and therefore incapable of addressing the underlying causal factors, and then turned around and suggested the use of path analyses—also a statistical tool! Aren't there several other approaches—field studies of various sorts, laboratory evolution experiments, and so forth,—that have been tried to study natural selection? What about those?

M&J: Funny you mentioned that; it hadn't occurred to us that our point could be missed so widely. Yes, path analysis is a statistical technique, and as such, it can be used as a more sophisticated version of, but essentially not differently from, multiple regression techniques (which in fact, mathematically speaking, are a subset of path analysis). However, as has been beautifully illustrated by Bill Shipley, path analysis is itself a special case of structural equation modeling, an approach that can be used to *model* causal paths and *predict* the statistical "shadows" they project. Such predictions can then be tested against actual empirical data. In one sense, path analysis is the quintessential tool of the biological detective, because it unifies statistical analysis (a must in the case of quantitative science) with causal hypothesis testing (a must for any science that wishes to move beyond pattern description). As for other approaches to studying selection, of course there are plenty, and we mentioned some examples throughout the book (a still valuable discussion of the many ways in which one can study selection can be found in the classic book by John Endler, by the way). But those other approaches fit in well with how we think selection ought to be studied, and they are part of the standard toolkit of evolutionary biology, so covering them in detail seemed unnecessary. Instead, we focused on one approach (well, the Lande-Arnold family of approaches) that is both widely used and believed to have particular power, and yet in our view is among the most problematic.

SKEPTIC: I was struck that when discussing selection, fitness, and the like, you kept using rather nonstandard terms such as "genetic resources" and "developmental resources" and referring to "developmental systems theory." First, it was never clear to me exactly what developmental systems theory was, or why I should care, and second, it was never clear what the language of "resources" was doing for you.

M&J: Taking the second question first, there are at least two things that the language of "resources" was supposed to do. First, we wanted to stress that genes are not the only source of heritable variation that can affect fitness. It has become increasingly clear that lots of other things are inherited reliably across generations, that those other things can (and do) vary between individuals, and that that variation can (and does) have phenotypic effects that can be associated with fitness differences. We hoped that our use of the language of developmental resources would keep these claims in the foreground. Second, when we wrote about "genetic resources," we did so because we didn't want to take a stand on what a "gene" is, how to identify genes, or what kinds of things "genes" can do. For example, a genetic resource may be a nucleotide sequence that, in the right cellular environment, "codes" for a protein. But it can also be a regulatory region, a binding site, or a number of other kinds of things. And all of these kinds of nucleotide regions are involved in and used in development, but none of them should be thought of as a privileged source of developmental control. Rather, they are all resources, constructively used during the development of organisms.

We tried in various places to give a bit of the flavor of developmental systems theory, but we recognize that we never presented a full summary or defense of the program. We think that's fine, actually—if you are interested in an extended defense of DST, there are a number of good books available (Oyama, Griffiths, and Gray's edited collection is a good place to start). This wasn't supposed to be another one. What this book is supposed to be is a defense of one way of making coherent the answers to a number of different conceptual difficulties and questions at the foundations of evolutionary biology. We believe that the insights of DST can be an important part of this coherent picture; its stress on the lack of a privileged site for the control of the developmental process, and other similar positions, resonate well with the views that we defend here.

SKEPTIC: Part of that makes me even less happy—namely, the idea that genes are a "developmental resource like any other." Surely that's a mistake. Are you really denying that genes carry most of the important information necessary for development? Or let me put this another way: We all know that all kinds of things are necessary for development, but, *pace* the weird cases you dug up, those things don't vary much between developing organisms. To use a hoary old metaphor, the genetic code is like a recipe. Yes, you need all the ingredients, and some of those ingredients are "information rich," but still, there is an important distinction between the ingredients and the recipe. When a recipe calls for a stick of butter, it's true that it doesn't contain the information about what a stick of butter is (or how to make one), and it is even true that different brands of butter (say) might give different results. But still, *most* of the important variation is variation in recipes, not in the raw materials. And the same relationship holds between genes and the other "resources" used in development. Is there really anything wrong with this picture?

M&J: Yes; in fact, there are a number of things wrong with it. On the empirical side, we think that the view you outline overestimates the importance of genetic variation and underestimates the importance of nongenetic (but still heritable) variation. This is, ultimately, an empirical point, but we believe that the evidence we cite, far from being a smattering of "weird" cases, is, as it were, the tip of the iceberg (many other cases are discussed in Jablonka and Lamb's book). On the more conceptual side, we want to reject the picture of genes as some stable repository of information (like a cookbook) that can turn raw ingredients into finished products; with the supporters of DST, or those people who stress various mechanisms of epigenetic inheritance, we think that the only ways in which genes can be said to carry information are the ways in which lots of other heritable resources can be said to carry information as well. But, again following the insights of DST, we think that the metaphor of information is misguided here; insofar as development involves information, that information is continually constructed by the process of development itself, which includes all the resources used, only some of which vary. Sometimes it is genetic variation that is associated with phenotypic differences, but sometimes the developmental system suppresses genetic variation; sometimes variation in other resources matters, but sometimes that variation is buffered. Neither is privileged in any particular way, and all must be explored on a case-by-case basis.

SKEPTIC: Well, I won't pretend I'm convinced, so let me just throw in another point that I think is going to be contentious: Do you really think that genetic drift doesn't exist? Are all these smart people who have published on this topic deluding themselves? Isn't this a rather extraordinary claim you are making?

M&J: It would be, if that's what we were saying. What we are actually suggesting is that "drift" isn't best *thought of* as the kind of thing that biologists usually think of when they talk about drift. It isn't, for example, a "force" (or whatever other metaphor one wishes to use) on the same level as the other phenomena that can throw a population off Hardy-Weinberg equilibrium. Selection, migration, mutation, recombination, are all—at bottom; that is, at the individual level—physical processes that cast a statistical shadow at the population level. But drift isn't a physical process at all, it is *only* a statistical shadow! That's why we suggest that it is best to think of drift as the (population size–dependent) variance around an expected outcome, where that outcome can be the predicted result of selection, migration, mutation, recombination, or any interaction among them. We know, you are about to fall back on your mantra: so what difference does this make for a practicing biologist? As we suggest in the final chapter, it does make a difference, because, for example, it means that random walks, not absence of change, should be the proper null hypothesis in population biology.

SKEPTIC: You realize, of course, that some of what you say here can and will be construed as a wholesale attack on the theory of evolution, which will help the public relations campaign of creationists and proponents of intelligent design. This is, as I'm sure you are aware, an even more serious concern than the actual scientific issues themselves, since it goes to the core of what science education means in an open society. Aren't you guys risking too much "collateral damage," so to speak? There are plenty of passages in the book that it wouldn't take a genius to quote only slightly out of context and make your arguments sound like music to a creationist's ears.

M&J: Perhaps. We have thought about this a lot. Indeed, one of us (Pigliucci) is, as you know, very active in the area of public outreach and teaching about the nature of science. There are two ways to respond to this criticism. First, creationists have always attempted to exploit legitimate disagreements among scientists as evidence of hopeless confusion within the field. But, frankly, this is either the result of extremely naive thinking about how complex activities like science actually work, or—worse—pure and simple intellectual dishonesty in the service of a narrow religious ideology. Having met several creationists, we can tell you that it is often hard to see where that line can be drawn, as self-deception is particularly easy for people who are absolutely convinced of being right. Second, and more important, we believe that the best way to eventually overcome the creationist threat is not to oversimplify science, but to present it to the public as the nuanced, provisional, and yet wonderful human undertaking that it really is. We are not suggesting that this book can do that for a general audience; it is too technical and is clearly aimed at graduate students and professional practitioners in biology and philosophy. But our hope is that it will trigger more appreciation of the provisional and messy methods of science among professionals themselves, who, after all, are the very same people who in turn train our science teachers and talk to journalists in public forums.

SKEPTIC: So, what's next? Where do you think Gould's, West-Eberhard's, Jablonka and Lamb's, and your attempts at conjuring a crisis for modern evolutionary theory are going to go? Do you guys see yourselves as the partial agents of a Kuhnian paradigm shift? Is evolutionary biology really mature for a new synthesis? And shouldn't that have come, if anything, from the so-called "evo-devo" (evolution of development) research program? What happened there?

M&J: Interesting language you use. We don't think we are trying to "conjure" a crisis: evolutionary theory isn't in crisis, but there is widespread (if not yet universal) acknowledgment that there is plenty of room for major improvement. There is also a feeling that such an improvement is about to mature, and of course, we and the other authors you mention are hoping to make our modest contribution to it. Evo-devo is an interesting beast on its own. It was hailed a few years ago as *the* long-awaited completion of the modern synthesis, the integration between developmental biology and the rest of evolutionary theory. What has actually happened is interesting: on the one hand, evo-devo research has produced, and continues to yield, fascinating insights about how animal and (to a lesser extent) plant body plans are put together during development, as well as information on the likely evolution of some of the underlying molecular mechanisms. However, it did not, and in fact cannot, deliver on its major promise, for an interesting reason: although we may say that we want to know how, say, major body plans evolved, or how evolution produced phenotypic novelties, a molecular study of master switches (which is most of what evo-devo has quickly become) can provide us with only a (very) partial answer. The remainder of the answer will have to come from knowledge of the selective pressures, ecological conditions, phylogenetic relationships, and even genotype-environment interactions of the plant and animal forms in question. These other pieces of the puzzle are much harder to come by, but without them we have only a highly incomplete integration of development into the main body of evolutionary theory. Development is much, much more than just the action of a series of master genes, *pace* modern genomics.

As for paradigm shifts, one can actually make a very good case that evolutionary theory has *never* gone through one of these (unlike, say, the shift from Ptolemaic to Copernican astronomy, or from Newtonian to relativistic physics, Kuhn's famous examples). That is because people have kept building on the foundation laid by Darwin, not replacing the foundation itself. We don't see this situation changing during the next synthetic phase (and neither did, for example, Stephen Jay Gould). Even after additional major empirical and conceptual advancements, evolutionary theory will remain essentially "Darwinian." The Master was (largely) right after all.

References

Abelson, R. P. 1995. *Statistics as Principled Argument.* Hillsdale, NJ: Lawrence Erlbaum Associates.

Alberch, P. 1991. From genes to phenotypes: Dynamical systems and evolvability. *Genetica* 84:5–11.

Anderson, D. R., K. P. Burnham, and W. L. Thompson. 2000. Null hypothesis testing: Problems, prevalence, and an alternative. *Journal of Wildlife Management* 64:912–23.

Ang, L. H., and X. W. Deng. 1994. Regulatory hierarchy of photomorphogenic loci: Allele-specific and light-dependent interaction between the *HY5* and *COP1* loci. *Plant Cell* 6: 613–28.

Antonovics, J., and P. H. van Tienderen. 1991. Ontoecogenophyloconstraints? The chaos of constraint terminology. *Trends in Ecology and Evolution* 6:166–68.

Ariew, A., and R. C. Lewontin. 2004. The confusions of fitness. *British Journal for the Philosophy of Science* 55:347–63.

Arnold, S. J., M. E. Pfrender, and A. G. Jones. 2001. The adaptive landscape as a conceptual bridge between micro- and macroevolution. *Genetica* 112/113:9–32.

Arnold, S. J., and P. C. Phillips. 1999. Hierarchical comparison of genetic variance-covariance matrices. II. Coastal-inland divergence in the garter snake, *Thamnophis elegans. Evolution* 53:1516–27.

Asins, M. J. 2002. Present and future of quantitative trait locus analysis in plant breeding. *Plant Breeding* 121:281–91.

Atchley, W. R., and B. K. Hall. 1991. A model for development and evolution of complex morphological structures. *Biological Review* 66:101–57.

Balter, M. 2005. Are humans still evolving? *Science* 309:234–37.

Barton, N. H., and M. Turelli. 1989. Evolutionary quantitative genetics: How little do we know? *Annual Review of Genetics* 23:337–70.

Bazylinski, D. A. 1999. Synthesis of the bacterial magnetosome: The making of a magnetic personality. *International Microbiology* 2:71–80.

Beatty, J. 1984. Chance and natural selection. *Philosophy of Science* 51:183–211.

Bégin, M., and D. A. Roff. 2001. An analysis of G matrix variation in two closely related cricket species, *Gryllus firmus* and *G. pennsylvanicus. Journal of Evolutionary Biology* 14: 1–13.

Behe, M. J. 1996. *Darwin's Black Box: The Biochemical Challenge to Evolution.* New York: Free Press.

Bell, G. 1997. *Selection: The Mechanism of Evolution.* New York: Chapman & Hall.

Bengtsson, B. O., and F. B. Christiansen. 1983. A two-locus mutation selection model and some of its evolutionary implications. *Theoretical Population Biology* 24:59–77.

Bennington, C. C., and J. B. McGraw. 1996. Environment-dependence of quantitative genetic parameters in *Impatiens pallida. Evolution* 50:1083–97.

Berger, J., and W. Jefferys. 1992. Ockham's razor and Bayesian analysis. *American Scientist* 80: 64–72.

Bjorklund, D. F., and A. D. Pellegrini. 2002. *The Origins of Human Nature: Evolutionary Developmental Psychology.* Washington, DC, American Psychological Association Press.

Block, N., O. Flanegan, and G. Guzeldere, eds. 1997. *The Nature of Consciousness.* Cambridge, MA: MIT Press.

Bohning-Gaese, K., and R. Oberrath. 1999. Phylogenetic effects on morphological, life-history, behavioural and ecological traits of birds. *Evolutionary Ecology Research* 1:347–64.

Bouchard, T. J. Jr., D. T. Lykken, M. McGue, N. L. Segal, and A. Tellegen. 1990. Sources of human psychological differences: The Minnesota study of twins reared apart. *Science* 250: 223–28.

Boyd, R. 1999. Homeostasis, species, and higher taxa. In *Species: New interdisciplinary essays,* ed. R. A. Wilson, 141–85. Cambridge, MA: MIT Press.

Bradshaw, A. D. 1965. Evolutionary significance of phenotypic plasticity in plants. *Advances in Genetics* 13:115–55.

Bradshaw, A. D., and K. Hardwick. 1989. Evolution and stress—genotypic and phenotypic components. *Biological Journal of the Linnean Society* 37:137–55.

Brandon, R. N. 1982. The levels of selection. In *PSA 1982: Proceedings of the 1982 Biennial Meetings of the Philosophy of Science Association,* vol. 1, ed. P. D. Asquith and T. Nickles, 315–23. East Lansing, MI: Philosophy of Science Association.

Brandon, R. N., and M. D. Rausher. 1996. Testing adaptationism: A comment on Orzack and Sober. *American Naturalist* 1481:189–201.

Brodie, E. D. III. 1993. Homogeneity of the genetic variance-covariance matrix for antipredator traits in two natural populations of the garter snake *Thamnophis ordinoides. Evolution* 47:844–54.

Brodie, E. D. III, A. J. Moore, and F. J. Janzen. 1995. Visualizing and quantifying natural selection. *Trends in Ecology and Evolution* 10:313–18.

Buller, D. 1998. Etiological theories of function: A geographical survey. *Biology and Philosophy* 13:505–27.

———. 2000. A guided tour of evolutionary psychology. In *A Field Guide to the Philosophy of Mind,* ed. M. Nani and M. Marraffa. An official electronic publication of the Department of Philosophy of University of Rome, http://host.uniroma3.it/progetti/kant/field/ep.htm.

———. 2005. *Adapting Minds: Evolutionary Psychology and the Persistent Quest for Human Nature.* Cambridge, MA: MIT Press.

Burian, R. M. 1997. Comments on complexity and experimentation in biology. *Philosophy of Science* 64:S279–91.

Buss, D. M. 1994. *The Evolution of Desire: Strategies of Human Mating.* New York: Basic Books.

———. 1995. Evolutionary psychology: A new paradigm for psychological science. *Psychological Inquiry* 6:1–30.

———. 1999. *Evolutionary Psychology: The New Science of the Mind.* Boston: Allyn & Bacon.

———. 2002. Human mating strategies. *Samfundsokonomen* 4:47–58.

Callahan, H. S., and M. Pigliucci. 2002. Shade-induced plasticity and its ecological significance in wild populations of *Arabidopsis thaliana. Ecology* 83:1965–80.

Callahan, H. S., C. L. Wells, and M. Pigliucci. 1999. Light-sensitive plasticity genes in *Arabidopsis thaliana:* Mutant analysis and ecological genetics. *Evolutionary Ecology Research* 1:731–51.

Camacho, J. P. M., T. F. Sharbel, and L. W. Beukeboom. 2000. B-chromosome evolution. *Philosophical Transactions of the Royal Society of London* B 355:163–78.

Carson, H. L., and A. R. Templeton. 1984. Genetic revolutions in relation to speciation phenomena: The founding of new populations. *Annual Review of Ecology and Systematics* 15:97–131.

Cartwright, N. 1983. *How the Laws of Physics Lie.* New York: Oxford University Press.

Casal, J. J. 1995. Coupling of phytochrome B to the control of hypocotyl growth in *Arabidopsis. Planta* 196:23–29.
Casal, J. J., and H. Boccalandro. 1995. Co-action between phytochrome B and HY4 in *Arabidopsis thaliana. Planta* 197:213–18.
Cavalier-Smith, T. 2000. Membrane heredity and early chloroplast evolution. *Trends in Plant Science* 5 (4): 174–82.
Cavalli-Sforza, L. L., and F. Cavalli-Sforza. 1995. *The Great Human Diasporas.* Trans. S. Thorne. Reading, MA: Addison-Wesley.
Chalmers, A. F. 1999. *What Is This Thing Called Science?* Milton Keynes: Open University Press.
Chamberlin, T. C. 1890. The method of multiple working hypotheses. *Science* (o.s.) 15:92–96. Reprinted 1965, 148:754–59.
———. 1897. The method of multiple working hypotheses. *Journal of Geology* 5:837–48.
Cheverud, J. M. 1984. Quantitative genetics and developmental constraints on evolution by selection. *Journal of Theoretical Biology* 110:155–71.
———. 1988. The evolution of genetic correlation and developmental constraints. In *Population Genetics and Evolution,* ed. G. de Jong, 94–101. Berlin: Springer-Verlag.
Cheverud, J. M., and E. J. Routman. 1995. Epistasis and its contribution to genetic variance components. *Genetics* 139:1455–61.
Chory, J., and J. Li. 1997. Gibberellins, brassinosteroids and light-regulated development. *Plant, Cell and Environment* 20:801–6.
Cipollini, D. F., and J. C. Schultz. 1999. Exploring cost constraints on stem elongation in plants using phenotypic manipulation. *American Naturalist* 153:236–42.
Claridge, M. F., H. A. Dawah, and M. R. Wilson, eds. 1997. *Species: The Units of Biodiversity.* Systematics Association Special Volume Series, 54. London: Chapman & Hall.
Cleland, C. E. 2001. Historical science, experimental science, and the scientific method. *Geology* 29:987–90.
———. 2002. Methodological and epistemic differences between historical science and experimental science. *Philosophy of Science* 69:474–96.
Clough, S. 2003. *Beyond Epistemology: A Pragmatic Approach to Feminist Science Studies.* Lanham, MD: Rowman and Littlefield.
Cohen, J. 1994. The earth is round ($p < .05$). *American Psychologist* 49:997–1003.,
Collard, M., and B. Wood. 2000. How reliable are human phylogenetic hypotheses? *Proceedings of the National Academy of Sciences, U.S.A.* 979:5003–6.
Collins, H. M., and T. J. Pinch. 1998. *The Golem: What You Should Know About Science.* 2nd ed. Cambridge: Cambridge University Press.
Colwell, R. K., and D. W. Winkler. 1984. A null model for null models in biogeography. In *Ecological Communities: Conceptual Issues and the Evidence,* ed. D. R. Strong Jr., D. Simberloff, L. G. Abele, and A. B. Thistle, 344–59. Princeton, NJ: Princeton University Press.
Conner, J. K. 2002. Genetic mechanisms of floral trait correlations in a natural population. *Nature* 420:407–10.
Cooper, R. M., and J. P. Zubek. 1958. Effects of enriched and restricted early environments on the learning ability of bright and dull rats. *Canadian Journal of Psychology* 12:159–64.
Cosmides, L., and J. Tooby. 1992. Cognitive adaptations for social exchange. In *The Adapted Mind: Evolutionary Psychology and the Generation of Culture,* ed. J. H. Barkow, L. Cosmides, and J. Tooby, 163–228. New York: Oxford University Press.
Cosmides, L., J. Tooby, and J. H. Barkow. 1992. Introduction: Evolutionary psychology and conceptual integration. In *The Adapted Mind: Evolutionary Psychology and the Generation of Culture,* ed. J. H. Barkow, L. Cosmides, and J. Tooby, 3–18. New York: Oxford University Press.
Cowley, D. E., and W. Atchley. 1992. Comparison of quantitative genetic parameters. *Evolution* 46:1965–66.
Coyne, J. A., N. H. Barton, and M. Turelli. 1997. A critique of Sewall Wright's shifting balance theory of evolution. *Evolution* 51:643–71.

Coyne, J. A., and H. A. Orr. 2004. *Speciation.* Sunderland, MA: Sinauer Associates.

Crespi, B. J. 1990. Measuring the effect of natural selection on phenotypic interaction systems. *American Naturalist* 135:32–47.

Crespi, B. J., and F. L. Bookstein. 1989. A path-analytic model for the measurement of selection on morphology. *Evolution* 43:18–28.

Cronbach, L. J. 1975. Beyond the two disciplines of scientific psychology. *American Psychologist* 30:116–27.

Crow, J. F. 2002. Here's to Fisher, additive genetic variance, and the fundamental theorem of natural selection. *Evolution* 56:1313–16.

Culp, S. 1997. Establishing genotype/phenotype relationships: Gene targeting as an experimental approach. *Philosophy of Science* 64:S268–78.

Cummins, R. 2002. Neo-teleology. In *Functions: New Essays in the Philosophy of Psychology and Biology,* ed. A. Ariew, R. Cummins, and M. Perlman, 157–72. New York: Oxford University Press.

Daday, H., F. E. Binet, A. Grassia, and J. W. Peak. 1973. The effect of environment on heritability and predicted selection response in *Medicago sativa. Heredity* 31:293–308.

Dahlberg, C., M. Bergström, and M. Hermansson. 1998. In situ detection of high levels of horizontal plasmid transfer in marine bacterial communities. *Applied and Environmental Microbiology* 64:2670–75.

Darwin, C. 1859. *On the Origin of Species by Means of Natural Selection: Or, the Preservation of Favored Races in the Struggle for Life.* New York: A. L. Burt, 1910.

———. 1868. *The Variation of Animals and Plants under Domestication.* New York: D. Appleton and Company, 1900.

———. 1871. *The Descent of Man.* Princeton, NJ: Princeton University Press, 1981.

———. 1872. *The Expression of the Emotions in Man and Animals.* New York: D. Appleton and Company.

Das, M. K. 1995. Sickle cell gene in central India: Kinship and geography. *American Journal of Human Biology* 7:565–73.

Davies, P. S. 1994. Troubles for direct proper functions. *Nous* 28 (3): 363–81.

———. 1995. "Defending" direct proper functions. *Analysis* 55 (4): 299–306.

———. 2000. Malfunctions. *Biology and Philosophy* 15:19–38.

Dawkins, R. 1976. *The Selfish Gene.* Oxford: Oxford University Press.

———. 1982. *The Extended Phenotype.* San Francisco: Freeman.

de Jong, G. 1994. The fitness of fitness concepts and the description of natural selection. *Quarterly Review of Biology* 69:3–29.

Den Boer, P. J. 1991. Seeing the trees for the wood: Random walks or bounded fluctuations of population size? *Oecologia* 86:484–91.

Dennett, D. 1995. *Darwin's Dangerous Idea: Evolution and the Meaning of Life.* New York: Simon & Schuster.

de Queiroz, K. 1992. Phylogenetic definitions and taxonomic philosophy. *Biology and Philosophy* 7:295–313.

———. 2005. Ernst Mayr and the modern concept of species. *Proceedings of the National Academy of Sciences, U.S.A.* 102 (supplement 1): 6600–6607.

de Quervain, D. J.-F., U. Fischbacher, V. Treyer, M. Schellhammer, U. Schnyder, A. Buck, and E. Fehr. 2004. The neural basis of altruistic punishment. *Science* 305:1254–58.

Dixon, P., and T. O'Reilly. 1999. Scientific versus statistical inference. *Canadian Journal of Experimental Psychology* 53 (2): 133–49.

Dobzhansky, T. 1937. *Genetics and the Origin of Species.* New York: Columbia University Press.

Dobzhansky, T., and B. Spassky. 1944. Genetics of natural populations. XI. Manifestation of genetic variants in *Drosophila pseudoobscura* in different environments. *Genetics* 29: 270–90.

Doebeli, M., and U. Dieckmann. 2000. Evolutionary branching and sympatric speciation caused by different types of ecological interactions. *American Naturalist* 156:S77–101.

Dudley, S., and J. Schmitt. 1995. Genetic differentiation in morphological responses to simulated foliage shade between populations of *Impatiens capensis* from open and woodland sites. *Functional Ecology* 9:655–66.

Dupré, J. 1993. *The Disorder of Things: Metaphysical Foundations of the Disunity of Science.* Cambridge, MA: Harvard University Press.

———. 1998. Normal people. *Social Research* 652:221–48.

Ebert, D., L. Yampolsky, and S. C. Stearns. 1993. Genetics of life history in *Daphnia magna.* I. Heritabilities at two food levels. *Heredity* 70:335–43.

The Economist. 1999. Come up and see my etchings. April 3, 71.

Eldredge, N., and S. J. Gould. 1972. Punctuated equilibria: An alternative to phyletic gradualism. In *Models in Paleobiology,* ed. T. J. M. Schopf, 82–115. San Francisco: Freeman, Cooper.

Endler, J. A. 1986. *Natural Selection in the Wild.* Princeton, NJ: Princeton University Press.

Ereshefsky, M., ed. 1991. *The Units of Evolution: Essays on the Nature of Species.* Cambridge, MA: MIT Press.

Etges, W. J., and M. A. Ahrens. 2001. Premating isolation is determined by larval rearing substrates in cactophilic *Drosophila mojavensis.* V. Deep geographic variation in epicuticular hydrocarbons among isolated populations. *American Naturalist* 158:585–98.

Falconer, D. S. 1981. *Introduction to Quantitative Genetics.* London: Longman.

Falconer, D. S., and T. F. C. Mackay. 1996. *Introduction to Quantitative Genetics.* 4th ed. Harlow, England: Prentice Hall.

Feldman, M. W., and R. C. Lewontin. 1990. The heritability hang-up. *Science* 190: 1163–68.

Fisher, R. A. 1930. *The Genetical Theory of Natural Selection.* Oxford: Clarendon.

Fitelson, B., C. Stephens, and E. R. Sober. 1999. How not to detect design—Critical notice: William A. Dembski, *The Design Inference. Philosophy of Science* 66:472–488.

Fontana, W., and P. Schuster. 1998a. Continuity in evolution: On the nature of transitions. *Science* 280:1451–55.

———. 1998b. Shaping space: The possible and the attainable in RNA genotype-phenotype mapping. *Journal of Theoretical Biology* 194:491–515.

Ford, M. J. 2002. Applications of selective neutrality tests to molecular ecology. *Molecular Ecology,* 11:1245–62.

Frank, S. A., and M. Slatkin. 1992. Fisher's fundamental theorem of natural selection. *Trends in Ecology and Evolution* 7:92–95.

Freeman, D. C., J. H. Graham, and J. M. Emlen. 1993. Developmental stability in plants: Symmetries, stress and epigenesis. *Genetica* 89:97–119.

Futuyma, D. J. 1998. *Evolutionary Biology.* 3rd ed. Sunderland, MA: Sinauer Associates.

Gabriel, W., and M. Lynch. 1992. The selective advantage of reaction norms for environmental tolerance. *Journal of Evolutionary Biology* 5:41–59.

Gagneux, P., C. Wills, U. Gerloff, D. Tautz, P. A. Morin, C. B. B. Fruth, G. Hohmann, O. Ryder, and D. S. Woodruff. 1999. Mitochondrial sequences show diverse evolutionary histories of African hominoids. *Proceedings of the National Academy of Sciences, U.S.A.* 96: 5077–82.

Ganfornina, M. D., and D. Sanchez. 1999. Generation of evolutionary novelty by functional shift. *BioEssays* 21:432–39.

Gavrilets, S. 1996. On phase three of the shifting balance. *Evolution* 50:1034–41.

———. 1997a. Evolution and speciation on holey adaptive landscapes. *Trends in Ecology and Evolution* 12:307–12.

———. 1997b. Hybrid zones with Dobzhansky-type epistatic selection. *Evolution* 51: 1027–35.

———. 1999. A dynamical theory of speciation on holey adaptive landscapes. *American Naturalist* 154:1–22.

———. 2003. Models of speciation: What have we learned in 40 years? *Evolution* 57:2197–2215.

———. 2004. *Fitness Landscapes and the Origin of Species.* Princeton, NJ: Princeton University Press.

Gavrilets, S., and A. Hastings. 1994. A quantitative-genetic model for selection on developmental noise. *Evolution* 48:1478–86.

———. 1995. Dynamics of polygenic variability under stabilizing selection, recombination, and drift. *Genetical Research* 65:63–74.

———. 1996. Founder effect speciation: A theoretical reassessment. *American Naturalist* 147:466–91.

Gavrilets, S., H. Li, and M. D. Vose. 1998. Rapid parapatric speciation on holey adaptive landscapes. *Proceedings of the Royal Society of London* B 265:1483–89.

Gazzaniga, M. S. 2000. Cerebral specialization and interhemispheric communication: Does the corpus callosum enable the human condition? *Brain* 123:1293–1326.

Giere, R. N. 1979. *Understanding Scientific Reasoning.* New York: Holt, Rinehart and Winston.

Gilchrist, G. W. 1995. Specialists and generalists in changing environments. I. Fitness landscapes of thermal sensitivity. *American Naturalist* 146:252–70.

Giray, E. 1976. An integrated biological approach to the species problem. *British Journal for the Philosophy of Science* 27:317–28.

Gleason, J. M., E. C. Griffith, and J. R. Powell. 1998. A molecular phylogeny of the *Drosophila willistoni* group: Conflicts between species concepts? *Evolution* 52:1093–1103.

Godfrey-Smith, P. 1994. A modern history theory of functions. *Nous* 28 (3): 344–62.

———. 2000. On the theoretical role of "genetic coding." *Philosophy of Science* 67:26–44.

———. 2001. Three kinds of adaptationism. In *Optimality and Adaptationism,* ed. E. Sober and S. Orzack, 335–57. Cambridge: Cambridge University Press.

Goldschmidt, R. 1940. *The Material Basis of Evolution.* New Haven, CT: Yale University Press.

Goodman, M., C. A. Porter, J. Czelusniak, S. L. Page, H. Schneider, J. Shoshani, G. Gunnell, and C. P. Groves. 1998. Toward a phylogenetic classification of primates based on DNA evidence complemented by fossil evidence. *Molecular Phylogenetics and Evolution* 93:585–98.

Gould, S. J. 1976. Darwin's untimely burial. *Natural History* 85 (8): 24–30.

———. 1980. The evolutionary biology of constraint. *Daedalus* 109:39–52.

———. 1989. A developmental constraint in *Cerion,* with comments on the definition and interpretation of constraint in evolution. *Evolution* 43:516–39.

———. 1996. *The Mismeasure of Man.* New York: Norton.

———. 2002. *The Structure of Evolutionary Theory.* Cambridge, MA: Harvard University Press.

Gould, S. J., and R. C. Lewontin. 1979. The spandrels of San Marco and the Panglossian paradigm: A critique of the adaptationist programme. *Proceedings of the Royal Society of London* B 205:581–98.

Gould, S. J., and E. S. Vrba. 1982. Exaptation—a missing term in the science of form. *Paleobiology* 8:4–15.

Grant, V. 1994. Evolution of the species concept. *Biologisches Zentralblatt* 113:401–15.

Gray, R. 1992. Death of the gene: Developmental systems strike back. In *Trees of Life,* ed. P. E. Griffiths, 165–209. Dordrecht: Kluwer Academic Publishers.

Gregson, R. A. M. 1997. Signs of obsolescence in psychological statistics: Significance versus contemporary theory. *Australian Journal of Psychology* 49:59–63.

Griesemer, J. R., and M. J. Wade. 1988. Laboratory models, casual explanation and group selection. *Biology and Philosophy* 3:67–96.

Griffiths, A. J. F., J. H. Miller, D. T. Suzuki, R. C. Lewontin, and W. M. Gelbart. 1996. *An Introduction to Genetic Analysis.* 6th ed. New York: Freeman.

Griffiths, P. E. 1992. Adaptive explanation and the concept of a vestige. In *Trees of Life,* ed. P. E. Griffiths, 111–32. Dordrecht: Kluwer Academic Publishers.

———. 1993. Functional analysis and proper functions. *British Journal for the Philosophy of Science* 44:409–22.
———. 1996. The historical turn in the study of adaptation. *British Journal for the Philosophy of Science* 47:511–32.
———. 1997. *What Emotions Really Are.* Chicago: University of Chicago Press.
———. 1999. Squaring the circle: Natural kinds with historical essences. In *Species: New Interdisciplinary Essays,* ed. R. Wilson, 209–28. Cambridge, MA: MIT Press.
Griffiths, P. E., and R. D. Gray. 1994. Developmental systems and evolutionary explanation. *Journal of Philosophy* 916:277–304.
———. 1997. Replicator II—Judgement Day. *Biology and Philosophy* 12:471–92.
Gromko, M. H. 1995. Unpredictability of correlated response to selection: Pleiotropy and sampling interact. *Evolution* 49:685–93.
Guasch, A., C. F. Zayas, J. R. Eckman, K. Muralidharan, W. Zhang, and L. J. Elsas. 1999. Evidence that microdeletions in the alpha globin gene protect against the development of sickle cell glomerulopathy in humans. *Journal of the American Society of Nephrology* 105: 1014–29.
Hacking, I. 1999. *The Social Construction of What?* Cambridge, MA: Harvard University Press.
Hadany, L. 2003. Adaptive peak shifts in a heterogeneous environment. *Theoretical Population Biology* 63:41–51.
Halliday, K. J., M. Koornneef, and G. C. Whitelam. 1994. Phytochrome B and at least one other phytochrome mediate the accelerated flowering response of *Arabidopsis thaliana* L. to low red/far-red ratio. *Plant Physiology* 104:1311–15.
Hamilton, W. D. 1963. The genetical evolution of social behavior. *American Naturalist* 97: 31–33.
Hansen, T. F. 1997. Stabilizing selection and the comparative analysis of adaptation. *Evolution* 51:1341–51.
Hardy, G. H. 1908. Mendelian proportions in a mixed population. *Science* 28:49–50.
Hartl, D. L., and A. G. Clark. 1989. *Principles of Population Genetics.* Sunderland, MA: Sinauer Associates.
Harvey, P. H., and A. Purvis. 1991. Comparative methods for explaining adaptations. *Nature* 351:619–25.
Hatcher, M. J. 2000. Persistence of selfish genetic elements: Population structure and conflict. *Trends in Ecology and Evolution* 15:271–77.
Hatfield, T., and D. Schluter. 1999. Ecological speciation in sticklebacks: Environment-dependent hybrid fitness. *Evolution* 53:866–73.
Hey, J. 1999. The neutralist, the fly and the selectionist. *Trends in Ecology and Evolution* 14: 35–37.
———. 2001a. The mind of the species problem. *Trends in Ecology and Evolution* 16:326–27.
———. 2001b. *Genes, Categories and Species.* Oxford: Oxford University Press.
Hilborn, R., and M. Mangel. 1997. *The Ecological Detective: Confronting Models with Data.* Princeton, NJ: Princeton University Press.
Hoffman, A., and W. H. Nitecki. 1987. Introduction. In *Neutral Models in Biology,* ed. W. H. Nitecki and A. Hoffman, 3–8. New York: Oxford University Press.
Hoffmann, A. A., R. Hallas, C. Sinclair, and L. Partridge. 2001. Rapid loss of stress resistance in *Drosophila melanogaster* under adaptation to laboratory culture. *Evolution* 55:436–38.
Hoffmann, A. A., and M. Schiffer. 1998. Changes in the heritability of five morphological traits under combined environmental stresses in *Drosophila melanogaster. Evolution* 52: 1207–12.
Horvath, C. D. 1997. Phylogenetic species concept: Pluralism, monism, and history. *Biology and Philosophy* 12:225–32.
Houle, D. 1991. Genetic covariance of fitness correlates: What genetic correlations are made of and why it matters. *Evolution* 45:630–48.
———. 1994. Adaptive distance and the genetic basis of heterosis. *Evolution* 48:1410–17.

Houston, A. I. 1990. Matching, maximizing and melloration as alternative descriptions of behaviour. In *From Animals to Animats: Proceedings of the First International Conference on Simulation of Adaptive Behavior,* ed. J.-A. Meyer and S. W. Wilson, 498–509. Cambridge, MA: MIT Press.

———. 1997. Are the spandrels of San Marco really Panglossian pendentives? *Trends in Ecology and Evolution* 123:125.

Howson, C., and P. Urbach. 1989. *Scientific Reasoning: The Bayesian Approach.* La Salle, IL: Open Court Press.

———. 1991. Bayesian reasoning in science. *Nature* 350:371–74.

Huelsenbeck, J. P., B. Rannala, and B. Larget. 2000. A Bayesian framework for the analysis of cospeciation. *Evolution* 54:353–64.

Hull, D. L. 1965. The effect of essentialism on taxonomy—Two thousand years of stasis. *British Journal for the Philosophy of Science* 15:314–26; 16:1–18.

———. 1978. A matter of individuality. *Philosophy of Science* 45:335–60.

Hurst, G. D. D., and J. H. Werren. 2001. The role of selfish genetic elements in eukaryotic evolution. *Nature Reviews* 2:597–606.

Hutchinson, G. E. 1965. *The Ecological Theater and the Evolutionary Play.* New Haven, CT: Yale University Press.

Jablonka, E. 2001. The systems of inheritance. In *Cycles of Contingency,* ed. S. Oyama, P. Griffith, and R. Gray, 99–116. Cambridge, MA: MIT Press.

Jablonka, E., and M. J. Lamb. 2005. *Evolution in Four Dimensions: Genetic, Epigenetic, Behavioral, and Symbolic Variation in the History of Life.* Cambridge, MA: MIT Press.

Jacob, F. 1977. Evolution and tinkering. *Science* 196:1161–66.

Jaenike, J. 1990. Host specialization in phytophagous insects. *Annual Review of Ecology and Systematics* 21:243–73.

Johanssen, W. 1911. The genotype conception of heredity. *American Naturalist* 45:129–59. In JStor: The Scholarly Journal Archive, http://www.jstor.org.

Johnson, N. A., and A. H. Porter. 2001. Toward a new synthesis: Population genetics and evolutionary developmental biology. *Genetica* 112/113:45–58.

Jones, A. G., S. J. Arnold, and R. Bürger. 2003. Stability of the **G**-matrix in a population experiencing pleiotropic mutation, stabilizing selection, and genetic drift. *Evolution* 57: 1747–60.

———. 2004. Evolution and stability of the **G**-matrix on a landscape with a moving optimum. *Evolution* 58:1639–54.

Jones, D. 1999. Evolutionary psychology. *Annual Review of Anthropology* 28:553–73.

Kaplan, D. T., and L. Glass. 1992. A direct test for determinism in a time series. *Physical Review Letters* 68:427–30.

Kaplan, J. M. 2002. Historical evidence and human adaptations. *Philosophy of Science* 69: S294–304.

Kaplan, J. M., and M. Pigliucci. 2001. Genes "for" phenotypes: A modern history view. *Biology and Philosophy* 16:189–213.

———. 2003. On the concept of biological race and its applicability to humans. *Philosophy of Science* 70:S1161–72.

Karoly, K., and J. K. Conner. 2000. Heritable variation in a family-diagnostic trait. *Evolution* 54:1433–38.

Kauffman, S. A., and S. Levin. 1987. Towards a general theory of adaptive walks on rugged landscapes. *Journal of Theoretical Biology* 128:11–45.

Kazazian, H. H. Jr., and J. V. Moran. 1998. The impact of L1 retrotransposons on the human genome. *Nature Genetics* 19:19–24.

Kearsey, M. J., and A. G. L. Farquhar. 1998. QTL analysis in plants: Where are we now? *Heredity* 80:137–42.

Kerr, B., and P. Godfrey-Smith. 2002. Individualist and multi-level perspectives on selection in structured populations. *Biology and Philosophy* 17:477–517.

Kevles, D. J. 1985. *In the Name of Eugenics.* Berkeley: University of California Press.

Kimura, D. 1999. *Sex and Cognition.* Cambridge, MA: MIT Press.
Kimura, M. 1968. Evolutionary rate at the molecular level. *Nature* 217:624–26.
———. 1983. *The Neutral Theory of Molecular Evolution.* Cambridge: Cambridge University Press.
Kimura, M., and J. F. Crow. 1964. The number of alleles that can be maintained in a finite population. *Genetics* 49:725–38.
Kimura, M., and T. Ohta. 1971. Protein polymorphism as a phase of molecular evolution. *Nature* 229:467–69.
Kingsolver, J. G., H. E. Hoekstra, J. M. Hoekstra, D. Berrigan, S. N. Vignieri, C. E. Hill, A. Hoang, P. Gibert, and P. Beerli. 2001. The strength of phenotypic selection in natural populations. *American Naturalist* 157:245–61.
Kingsolver, J. G., and D. W. Schemske. 1991. Path analyses of selection. *Trends in Ecology and Evolution* 6:276–80.
Kirk, K. M., S. P. Blomberg, D. L. Duffy, A. C. Heath, I. P. Owens, and N. G. Martin. 2001. Natural selection and quantitative genetics of life-history traits in Western women: A twin study. *Evolution* 55:423–35.
Kirkpatrick, M. 1982. Quantum evolution and punctuated equilibria in continuous genetic characters. *American Naturalist* 119:833–48.
Kitcher, P. 1984. Species. *Philosophy of Science* 51:308–33.
———. 1985. *Vaulting Ambition: Sociobiology and the Quest for Human Nature.* Cambridge, MA: MIT Press.
———. 2001. Battling the undead: How and how not to resist genetic determinism. In *Thinking about Evolution: Historical, Philosophical and Political Perspectives,* ed. R. Singh, K. Krimbas, J. Beatty, and D. Paul, 396–414. Cambridge: Cambridge University Press.
Kohler, R. E. 1994. *The Lords of the Fly.* Chicago: University of Chicago Press.
Komeda, Y. 1993. The use of transgenic *Arabidopsis thaliana* for studies of the regulation of genes for heat-shock proteins. *Journal of Plant Research* 3:213–19.
Korol, A. B., and K. G. Iliadi. 1994. Increased recombination frequencies resulting from directional selection for geotaxis in *Drosophila. Heredity* 72:64–68.
Krebs, R. A., and V. Loeschcke. 1994. Costs and benefits of activation of the heat-shock response in *Drosophila melanogaster. Functional Ecology* 8:730–37.
Kurbatova, O. L., O. K. Botvinyev, and Y. P. Altukhov. 1990. Adaptive norm and stabilizing selection for anthropometric characters at birth. *Genetika* 27:1229–40.
Lakatos, I. 1977. *The Methodology of Scientific Research Programmes.* Vol. 1 of *Philosophical Papers,* ed. J. Worrall and G. Currie. Cambridge: Cambridge University Press.
Lande, R. 1976. Natural selection and random genetic drift in phenotypic evolution. *Evolution* 30:314–34.
———. 1979. Quantitative genetic analysis of multivariate evolution, applied to brain:body size allometry. *Evolution* 33:402–16.
Lande, R., and S. J. Arnold. 1983. The measurement of selection on correlated characters. *Evolution* 37:1210–26.
Lander, E. S., and D. Botstein. 1989. Mapping Mendelian factors underlying quantitative traits using RFLP linkage maps. *Genetics* 121:185–99.
Landman, O. E. 1991. The inheritance of acquired characteristics. *Annual Review of Genetics* 25:1–20.
Larson, A., and J. B. Losos. 1996. Phylogenetic systematics of evolution. In *Adaptation,* ed. M. R. Rose and G. V. Lauder, 187–220. San Diego, CA: Academic Press.
Lasceve, G., J. Leymarie, M. A. Olney, E. Liscum, J. M. Christie, A. Vavasseur, and W. Briggs. 1999. *Arabidopsis* contains at least four independent blue-light-activated signal transduction pathways. *Plant Physiology* 120:605–14.
Latour, B., and S. Woolgar. 1979. *Laboratory Life: The Social Construction of Scientific Facts.* Cambridge, MA: Harvard University Press.
Lauder, G. V. 1996. The argument from design. In *Adaptation,* ed. M. R. Rose and G. V. Lauder, 55–92. San Diego, CA: Academic Press.

Leroi, A. M., M. R. Rose, and G. V. Lauder. 1994. What does the comparative method reveal about adaptation? *American Naturalist* 1433:381–402.

Levene, H. 1953. Genetic equilibrium when more than one ecological niche is available. *American Naturalist* 87:331–33.

Levins, R., and R. C. Lewontin. 1985. *The Dialectical Biologist.* Cambridge, MA: Harvard University Press.

Levitt, N. 1998. Why professors believe weird things: Sex, race, and the trials of the new left. *Skeptic* 6 (3): 28–35.

Lewontin, R. C. 1970. The units of selection. *Annual Review of Ecology and Systematics* 1: 1–14.

———. 1974. The analysis of variance and the analysis of causes. *American Journal of Human Genetics* 26:400–411.

———. 1978. Adaptation. *Scientific American,* September, 213–30.

———. 1979. Sociobiology as an adaptationist program. *Behavioral Science* 241:5–14.

———. 1983. The organism as the subject and object of evolution. *Scientia* 118:63–82.

———. 1985. Evolution as theory and ideology. In *The Dialectical Biologist,* ed. R. Levins and R. C. Lewontin, 9–64. Cambridge, MA: Harvard University Press. (Originally published in *Enciclopedia Einaudi,* vol. 3, ed. G. Einaudi, 1977.)

———. 1992. *Biology as Ideology: The Doctrine of DNA.* New York: HarperPerennial.

———. 1995. *Human Diversity.* New York: Scientific American Library.

———. 1998. The evolution of cognition: Questions we will never answer. In *Methods, Models, and Conceptual Issues: An Invitation to Cognitive Science,* ed. D. Scarborough and S. Sternberg, 107–32. Cambridge, MA: MIT Press.

Lewontin, R. C., and C. C. Cockerham. 1959. The goodness-of-fit test for detecting natural selection in random mating populations. *Evolution* 13:561–64.

Lewontin, R. C., and J. L. Hubby. 1966. A molecular approach to the study of genic heterozygosity in natural populations. II. Amount of variation and degree of heterozygosity in natural populations of *Drosophila pseudoobscura. Genetics* 54:595–609.

Lloyd, E. A. 1999. Evolutionary psychology: The burdens of proof. *Biology and Philosophy* 14:211–33.

Loeschcke, V., J. Bundgaard, and J. S. F. Barker. 1999. Reaction norms across and genetic parameters at different temperatures for thorax and wing size traits in *Drosophila aldrichi* and *D. buzzatii. Journal of Evolutionary Biology* 12:605–23.

Loeschcke, V., and R. A. Krebs. 1994. Genetic variation for resistance and acclimation to high temperature stress in *Drosophila buzzatii. Biological Journal of the Linnean Society* 52: 83–92.

Loftus, G. R. 1993. A picture is worth a thousand rho values: On the irrelevance of hypothesis testing in the microcomputer age. *Behavioral Research Methods, Instruments, and Computers* 25:250–56.

Lopez-Juez, E., M. Kobayashi, A. Sakurai, Y. Kamiya, and R. E. Kendrick. 1995. Phytochrome, gibberellins, and hypocotyl growth. *Plant Physiology* 107:131–40.

Lorenz, K. 1963. *On Aggression.* Trans. M. K. Wilson. New York: Harcourt, Brace and World.

Lynch, M. 1999. Estimating genetic correlations in natural populations. *Genetical Research* 74:255–64.

Macnair, M. R., V. E. Macnair, and B. E. Martin. 1989. Adaptive speciation in *Mimulus:* An ecological comparison of *M. cupriphilus* with its presumed progenitor, *M. guttatus. New Phytologist* 112:269–79.

Maestripieri, D., and J. R. Roney. In press. Evolutionary developmental psychology: Contributions for comparative research with nonhuman primates. *Developmental Review.*

Magnus, D. 1998. Evolution without change in gene frequencies. *Biology and Philosophy* 13: 255–61.

Magwene, P. M. 2001. New tools for studying integration and modularity. *Evolution* 55: 1734–45.

Malakoff, D. 1999. Bayes offers a "new" way to make sense of numbers. *Science* 286:1460–64.
Maley, C. 1997. Mutation rates as adaptations. *Journal of Theoretical Biology* 186:339–48.
Mallet, J. 2001. The speciation revolution. *Journal of Evolutionary Biology* 14:887–88.
Manly, B. F. J. 1985. *The Statistics of Natural Selection.* London: Chapman & Hall.
Manning, R. N. 1997. Biological function, selection, and reduction. *British Journal for the Philosophy of Science* 48:69–82.
Maresca, B., E. Patriarca, C. Goldenberg, and M. Sacco. 1988. Heat shock and cold adaptation in Antarctic fishes: A molecular approach. *Comparative Biochemistry and Physiology* 90B:623–29.
Mark, R. 1996. Architecture and evolution. *American Scientist* 84:383–89.
Marks, J. 2005. Phylogenetic trees and evolutionary forests. *Evolutionary Anthropology* 14: 49–53.
Marrow, P., and R. A. Johnstone. 1996. Riding the evolutionary streetcar: Where population genetics and game theory meet. *Trends in Ecology and Evolution* 11:445–46.
Martins, E. P. 2000. Adaptation and the comparative method. *Trends in Ecology and Evolution* 15:295–99.
Mathews, S., M. Lavin, and R. A. Sharrock. 1995. Evolution of the phytochrome gene family and its utility for phylogenetic analyses of angiosperms. *Annals of the Missouri Botanical Gardens* 82:296–321.
Matloff, N. S. 1991. Statistical hypothesis testing: Problems and alternatives. *Environmental Entomology* 20:1246–50.
Matthen, M., and A. Ariew. 2002. Two ways of thinking about fitness and natural selection. *Journal of Philosophy* 992:55–83.
Mayden, R. L. 1997. A hierarchy of species concepts: The denouement in the saga of the species problem. In *Species: The Units of Biodiversity,* ed. M. F. Claridge, H. A. Dawah, and M. R. Wilson, 381–424. Systematics Association Special Volume Series, 54. London: Chapman & Hall.
Maynard Smith, J., R. Burian, S. Kauffman, P. Alberch, J. Campbell, B. Goodwin, R. Lande, D. Raup, and L. Wolpert. 1985. Developmental constraints and evolution. *Quarterly Review of Biology* 60:265–87.
Mayr, E. 1980. Some thoughts on the history of the evolutionary synthesis. In *The Evolutionary Synthesis: Perspectives on the Unification of Biology,* ed. E. Mayr and W. B. Provine, 1–48. Cambridge, MA: Harvard University Press.
———. 1982. *The Growth of Biological Thought.* Cambridge, MA: Harvard University Press.
———. 1996. What is a species, and what is not? *Philosophy of Science* 63:262–77.
Mayr, E., and W. B. Provine, eds. 1980. *The Evolutionary Synthesis: Perspectives on the Unification of Biology.* Cambridge, MA: Harvard University Press.
Mazer, S. J., and C. T. Schick. 1991. Constancy of population parameters for life-history and floral traits in *Raphanus sativus* L. II. Effects of planting density on phenotype and heritability estimates. *Evolution* 45:1888–1907.
McBurney, D. H., S. J. C. Gaulin, T. Devineni, and C. Adams. 1997. Superior spatial memory of women: Stronger evidence for the gathering hypothesis. *Evolution and Human Behavior* 18:165–74.
McClamrock, R. 1995. Screening-off and the levels of selection. *Erkenntnis* 42:107–12.
McCloskey, D. N., and S. T. Ziliak. 1996. The standard error of regressions. *Journal of Economic Literature* 34:97–114.
McCormac, A. C., D. Wagner, M. T. Boylan, P. H. Quail, H. Smith, and G. C. Whitelam. 1993. Photoresponses of transgenic *Arabidopsis* seedlings expressing introduced phytochrome B-encoding cDNAs: Evidence that phytochrome A and phytochrome B have distinct photoregulatory functions. *Plant Journal* 4:19–27.
Meehl, P. 1978. Theoretical risks and tabular asterisks: Sir Karl, Sir Ronald, and the slow progress of soft psychology. *Journal of Consulting and Clinical Psychology* 46 (4): 806–34.
Mezey, J. G., J. M. Cheverud, and G. P. Wagner. 2000. Is the genotype-phenotype map

modular? A statistical approach using mouse Quantitative Trait Loci data. *Genetics* 156: 305–11.

Miller, G. F. 1998. How mate choice shaped human nature: A review of sexual selection and human evolution. In *Handbook of Evolutionary Psychology: Ideas, Issues, and Applications,* ed. C. Crawford and D. R. Krebs, 87–130. Mahwah, NJ: Lawrence Erlbaum Associates.

———. 2000. *The Mating Mind: How Sexual Choice Shaped the Evolution of Human Nature.* New York: Doubleday.

Miller, G. F., and P. M. Todd. 1998. Mate choice turns cognitive. *Trends in Cognitive Science* 2 (5): 190–98.

Millikan, R. G. 1984. *Language, Thought, and Other Biological Categories.* Cambridge, MA: MIT Press.

Mills, S., and J. Beatty. 1979. The propensity interpretation of fitness. *Philosophy of Science* 46:263–88.

Millstein, R. L. 2002. Are random drift and natural selection conceptually distinct? *Biology and Philosophy* 17:33–53

Mishler, B. D., and R. N. Brandon. 1987. Individuality, pluralism, and the phylogenetic species concept. *Biology and Philosophy* 2:397–414.

Mishler, B. D., and M. J. Donoghue. 1982. Species concepts: A case for pluralism. *Systematic Zoology* 31:491–503.

Mitchell, R. J. 1992. Testing evolutionary and ecological hypotheses using path analysis and structural equation modelling. *Functional Ecology* 6:123–29.

Mitchell-Olds, T. 1995. The molecular basis of quantitative genetic variation in natural populations. *Trends in Ecology and Evolution* 10:324–27.

Mitchell-Olds, T., and J. Bergelson. 1990. Statistical genetics of an annual plant, *Impatiens capensis.* I. Genetic basis of quantitative variation. *Genetics* 124:407–15.

Mitchell-Olds, T., and R. G. Shaw. 1987. Regression analysis of natural selection: Statistical inference and biological interpretation. *Evolution* 41:1149–61.

Moller, A. P., and M. D. Jennions. 2001. Testing and adjusting for publication bias. *Trends in Ecology and Evolution* 16:580–86.

Monod, J. 1971. *Chance and Necessity: An Essay on the Natural Philosophy of Modern Biology.* New York: Knopf.

Morris, D. 1962. *The Biology of Art: A Study of the Picture-Making Behaviour of the Great Apes and its Relationship to Human Art.* New York: Knopf.

Moss, L. 2003. *What Genes Can't Do.* MIT Press. Cambridge, MA.

Müller, G. B., and S. A. Newman. 2003. Origination of organismal form: The forgotten cause in evolutionary theory. In *Origination of Organismal Form,* ed. G. B. Müller and S. A. Newman, 3–10. Cambridge, MA: MIT Press.

Muller, H. J. 1942. Isolating mechanisms, evolution and temperature. *Biological Symposia* 6: 71–125.

Neff, M. M., and E. V. Volkenburgh. 1994. Light-stimulated cotyledon expansion in *Arabidopsis* seedlings: The role of phytochrome B. *Plant Physiology* 104:1027–32.

Nei, M., T. Maruyama, and C.-I. Wu. 1983. Models of evolution of reproductive isolation. *Genetics* 103:557–79.

Nice, C. C., J. A. Fordyce, A. M. Shapiro, and R. Ffrench-Constant. 2002. Lack of evidence for reproductive isolation among ecologically specialized lycaenid butterflies. *Ecological Entomology* 27:702–12.

Niklas, K. J. 1994. Morphological evolution through complex domains of fitness. *Proceedings of the National Academy of Sciences, U.S.A.* 91:6772–79.

———. 1997. Adaptive walks through fitness landscapes for early vascular land plants. *American Journal of Botany* 841:16–25.

Noor, M. A. F. 2002. Is the biological species concept showing its age? *Trends in Ecology and Evolution* 17:153–54.

Odling-Smee, F. J. 1988. Niche-constructing phenotypes. In *The Role of Behavior in Evolution,* ed. H. C. Plotkin, 73–132. Cambridge, MA: MIT Press.

Odling-Smee, F. J., K. N. Laland, and M. W. Feldman. 2003. *Niche Construction: The Neglected Process in Evolution.* Princeton, NJ: Princeton University Press.
Okasha, S. 2000. The underdetermination of theory by data and the "strong programme" in the sociology of knowledge. *International Studies in the Philosophy of Science* 14:283–97.
———. 2003. Recent work on the levels of selection problem. *Human Nature Review* 3: 349–56.
Orr, H. A. 1998. Testing natural selection vs. genetic drift in phenotypic evolution using quantitative trait locus data. *Genetics* 149:2099–2104.
Orzack, S. H., and E. Sober. 1994. Optimality models and the test of adaptationism. *American Naturalist* 143:361–80.
Oster, G. F., and P. Alberch. 1982. Evolution and bifurcation of developmental programs. *Evolution* 36:444–59.
Otte, D., and J. A. Endler. 1989. *Speciation and Its Consequences.* Sunderland, MA: Sinauer Associates.
Otto, S. P., and Y. Michalakis. 1998. The evolution of recombination in changing environments. *Trends in Ecology and Evolution* 13:145–50.
Oyama, S. 1992. Ontogeny and phylogeny: A case of metarecapitulation? In *Trees of Life,* ed. P. E. Griffiths, 211–39. Dordrecht: Kluwer Academic Publishers.
———. 2000. *The Ontogeny of Information: Developmental Systems and Evolution.* 2nd ed. Durham, NC: Duke University Press.
Oyama, S., P. E. Griffiths, and R. D. Gray, eds. 2001. *Cycles of Contingency.* Cambridge, MA: MIT Press.
Pal, C. 1998. Plasticity, memory, and the adaptive landscape of the genotype. *Proceedings of the Royal Society of London* B 265:1319–23.
Paulsen, S. M. 1996. Quantitative genetics of the wing color pattern in the buckeye butterfly *Precis coenia* and *Precis evarete. Evolution* 50:1585–97.
Paulson, S. G., C. W. Roberts, and L. M. Staley. 1973. The effect of environment on body weight heritability estimates. *Poultry Science* 52:1557–63.
Phillips, P. C. 1999. From complex traits to complex alleles. *Trends in Genetics* 15:6–8.
Phillips, P. C., and S. J. Arnold. 1989. Visualizing multivariate selection. *Evolution* 43: 1209–22.
———. 1999. Hierarchical comparison of genetic variance-covariance matrices. I. Using the Flury hierarchy. *Evolution* 53:1506–15.
Phillips, P. C., M. C. Whitlock, and K. Fowler. 2001. Inbreeding changes the shape of the genetic covariance matrix in *Drosophila melanogaster. Genetics* 158:1137–45.
Pickering, A. 1984. *Constructing Quarks: A Sociological History of Particle Physics.* Edinburgh: Edinburgh University Press.
Pier, G. B., M. Grout, T. Zaide, G. Meluleni, S. S. Mueschenborn, G. Banting, R. Ratcliff, M. J. Evans, and W. H. Colledge. 1998. *Salmonella typhi* uses CFTR to enter intestinal epithelial cells. *Nature* 393:79–82.
Pigliucci, M. 1996a. How organisms respond to environmental changes: From phenotypes to molecules and vice versa. *Trends in Ecology and Evolution* 11:168–73.
———. 1996b. Modelling phenotypic plasticity. II. Do genetic correlations matter? *Heredity* 77:453–60.
———. 2001. *Phenotypic Plasticity: Beyond Nature and Nurture.* Baltimore: Johns Hopkins University Press.
———. 2002. Are ecology and evolutionary biology "soft" sciences? *Annales Zoologici Fennici* 39:87–98.
———. 2003a. Design yes, intelligent no: A critique of intelligent design theory and neocreationism. In *Science and Religion: Are They Compatible?* ed. P. Kurtz, B. Karr, and R. Sandhu, 99–110. Amherst, NY: Prometheus.
———. 2003b. Epigenetics is back! *Cell Cycle* 2 (1): 33–35.
———. 2003c. From molecules to phenotypes? The promise and limits of integrative biology. *Basic and Applied Ecology* 4:297–306.

———. 2003d. Species as family resemblance concepts: The dissolution of the species problem? *BioEssays* 25:596–602.

Pigliucci, M., and J. M. Kaplan. 2000. The fall and rise of Dr. Pangloss: Adaptationism and the *Spandrels* paper 20 years later. *Trends in Ecology and Evolution* 152:66–70.

Pigliucci, M., and C. Murren. 2003. Genetic assimilation and a possible evolutionary paradox: Can macroevolution sometimes be so fast as to pass us by? *Evolution* 57:1455–64.

Pigliucci, M., and C. D. Schlichting. 1996. Reaction norms of *Arabidopsis*. IV. Relationships between plasticity and fitness. *Heredity* 76:427–36.

———. 1997. On the limits of quantitative genetics for the study of phenotypic evolution. *Acta Biotheoretica* 45:143–60.

Pigliucci, M., and J. Schmitt. 1999. Genes affecting phenotypic plasticity in *Arabidopsis:* Pleiotropic effects and reproductive fitness of photomorphogenic mutants. *Journal of Evolutionary Biology* 12:551–62.

———. 2004. Phenotypic plasticity in response to foliar and neutral shade in gibberellin mutants of *Arabidopsis thaliana. Evolutionary Ecology Research* 6:243–59.

Pinker, S. 1997. *How the Mind Works.* New York: Norton.

Platt, J. R. 1964. Strong inference: Certain systematic methods of scientific thinking may produce much more rapid progress than others. *Science* 146:347–53.

Popper, K. R. 1963. *Conjectures and Refutations: The Growth of Scientific Knowledge.* London: Routledge.

Prandl, R., E. Kloske, and F. Schoffl. 1995. Developmental regulation and tissue-specific differences of heat shock gene expression in transgenic tobacco and *Arabidopsis* plants. *Plant Molecular Biology* 28:73–82.

Preziosi, R. F., W. E. Snyder, C. P. Grill, and A. J. Moore. 1999. The fitness of manipulating phenotypes: Implications for studies of fluctuating asymmetry and multivariate selection. *Evolution* 53:1312–18.

Price, P. W. 1996. *Biological Evolution.* Fort Worth, TX: Saunders.

Quail, P. H. 1997a. An emerging molecular map of the phytochromes. *Plant, Cell and Environment* 20:657–65.

———. 1997b. The phytochromes: A biochemical mechanism of signaling in sight? *BioEssays* 19:571–79.

Quail, P. H., M. T. Boylan, B. M. Parks, T. W. Short, Y. Xu, and D. Wagner. 1995. Phytochromes: Photosensory perception and signal transduction. *Science* 268:675–80.

Queitsch, C., T. A. Sangster, and S. Lindquist. 2002. Hsp90 buffers genetic variation, environmental responses, and maintains developmental stability. *Nature* 417:618–24.

Quine, W. V. O. 1960. *Word and Object.* Cambridge, MA: MIT Press.

Rausher, M. D. 1992. The measurement of selection on quantitative traits: Biases due to environmental covariances between traits and fitness. *Evolution* 46:616–26.

Reed, J. W., A. Nagatani, T. D. Elich, M. Fagan, and J. Chory. 1994. Phytochrome A and phytochrome B have overlapping but distinct functions in *Arabidopsis* development. *Plant Physiology* 104:1139–49.

Reed, J. W., P. Nagpal, D. S. Poole, M. Furuya, and J. Chory. 1993. Mutations in the gene for the Red/Far-Red light receptor phytochrome B alter cell elongation and physiological responses throughout *Arabidopsis* development. *Plant Cell* 5:147–57.

Reidys, C., P. F. Stadler, and P. Schuster. 1997. Generic properties of combinatory maps—neutral networks of RNA secondary structure. *Bulletin of Mathematical Biology* 59: 339–97.

Rice, W. R., and E. E. Hostert. 1993. Laboratory experiments on speciation: What have we learned in forty years? *Evolution* 47:1637–1753.

Ridley, M. 1989. The cladistic solution to the species problem. *Biology and Philosophy* 4:1–16.

Rieppel, O. 1990. Structuralism, functionalism. and the four Aristotelian causes. *Journal of the History of Biology* 23:291–320.

Robert, J. S. 2004. *Embryology, Epigenesis, and Evolution: Taking Development Seriously.* Cambridge: Cambridge University Press.

Robert, J. S., B. K. Hall, and W. M. Olson. 2001. Bridging the gap between developmental systems theory and evolutionary developmental biology. *BioEssays* 23:954–62.

Roff, D. A. 2000. The evolution of the G matrix: Selection or drift? *Heredity* 84:135–42.

———. 2002. Comparing G matrices: A MANOVA approach. *Evolution* 56:1286–91.

Roff, D. A., and T. A. Mousseau. 1999. Does natural selection alter genetic architecture? An evaluation of quantitative genetic variation among populations of *Allenomobius socius* and *A. fasciatus*. *Journal of Evolutionary Biology* 12:361–69.

Rohlf, F. J., and L. F. Marcus. 1993. A revolution in morphometrics. *Trends in Ecology and Evolution* 8:129–32.

Rollo, C. D. 1995. *Phenotypes: Their Epigenetics, Ecology and Evolution.* New York: Chapman & Hall.

Roopnarine, P. D., G. Byars, and P. Fitzgerald. 1999. Anagenetic evolution, stratophenetic patterns, and random walk models. *Paleobiology* 25:41–57.

Rose, M. R., and G. V. Lauder. 1996. Post-spandrel adaptationism. In *Adaptation,* ed. M. R. Rose and G. V. Lauder, 1–10. San Diego, CA: Academic Press.

Rose, M. R., T. J. Nusbaum, and A. K. Chippindale. 1996. Laboratory evolution: The experimental wonderland and the Cheshire Cat syndrome. In *Adaptation,* ed M. R. Rose and G. V. Lauder, 221–42. San Diego, CA: Academic Press.

Rosenberg, A. 1978. The supervenience of biological concepts. *Philosophy of Science* 45: 368–86.

———. 1983. Fitness. *Journal of Philosophy* 80:457–73.

Roth, V. L. 1988. The biological basis of homology. In *Ontogeny and Systematics,* ed. C. J. Humphries, 1–26. New York: Columbia University Press.

Rudge, D. 1998. A Bayesian analysis of strategies in evolutionary biology. *Perspectives on Science* 6 (4): 341–60.

Ruse, M. 1969. Definitions of species in biology. *British Journal for the Philosophy of Science* 20:97–119.

———. 1996. Are pictures really necessary? The case of Sewall Wright's "adaptive landscapes." In *Picturing Knowledge: Historical and Philosophical Programs Concerning the Use of Art in Science,* ed. B. S. Baigrie, 303–37. Toronto: University of Toronto Press.

Rutherford, S. L., and S. Lindquist. 1998. *Hsp90* as a capacitor for morphological evolution. *Nature* 396:336–42.

Sarkar, S. 1996. Decoding "coding"—Information and DNA. *BioScience* 46:857–64.

Scheiner, S. M. 1993. Genetics and evolution of phenotypic plasticity. *Annual Review of Ecology and Systematics* 24:35–68.

Scheiner, S. M., and H. S. Callahan. 1999. Measuring natural selection on phenotypic plasticity. *Evolution* 53:1704–13.

Scheiner, S. M., and T. J. DeWitt. 2004. Future research directions. In *Phenotypic Plasticity: Functional and Conceptual Approaches,* ed. T. J. DeWitt and S. M. Scheiner, 201–6. New York: Oxford University Press.

Scheiner, S. M., K. Donohue, L. A. Dorn, S. J. Mazer, and L. M. Wolfe. 2002. Reducing environmental bias when measuring natural selection. *Evolution* 56:2156–67.

Scheiner, S. M., R. J. Mitchell, and H. S. Callahan. 2000. Using path analysis to measure natural selection. *Journal of Evolutionary Biology* 13:423–33.

Schlichting, C. D. 1986. The evolution of phenotypic plasticity in plants. *Annual Review of Ecology and Systematics* 17:667–93.

Schlichting, C. D., and M. Pigliucci. 1995. Gene regulation, quantitative genetics and the evolution of reaction norms. *Evolutionary Ecology* 9:154–68.

———. 1998. *Phenotypic Evolution: A Reaction Norm Perspective.* Sunderland, MA: Sinauer Associates.

Schluter, D., and D. Nychka. 1994. Exploring fitness surfaces. *American Naturalist* 143:597–616.

Schmitt, J. 1997. Is photomorphogenic shade avoidance adaptive? Perspectives from population biology. *Plant, Cell and Environment* 20:826–30.

Schmitt, J., S. Dudley, and M. Pigliucci. 1999. Manipulative approaches to testing adaptive plasticity: Phytochrome-mediated shade avoidance responses in plants. *American Naturalist* 154:S43–54.

Schmitt, J., and R. D. Wulff. 1993. Light spectral quality, phytochrome and plant competition. *Trends in Ecology and Evolution* 8:47–50.

Schuster, P., W. Fontana, P. F. Stadler, and I. Hofacker. 1994. From sequences to shapes and back: A case study in RNA secondary structure. *Proceedings of the Royal Society of London* B 255:279–84.

Seger, J., and J. W. Stubblefield. 1996. Optimization and adaptation. In *Adaptation,* ed. M. R. Rose and G. V. Lauder, 93–124. San Diego, CA: Academic Press.

Sell, A., E. Hagen, L. Cosmides, and J. Tooby. 2003. Evolutionary psychology: Applications and criticisms. In *Encyclopedia of Cognitive Science,* 47–53. London: Macmillan.

Sgrò, C. M., and A. A. Hoffmann. 1998. Effects of temperature extremes on genetic variances for life history traits in *Drosophila melanogaster* as determined from parent-offspring comparisons. *Journal of Evolutionary Biology* 11:1–20.

Sgrò, C. M., and L. Partridge. 2000. Evolutionary responses of the life history of wild-caught *Drosophila melanogaster* to two standard methods of laboratory culture. *American Naturalist* 156:341–53.

Shanks, N., and K. Joplin. 1999. Redundant complexity: A critical analysis of intelligent design in biochemistry. *Philosophy of Science* 66:268–82.

Shaw, F. H., R. G. Shaw, G. S. Wilkinson, and M. Turelli. 1995. Changes in genetic variances and covariances: **G** whiz! *Evolution* 49:1260–67.

Shaw, R. G. 1992. Comparison of quantitative genetic parameters: Reply to Cowley and Atchley. *Evolution* 46:1967–68.

Sheets, H. D., and C. E. Mitchell. 2001. Why the null matters: Statistical tests, random walks and evolution. *Genetica* 112/113:105–25.

Shinomura, T., A. Nagatani, J. Chory, and M. Furuya. 1994. The induction of seed germination in *Arabidopsis thaliana* is regulated principally by phytochrome B and secondarily by phytochrome A. *Plant Physiology* 104:363–71.

Shipley, B. 2000. *Cause and Correlation in Biology: A User's Guide to Path Analysis, Structural Equations and Causal Inference.* Cambridge: Cambridge University Press.

Shoemaker, J. S., I. S. Painter, and B. S. Weir. 1999. Bayesian statistics in genetics: A guide for the uninitiated. *Trends in Genetics* 15:354–58.

Simon, M. A. 1969. When is a resemblance a family resemblance? *Mind* 78:408–16.

Simons, A. M., and D. A. Roff. 1994. The effect of environmental variability on the heritabilities of traits of a field cricket. *Evolution* 48:1637–49.

Simpson, G. G. 1944. *Tempo and Mode in Evolution.* New York: Columbia University Press.

Sinervo, B., and A. L. Basolo. 1996. Testing adaptation using phenotypic manipulations. In *Adaptation,* ed M. R. Rose and G. V. Lauder, 149–86. San Diego, CA: Academic Press.

Sinervo, B., and E. Svensson. 1998. Mechanistic and selective causes of life history trade-offs and plasticity. *Oikos* 83:432–42.

Sinsheimer, J., J. Lake, and R. J. A. Little. 1996. Bayesian hypothesis testing for four-taxon topologies using molecular sequence data. *Biometrics* 52:193–210.

Skipper, R. A. Jr. 2004. The heuristic role of Sewall Wright's 1932 adaptive landscape diagram. *Philosophy of Science* 71:1176–88.

Slobodkin, L. B. 1987. How to be objective in community studies. In *Neutral Models in Biology,* ed. M. H. Nitecki and A. Hoffman, 93–108. New York: Oxford University Press.

Smith, H. 1982. Light quality, photoreception, and plant strategy. *Annual Review of Plant Physiology* 33:481–518.

Sniegowski, P. D., P. J. Gerrish, T. Johnson, and A. Shaver. 2000. The evolution of mutation rates: Separating causes from consequences. *BioEssays* 22:1057–66.

Sniegowski, P. D., P. J. Gerrish, and R. E. Lenski. 1997. Evolution of high mutation rates in experimental populations of *E. coli. Nature* 387:703–5.

Sober, E. 1984. Fact, fiction, and fitness. *Journal of Philosophy* 81:372–84.

———. 1992. Screening-off and the units of selection. *Philosophy of Science* 59:142–52.
———. 1993. *Philosophy of Biology*. Boulder, CO: Westview Press.
———. 1996. Evolution and optimality: Feathers, bowling balls, and the thesis of adaptation. *Philosophic Exchange* 26:41–57.
———. 2002. Instrumentalism, parsimony, and the Akaike framework. *Philosophy of Science* 69:S112–23.
Sober, E., and R. C. Lewontin. 1982. Artifact, cause and genic selection. *Philosophy of Science* 49:157–80.
Sober, E., and M. Steel. 2002. Testing the hypothesis of common ancestry. *Journal of Theoretical Biology* 218:395–408.
Sober, E., and D. S. Wilson. 1998. *Unto Others: The Evolution and Psychology of Unselfish Behavior.* Cambridge, MA: Harvard University Press.
Sokal, R. R., and F. J. Rohlf. 1995. *Biometry.* New York: Freeman.
Sollars, V., X. Lu, X. Wang, M. D. Garfinkel, and D. M. Ruden. 2003. Evidence for an epigenetic mechanism by which Hsp90 acts as a capacitor for morphological evolution. *Nature Genetics* 33:70–74.
Sorell, T. 1991. *Scientism: Philosophy and the Infatuation with Science.* London: Routledge.
Sorenson, M. D., K. M. Sefc, and R. B. Payne. 2003. Speciation by host switch in brood parasitic indigobirds. *Nature* 424:928–31.
Splitter, L. J. 1988. Species and identity. *Philosophy of Science* 55:323–48.
Stadler, B. M. R., P. F. Stadler, G. P. Wagner, and W. Fontana. 2001. The topology of the possible: Formal spaces underlying patterns of evolutionary change. *Journal of Theoretical Biology* 212:241–74.
Stanley, S. M. 1975. A theory of evolution above the species level. *Proceedings of the National Academy of Sciences, U.S.A.* 72:646–50.
Stephens, J. C., D. E. Reich, D. B. Goldstein, et al. 1998. Dating the origin of the CCR5-Delta32 AIDS-resistance allele by the coalescence of haplotypes. *American Journal of Human Genetics* 626:1507–15.
Steppan, S. J. 1997a. Phylogenetic analysis of phenotypic covariance structure. I. Contrasting results from matrix correlation and common principal component analyses. *Evolution* 51:571–86.
———. 1997b. Phylogenetic analysis of phenotypic covariance structure. II. Reconstructing matrix evolution. *Evolution* 51:587–94.
Steppan, S. J., P. C. Phillips, and D. Houle. 2002. Comparative quantitative genetics: Evolution of the G matrix. *Trends in Ecology and Evolution* 17:320–27.
Sterelny, K. 1994. The nature of species. *Philosophical Books* 35:9–20.
Sterelny,. K., and P. E. Griffiths. 1999. *Sex and Death.* Chicago: University of Chicago Press.
Sterelny, K., and P. Kitcher. 1988. The return of the gene. *Journal of Philosophy* 85:339–61.
Stern, C., and E. R. Sherwood. 1966. *The Origin of Genetics: A Mendel Source Book.* New York: Freeman.
Stigler, S. M. 1987. Testing hypotheses or fitting models? Another look at mass extinctions. In *Neutral Models in Biology,* ed. M. H. Nitecki and A. Hoffman, 147–59. New York: Oxford University Press.
Stinchcombe, J. R., M. T. Rutter, D. S. Burdick, P. Tiffin, M. D. Rausher, and R. Mauricio. 2002. Testing for environmentally induced bias in phenotypic estimates of natural selection: Theory and practice. *American Naturalist* 160:511–23.
Stone, J. R. 1996. Computer-simulated shell size and shape variation in the Caribbean land snail genus *Cerion:* A test of geometrical constraints. *Evolution* 50:341–47.
Stotz, K., and P. E. Griffiths. 2004. Genes: Philosophical analyses put to the test. In *Genes, Genomes and Genetic Elements,* ed. by K. Stotz, special issue, *History and Philosophy of the Life Sciences,* 26:5–28.
Su, W., and S. H. Howell. 1995. The effects of cytokinin and light on hypocotyl elongation in *Arabidopsis* seedlings are independent and additive. *Plant Physiology* 108:1423–30.

Sultan, S. E. 1987. Evolutionary implications of phenotypic plasticity in plants. *Evolutionary Biology* 21:127–78.

———. 1995. Phenotypic plasticity and plant adaptation. *Acta Botanica Neederlandica* 44:363–83.

Svensson, E., and B. Sinervo. 2000. Experimental excursions on adaptive landscapes: Density-dependent selection on egg size. *Evolution* 54:1396–1403.

Symons, D. 1995. Beauty is in the adaptations of the beholder: The evolutionary psychology of human female sexual attractiveness. In *Sexual Nature, Sexual Culture,* ed. P. R. Abramson and S. D. Pinkerton, 80–118. Chicago: University of Chicago Press.

Tattersall, I. 1995. *The Fossil Trail: How We Know What We Think We Know about Human Evolution.* New York: Oxford University Press.

———. 2000. Once we were not alone. *Scientific American,* January, 56–62.

Templeton, A. R. 1981. Mechanisms of speciation—a population genetic approach. *Annual Review of Ecology and Systematics* 12:23–48.

———. 1989. The meaning of species and speciation: A genetic perspective. In *Speciation and Its Consequences,* ed. D. Otte and J. A. Endler, 3–27. Sunderland, MA: Sinauer Associates.

———. 1996. Experimental evidence for the genetic-transilience model of speciation. *Evolution* 50:909–15.

———. 1999. Human races: A genetic and evolutionary perspective. *American Anthropologist* 100:632–50.

Templeton, A. R., S. D. Maskas, and M. B. Cruzan. 2000. Gene trees: A powerful tool for exploring the evolutionary biology of species and speciation. *Plant Species Biology* 15:211–22.

Thornhill, R., and C. T. Palmer. 2000. *A Natural History of Rape: Biological Bases of Sexual Coercion.* Cambridge, MA: MIT Press.

Tooby, J., and L. Cosmides. 1989. The innate versus the manifest: How universal does universal have to be? *Behavioral and Brain Sciences* 12:36–37.

———. 1990. On the universality of human nature and the uniqueness of the individual: The role of genetics and adaptation. *Journal of Personality* 58:17–67.

———. 1992. The psychological foundations of culture. In *The Adapted Mind: Evolutionary Psychology and the Generation of Culture,* ed. J. H. Barkow, L. Cosmides, and J. Tooby, 19–136. New York: Oxford University Press.

———. 2000. Toward mapping the evolved functional organization of the mind and brain. In *The New Cognitive Neurosciences,* ed. M. S. Gazzaniga, 1167–78. Cambridge, MA: MIT Press.

Travisano, M., J. A. Mongold, A. F. Bennett, and R. E. Lenski. 1995. Experimental tests of the roles of adaptation, chance, and history in evolution. *Science* 267:87–90.

Trickett, A. J., and R. K. Butlin. 1994. Recombination suppressors and the evolution of new species. *Heredity* 73:339–45.

Trout, J. D. 2002. Scientific explanation and the sense of understanding. *Philosophy of Science* 69:212–33.

Turelli, M. 1988. Phenotypic evolution, constant covariances, and the maintenance of additive variance. *Evolution* 42:1342–47.

Upton, G. J. G., and I. T. Cook. 2002. *The Oxford Dictionary of Statistics,* s.v. "random walk." Oxford: Oxford University Press.

van Tienderen, P. H., and G. de Jong. 1994. A general model of the relation between phenotypic selection and genetic response. *Journal of Evolutionary Biology* 7:1–12.

Van Valen, L. 1973. A new evolutionary law. *Evolutionary Theory* 1:1–30.

Van Valen, L. M. 1988. Species, sets, and the derivative nature of philosophy. *Biology and Philosophy* 3:49–66.

Vela-Cardenas, M., and K. J. Frey. 1972. Optimum environment for maximizing heritability and genetic gain from selection. *Iowa State Journal of Science* 46:381–94.

Via, S. 1987. Genetic constraints on the evolution of phenotypic plasticity. In *Genetic Constraints on Adaptive Evolution,* ed. V. Loeschcke, 47–71. Berlin: Springer-Verlag.

Wade, M. J. 1977. An experimental study of group selection. *Evolution* 31:134–53.

Wade, M. J., and C. J. Goodnight. 1998. The theories of Fisher and Wright in the context of metapopulations: When nature does many small experiments. *Evolution* 52:1537–53.

Wade, P. R. 2000. Bayesian methods in conservation biology. *Conservation Biology* 14: 1308–16.

Wagner, A., G. P. Wagner, and P. Similion. 1994. Epistasis can facilitate the evolution of reproductive isolation by peak shifts: A two-locus two-allele model. *Genetics* 138:533–45.

Wagner, G. 2001. *The Character Concept in Evolutionary Biology.* San Diego, CA: Academic Press.

Wagner, G., and M. D. Laubichler. 2001. Character identification: The role of the organism. In *The Character Concept in Evolutionary Biology,* ed. G. P. Wagner, 141–63. San Diego, CA: Academic Press.

Waldmann, P., and S. Andersson. 2000. Comparison of genetic covariance matrices within and between *Scabiosa canescens* and *S. columbaria. Journal of Evolutionary Biology* 13: 826–35.

Walsh, D. M. 2003. Fit and diversity: Explaining adaptive evolution. *Philosophy of Science* 70:280–301.

Walsh, D. M., T. Lewens, and A. Ariew. 2002. The trials of life: Natural selection and random drift. *Philosophy of Science* 693:429–46.

Ward, D. 1992. The role of satisficing in foraging theory. *Oikos* 63:312–17.

Waters, C. K. 1991. Tempered realism about the force of selection. *Philosophy of Science* 58: 553–73.

———. 1994. Genes made molecular. *Philosophy of Science* 61:163–85.

Wayne, M. L., and K. Simonsen. 1998. Statistical tests of neutrality in the age of weak selection. *Trends in Ecology and Evolution* 13:236–40.

Weber, M. 2005. *Philosophy of Experimental Biology.* New York: Cambridge University Press.

Weigensberg, I., and D. A. Roff. 1996. Natural heritabilities: Can they be reliably estimated in the laboratory? *Evolution* 50:2149–57.

Weinberg, S. 1992. Against philosophy. In *Dreams of a Final Theory,* 166–90. New York: Pantheon.

Weinberg, W. 1908. On the demonstration of heredity in man. In *Papers on Human Genetics,* trans. S. H. Boyer. Englewood Cliffs, NJ: Prentice Hall, 1963.

Weiner, J. 1999. *Time, Love, Memory: A Great Biologist and His Quest for the Origins of Behavior.* New York: Knopf.

Wells, C., and M. Pigliucci. 2000. Heterophylly in aquatic plants: Considering the evidence for adaptive plasticity. *Perspectives in Plant Ecology, Evolution and Systematics* 3:1–18.

Wen, J. 1999. Evolution of eastern Asian and eastern North American disjunct distributions in flowering plants. *Annual Review of Ecology and Systematics* 30:421–55.

West-Eberhard, M. J. 2003. *Developmental Plasticity and Evolution.* Oxford: Oxford University Press.

Whitlock, M. C. 1995. Variance-induced peak shifts. *Evolution* 49:252–59.

———. 1997. Founder effects and peak shifts without genetic drift: Adaptive peak shifts occur easily when environments fluctuate slightly. *Evolution* 51:1044–48.

Wilke, C. O., J. L. Wang, C. Ofria, R. E. Lenski, and C. Adami. 2001. Evolution of digital organisms at high mutation rates leads to survival of the flattest. *Nature* 412:331–33.

Wilkins, A. S. 2002. *The Evolution of Developmental Pathways.* Sunderland, MA: Sinauer Associates.

Williams, G. C. 1966. *Adaptation and Natural Selection.* Princeton, NJ: Princeton University Press. Reissued with new Introduction, 1996.

———. 1992. *Natural Selection: Domains, Levels, and Challenges.* New York: Oxford University Press.

Wilson, D. S. 1994. Adaptive genetic variation and human evolutionary psychology. *Ethology and Sociobiology* 15:219–35.
Wilson, E. O. 1975. *Sociobiology: The New Synthesis.* Harvard University Press
———. 1998. *Consilience: The Unity of Knowledge.* New York: Knopf.
Wilson, R. 2003. Pluralism, entwinement, and the levels of selection. *Philosophy of Science* 70:531–52.
Wimsatt, W. C. 1987. False models as means to truer theories. In *Neutral Models in Biology,* ed. W. H. Nitecki and A. Hoffman, 23–55. New York: Oxford University Press.
Wittgenstein, L. 1958. *Philosophical Investigations.* Trans. G. E. M. Anscombe. New York: Macmillan, 1973.
Wolpert, L. 2002. *Principles of Development.* 2nd ed. Oxford: Oxford University Press.
Woodward, B., and S. Armstrong. 1979. *The Brethren: Inside the Supreme Court.* New York: Simon & Schuster.
Wray, G. A. 1994. Developmental evolution: New paradigms and paradoxes. *Developmental Genetics* 15:1–6.
Wright, S. 1932a. Evolution in Mendelian populations. *Genetics* 16:97–159.
———. 1932b. The roles of mutation, inbreeding, crossbreeding and selection in evolution. *Proceedings of the Sixth International Congress of Genetics* 1:356–66.
Wu, S.-H., and J. C. Lagarias. 1997. The phytochrome photoreceptor in the green alga *Mesotaenium caldariorum:* Implication for a conserved mechanism of phytochrome action. *Plant, Cell and Environment* 20:691–99.
Wynne-Edwards, V. C. 1962. *Animal Dispersion in Relation to Social Behaviour.* New York: Hafner.
Yang, Y.-Y., A. Nagatani, Y. J. Zhao, B. J. Kang, R. E. Kendrick, and Y. Kamiya. 1995. Effects of gibberellins on seed germination of phytochrome-deficient mutants of *Arabidopsis thaliana. Plant Cell Physiology* 36:1205–11.
Yanovsky, M. J., J. J. Casal, and G. C. Whitelam. 1995. Phytochrome A, phytochrome B and HY4 are involved in hypocotyl growth responses to natural radiation in *Arabidopsis:* Weak de-etiolation of the *phyA* mutant under dense canopies. *Plant, Cell and Environment* 18:788–94.
Yoshimura, J., and C. W. Clark, eds. 1993. *Adaptation in Stochastic Environments.* Berlin: Springer.
Zhivotovsky, L. A., M. W. Feldman, and A. Bergman. 1996. Fitness patterns and phenotypic plasticity in a spatially heterogeneous environment. *Genetical Research* 68:241–48.
Ziliak, S. T., and D. N. McCloskey. 2004. Size matters: The standard error of regressions in the *American Economic Review. Journal of Socio-Economics* 33:527–46.

Index

Page references to tables, figures, and boxes appear in italics.